Advances in
PARASITOLOGY

VOLUME 65

Advances in PARASITOLOGY

VOLUME 65

Edited by

R. MULLER
London School of Hygiene and Tropical Medicine, London, UK

D. ROLLINSON
Department of Zoology
The Natural History Museum
Cromwell Road, London, UK

S. I. HAY
Senior Research Fellow
Malaria Public Health & Epidemiology Group
Centre for Geographic Medicine
KEMRI/University of Oxford Wellcome
Trust Collaborative Programme, Nairobi, Kenya

ELSEVIER

AMSTERDAM • BOSTON • HEIDELBERG • LONDON
NEW YORK • OXFORD • PARIS • SAN DIEGO
SAN FRANCISCO • SINGAPORE • SYDNEY • TOKYO
Academic Press is an imprint of Elsevier

Academic Press is an imprint of Elsevier
84 Theobald's Road, London WC1X 8RR, UK
Radarweg 29, PO Box 211, 1000 AE Amsterdam, The Netherlands
The Boulevard, Langford Lane, Kidlington, Oxford OX5 1GB, UK
30 Corporate Drive, Suite 400, Burlington, MA 01803, USA
525 B Street, Suite 1900, San Diego, CA 92101-4495, USA

First edition 2007

ISBN: 978-0-12-374166-0
ISSN: 0065-308X

Printed and bound in The Netherlands
07 08 09 10 11 10 9 8 7 6 5 4 3 2 1

CONTENTS

CONTRIBUTORS

J. P. Ackers
Department of Infectious and Tropical Diseases, London School of Hygiene and Tropical Medicine, London WC1E 7HT, United Kingdom.

V. Ali
Department of Parasitology, Gunma University Graduate School of Medicine, Maebashi, Japan.

U. C. M. Alsmark
Division of Biology, Newcastle University, Newcastle NE1 7RU, United Kingdom.

A. Bhattacharya
School of Life Sciences and Information Technology, Jawaharlal Nehru University, New Delhi 110067, India.

S. Bhattacharya
School of Environmental Sciences, Jawaharlal Nehru University, New Delhi 110067, India.

Xavier Bosch-Capblanch
International Health Group, Liverpool School of Tropical Medicine, Liverpool L3 5QA, United Kingdom.

Bernard J. Brabin
Department of Community Child Health, Royal Liverpool Children's Hospital NHS Trust, Alder Hey, West Derby, Liverpool L12 2AP, United Kingdom, and Child and Reproductive Health Group, Liverpool School of Tropical Medicine, Liverpool L3 5QA, United Kingdom, and Emma Kinderziekenhuis, Academic Medical Centre, University of Amsterdam, 1100 DD Amsterdam, The Netherlands.

James E. Bron
Institute of Aquaculture, University of Stirling, Stirling, Scotland FK9 4LA, United Kingdom.

Adam J. Brooker
Institute of Aquaculture, University of Stirling, Stirling, Scotland FK9 4LA, United Kingdom.

I. Bruchhaus
Bernhard Nocht Institute for Tropical Medicine, D-20359 Hamburg, Germany.

James Bunn
Department of Paediatrics, College of Medicine, University of Malawi, Blantyre, Malawi, and Child and Reproductive Health Group, Liverpool School of Tropical Medicine, Liverpool L3 5QA, United Kingdom.

C. G. Clark
Department of Infectious and Tropical Diseases, London School of Hygiene and Tropical Medicine, London WC1E 7HT, United Kingdom.

M. Duchêne
Department of Specific Prophylaxis and Tropical Medicine, Center for Physiology and Pathophysiology, Medical University of Vienna, A-1090 Vienna, Austria.

T. M. Embley
Division of Biology, Newcastle University, Newcastle NE1 7RU, United Kingdom.

P. G. Foster
Department of Zoology, Natural History Museum, London, SW7 5BD, United Kingdom.

Bernard Fried
Department of Biology, Lafayette College, Easton, Pennsylvania 10842, USA.

C. A. Gilchrist
Department of Medicine, Division of Infectious Diseases, University of Virginia Health Sciences Center, Charlottesville, Virginia 22908, USA.

N. Guillén
Institut Pasteur, Unité Biologie Cellulaire du Parasitisme and INSERM U786, F-75015 Paris, France.

N. Hall
The Institute for Genomic Research, Rockville, Maryland 20850, and School of Biological Sciences, University of Liverpool, Liverpool L69 7ZB, United Kingdom.

R. P. Hirt
Division of Biology, Newcastle University, Newcastle NE1 7RU, United Kingdom.

M. Tazreiter
Department of Specific Prophylaxis and Tropical Medicine, Center for Physiology and Pathophysiology, Medical University of Vienna, A-1090 Vienna, Austria.

James W. Kazura
Center for Global Health & Diseases, Case Western Reserve University, Cleveland, Ohio 44106–7286, USA.

M. Leippe
Zoologisches Institut der Universität Kiel, D-24098 Kiel, Germany.

A. Lohia
Department of Biochemistry, Bose Institute, Kolkata 700054, India.

María-Paz Loscertales
Child and Reproductive Health Group, Liverpool School of Tropical Medicine, Liverpool L3 5QA, United Kingdom.

Mwele N. Malecela-Lazaro
National Institute for Medical Research, Dar es Salaam, Tanzania.

B. J. Mann
Department of Medicine, Division of Infectious Diseases, University of Virginia Health Sciences Center, Charlottesville, Virginia 22908, USA.

S. Marion
Institut Pasteur, Unité Biologie Cellulaire du Parasitisme and INSERM U786, F-75015 Paris, France, and Cell Biology and Biophysics Program, European Molecular Biology Laboratory, 69117 Heidelberg, Germany.

David A. Mayer
Department of Surgery, New York Medical College, Valhalla, New York 10595, USA.

Edwin Michael
Department of Infectious Disease Epidemiology, Imperial College London, Norfolk Place, London W2 1PG, United Kingdom.

C. Mukherjee
Department of Biochemistry, Bose Institute, Kolkata 700054, India.

C. J. Noël
Division of Biology, Newcastle University, Newcastle NE1 7RU, United Kingdom.

T. Nozaki
Department of Parasitology, Gunma University Graduate School of Medicine, Maebashi, Japan.

James O'Donnell
Haemostasis Research Group, Institute of Molecular Medicine, Trinity College Dublin, St. James's Hospital, Dublin 8, Republic of Ireland.

Stephen Owens
Medical Research Council Laboratories, Banjul, Fajara, The Gambia, and Child and Reproductive Health Group, Liverpool School of Tropical Medicine, Liverpool L3 5QA, United Kingdom.

Y. Saito-Nakano
Department of Parasitology, National Institute of Infectious Diseases, Tokyo, Japan.

J. Samuelson
Department of Molecular and Cell Biology, Boston University Goldman School of Dental Medicine, Boston, Massachusetts 02118, USA.

Andrew P. Shinn
Institute of Aquaculture, University of Stirling, Stirling, Scotland FK9 4LA, United Kingdom.

T. Sicheritz-Ponten
Center for Biological Sequence Analysis, BioCentrum-DTU, Technical University of Denmark, DK-2800 Lyngby, Denmark.

U. Singh
Departments of Internal Medicine, Microbiology, and Immunology, Stanford University School of Medicine, Stanford, California 94305, USA.

E. Tannich
Bernhard Nocht Institute for Tropical Medicine, D-20359 Hamburg, Germany.

C. Weber
Institut Pasteur, Unité Biologie Cellulaire du Parasitisme and INSERM U786, F-75015 Paris, France.

PREFACE

María-Paz Loscertales and colleagues from the Liverpool School of Tropical Medicine open this volume with a review of the mechanisms behind the protection conferred by ABO blood group phenotypes on *Plasmodium falciparum* malaria. Given the long pedigree of research in this area, it is surprising that such a review had not already been undertaken. The authors provide an authoritative overview of the subject, ranging from the biological basis of the protective mechanisms to the population genetics of these blood groups. Moreover, they conduct detailed meta-analyses with regard to the association of ABO phenotypes and uncomplicated and severe *P. falciparum* malaria outcomes. No consistent summary of the evidence could be provided for uncomplicated malaria. Within the severe disease examples, however, they did find that group A was associated with more severe and group O with milder disease. These analyses were further extended to explore blood group type as risk factors for *P. falciparum* during pregnancy, the association of which was found to vary according to pregnancy stage. All the results are then subject to a detailed discussion of the potential mechanisms for such interactions.

Graham Clark from the Department of Infectious and Tropical Diseases, London School of Hygiene and Tropical Medicine leads an international authorship in explaining many previously unreported aspects of the genome of the intestinal parasite *Entamoeba histolytica*, with an emphasis on how this allows a better understanding of the biology of the parasite. This predatory protist has an expanded repertoire of gene families involved with signalling and trafficking that allow the parasite to interact with the environment. Conversely, it has lost many other metabolic pathways leading to significant compaction of the genome in other areas. This is a vast review and illustrative of the huge amounts of information being generated for genome sequence initiatives. The authors emphasise, however, that this is a work in progress and a full understanding of the biology at the macro level will not be possible until the genome is complete.

The next paper concerns monitoring and evaluation of lymphatic filariasis control and complements papers in Volume 61 on the Control of Human Parasitic Diseases. Edwin Michael (Imperial College School of Medicine, London, UK), Mwela Malecela-Lazaro (National Institute for Medical Research, Dar es Salaam, Tanzania) and James Kazura (Cape

Western Reserve University, Cleveland, USA) join forces to present cogent arguments for the need to monitor parasite control programmes. The development and design of mathematical models of parasite transmission can contribute to parasite control monitoring programmes. The ability to estimate endpoints, understand trends and determine reliable indicators can all contribute to effective control. The authors stress that with increasing emphasis being given to the control of filariasis and other parasitic diseases, it is essential to design and implement rational monitoring plans backed by a flexible management approach.

David A. Mayer of The New York Medical College, Valhalla, New York, USA, and Bernard Fried of the Department of Biology, Lafayette, Easton, USA, have found in their review that although many helminth infections have been implicated in carcinogenesis, only two, *Schistosoma haematobium* and *Opisthorchis viverrini*, have been definitely proven to cause cancer in humans. The authors provide a comprehensive overview of a wide range of trematodes, cestodes and nematodes associated with carcinogenesis. Recent researches on the mechanisms are evaluated, including chronic inflammation, modulation of the host immune system, impairment of immunological surveillance, inhibition of cell–cell communication, disruption of proliferation–antiproliferation pathways, induction of genomic instability and stimulation of malignant stem cell progeny. The authors point out that in 1926, Johanes Fibiger won the only Nobel Prize awarded in helminthology for his work on the causation of cancer by a helminth; and, although doubt has been cast on his conclusions over the years, recent research appears to validate his results.

The final review in this volume by Adam Brooker, Andrew Shinn and James Bron from the Institute of Aquaculture, University of Stirling, Scotland, deals with the parasitic copepod *Lernaeocerca branchialis*. There is increasing pressure on many fish stocks from over fishing and environmental changes. *L. branchialis* poses a significant threat to the health of both wild and cultured gadoid fish, with cod, whiting and haddock being the most common hosts. The review is timely given the increasing interest and development of fish aquaculture. This paper considers many aspects of the copepod's biology, including taxonomy, morphology, reproduction, pathogenicity and distribution. The paper highlights many gaps in knowledge and suggests areas that require more detailed examination and research. Special attention is drawn to the need to anticipate problems likely to be caused by this parasitic copepod as fish culture sites are developed in coastal areas where most transmission takes place.

R. Muller
D. Rollinson
S. I. Hay

CHAPTER 1

ABO Blood Group Phenotypes and *Plasmodium falciparum* Malaria: Unlocking a Pivotal Mechanism

María-Paz Loscertales,* **Stephen Owens,*,†** **James O'Donnell,‡** **James Bunn,*,§** **Xavier Bosch-Capblanch,¶** and **Bernard J. Brabin*,||,#**

Contents

* Child and Reproductive Health Group, Liverpool School of Tropical Medicine, Liverpool L3 5QA, United Kingdom
† Medical Research Council Laboratories, Banjul, Fajara, The Gambia
‡ Haemostasis Research Group, Institute of Molecular Medicine, Trinity College Dublin, St. James's Hospital, Dublin 8, Republic of Ireland
§ Department of Paediatrics, College of Medicine, University of Malawi, Blantyre, Malawi
¶ International Health Group, Liverpool School of Tropical Medicine, Liverpool L3 5QA, United Kingdom
|| Emma Kinderziekenhuis, Academic Medical Centre, University of Amsterdam, 1100 DD Amsterdam, The Netherlands
Department of Community Child Health, Royal Liverpool Children's Hospital NHS Trust, Alder Hey, West Derby, Liverpool L12 2AP, United Kingdom

Advances in Parasitology, Volume 65
ISSN 0065-308X, DOI: 10.1016/S0065-308X(07)65001-5

Abstract Host susceptibility to *Plasmodium falciparum* infection is central for improved understanding of malaria in human populations. Red blood cell (RBC) polymorphisms have been proposed as factors associated with malaria infection or its severity, although no systematic appraisal of ABO phenotypes and malaria risk has been undertaken. This analysis summarises epidemiological, clinical and immunological evidence on the nature of ABO histo-blood antigens and their interaction with malaria in terms of population genetics, infection risk, severe malaria and placental malaria. In non-pregnant subjects, a meta-analysis showed no conclusive evidence associating ABO phenotypes with risk of uncomplicated malaria. There was stronger evidence that ABO phenotype modulates severity of *P. falciparum* malaria, with group A associated with severe disease and blood group O with milder disease. Among pregnant subjects, group O was associated with increased risk of placental malaria in primigravidae and reduced risk in multigravidae. The biological basis for ABO-related susceptibility to malaria is reviewed. Several mechanisms relate to these associations including affinity for *Anopheles gambiae*; shared ABO antigens with *P. falciparum*; impairment of merozoite penetration of RBCs; and cytoadherence, endothelial activation and rosetting. ABO phenotypic associations with malaria are related to its pathogenesis and improved understanding of these interactions is required for understanding the glycobiology of malaria infection.

1. INTRODUCTION

The ABO blood group system of carbohydrate antigen expression on the surface of human red blood cells (RBCs) was first described by Karl Landsteiner in 1900 (Lansdsteiner, 1900, 1991) and represented an important step towards development of safer blood transfusions (Owen, 2000). Later in 1941, Hartmann (1941 and 1970) demonstrated that expression of ABO antigens was not confined to RBCs, which led to the more accurate name of histo-blood group antigens. The system was first described with three antigens, six genotypes and four phenotypes, but has now been shown to be highly polymorphic (Chester and Olsson, 2001). Histo-blood group antigens of ABO specificity have been detected in many other species (Kominato *et al.*, 1992). Although ABO antigens have not been shown to have any specific physiological function, individual phenotypes have been reported to be associated with a variety of conditions, including infection and cancer (Greenwell, 1997). The hypothesis that ABO groups may be of importance is supported by the observation that their geographical distribution varies markedly, suggesting that positive selective factors may have influenced gene spread.

In regions highly endemic for *Plasmodium falciparum* malaria, it is well recognised that a range of RBC polymorphisms associated with resistance to severe disease have undergone positive selection (Min-Oo and Gros, 2005). Tolerance to malaria based on inherited factors often occurs amongst Africans or persons of African descent presenting haemoglobins S, C and E, α- or β-thalassaemias, glucose-6-phosphate dehydrogenase deficiency, southern Asian ovalocytosis and glycophorins A, B and C variants (Pasvol, 2003). In addition, a number of studies have analysed whether ABO blood groups may be associated with malaria risk and/or severity. Some early studies were collected by Mourant *et al.* (1978), and more recently Uneke (2007) has summarised some of the available evidences. For a number of reasons, these often small studies produced conflicting conclusions, so that the relationship between ABO blood group and malaria has not been clearly defined. In this analysis, we review the relationship between ABO blood group and malaria, in terms of the host's immune response, geographic distribution in relation to malaria epidemiology, clinical disease severity and the nature of underlying biological mechanisms. The aims of the review were (a) to estimate the association or lack of association of ABO blood groups and malaria (clinical or parasitological) and to determine the magnitude of these associations; and (b) to provide an overview of the mechanisms that underlie the contribution of ABO blood groups to malaria patterns and pathogenesis. The nature of ABO histo-blood group antigens is also considered in order to provide an overview as a basis for assessing their role in malaria.

2. METHODS

To estimate the association between blood groups and malaria [aim (a)], we searched the literature for geographical and population genetics and clinical or epidemiological data on ABO phenotypes and malaria; we included all study designs, including any type of participants, where data on at least two blood groups and on *falciparum* malaria (whether clinical or parasitological) were present. To describe the biological mechanisms underlying this association [aim (b)], we searched studies on malaria pathogenesis that included information on ABO histo-blood group antigens.

2.1. Inclusion criteria

For aim (a), the literature was searched for experimental or quasi-experimental, observational and case studies which assessed possible associations between ABO phenotypes and *P. falciparum* malaria. Participants could include children or adults exposed to or experiencing *falciparum* malaria. Epidemiological outcomes of interest included malaria parasite prevalence or parasite density, malaria severity defined by clinical parameters (cerebral malaria, severe anaemia, acidotic breathing) or analytical parameters (lactic acidosis, hypoglycaemia) and death associated or attributed to malaria. For aim (b), experimental studies included placental or necropsy samples, or *in vitro* models as well as concept papers.

We searched studies in any language or publication status in the following data bases: The Cochrane library and MEDLINE (1966 to November 2006) and the reference lists of identified studies. We also performed a manual search prior to 1966, from book indices and early publication references and contacted researchers if necessary for unpublished data. The search strategy for aim (a) included the terms 'malaria' and 'blood groups' and their variants, using the appropriate syntaxes for each data base. For aim (b), the terms malaria and blood group were used together with 'rosetting' or 'cytoadherence'. We assessed eligibility using pre-established selection criteria. We followed the recommendations of The Cochrane Collaborative Review Group on HIV Infection and AIDS [CCRG HIV/AIDS. Editorial Policy: Inclusion and Appraisal of Experimental and Non-experimental (Observational) Studies].

From the search results, we identified further potentially relevant studies. The inclusion criteria were subsequently applied. We extracted the following data items: year, country, types of participants, health care setting, urban or rural context and number of subjects with and without malaria and with each blood group phenotype.

2.2. Data analysis

Data were entered into an MS Excel worksheet. Studies were grouped according to the study design (case-control and cross-sectional) and case-control studies were divided in two categories: those in which cases had uncomplicated malaria and those with severe malaria. Data were cross-checked and entered into Review Manager version 4.2 (The Nordic Cochrane Centre, Rigshospitalet 2003) to estimate odds ratios (ORs) and confidence intervals (CIs) for the malaria and blood group antigen associations and to produce forest plots. Relative risks (RR) and their 95% CIs were computed for dichotomous data. Studies were grouped into those assessing infection risk (or prevalence) and those with data severity. Studies were sub-grouped as descriptive or case-control studies. In most case-control studies, controls were not paired with cases, but selected as a group. Heterogeneity was assessed by visually examining forest plots and using the chi-square test for heterogeneity with a 10% level of statistical significance. If heterogeneity was present and it was considered appropriate to combine trial data then a random-effects model (REM) was used instead of a fixed-effect model.

Of 319 potentially relevant articles, 34 studies met the inclusion criteria. We also located one on-going study. In the manual bibliographic search, three early studies and one early data summary were identified. One study was included with complete unpublished data provided by the researcher. Most studies were undertaken between 1979 and 1999. In the initial analysis, there were no significant differences between A and B phenotypes which showed the same trend in relation to malaria infection risk and severity. For this reason, the data and OR estimates in the tables were presented with comparison of A, B or AB phenotypes with the O blood group.

3. THE NATURE OF ABO HISTO-BLOOD ANTIGENS

Blood group antigens are stable characteristics controlled by genes inherited in a simple Mendelian manner. The ABO blood group locus is located on chromosome 9, with co-dominant A and B alleles that express glycosyltransferases, as well as silent alleles with no observable expression (O alleles) (Watkins and Morgan, 1959; Yamamoto, 2004). The A and B glycosyltransferase enzymes catalyse the addition of specific sugar moieties onto a precursor carbohydrate chain. The ABO antigens are carbohydrate in nature and differ only with respect to their terminal sugar (Fig. 1.1). The precursor acceptor structure is known as the H antigen, and is synthesised by an α1-2-fucosyltransferase encoded by the H gene locus on chromosome 19. The H antigen can also be produced by the action of other α1-2-fucosyltransferase, encoded by another locus (the secretor gene Se), which has a different affinity for carrier carbohydrate chains and determines the presence of ABH antigens in many tissues. The biosynthetic pathway of ABO is intimately related to the synthesis of other carbohydrate blood group antigens, including Le, P and Ii.

In clinical blood transfusion practise, human erythrocytes are typically grouped into six main ABO phenotypes, O, A_1, A_2, B, A_1B and A_2B, using serological methods. The difference between the A_1 and A_2 subgroups is

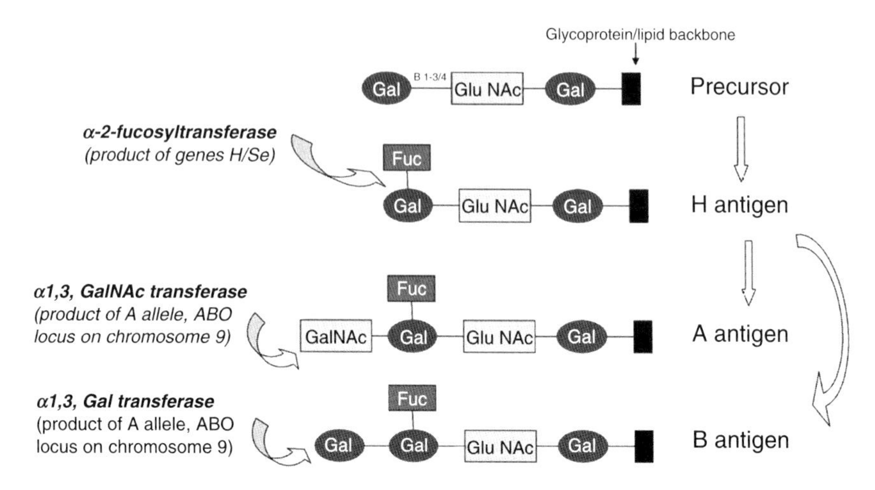

FIGURE 1.1 Structure and synthesis of ABO blood groups.

partly quantitative and only a quarter of the A or AB phenotypes possess the A_2 antigen (Clausen and Hakomori, 1989). DNA analysis is necessary to discriminate between different underlying genotypes (e.g., AO vs AA, BB vs BO). In blood group O individuals, the RBCs express H antigen, but fail to express either A or B antigenic determinants. Thus, H antigen is present on all RBCs, except in the rare Bombay phenotype. However, quantitative H antigen expression varies significantly between different groups ($O > A_2 > A_2B > B > A_1 > A_1B$) (Contreras and Lubenko, 1999).

Although RBCs are the main carriers of ABH antigens, these antigens are actually also expressed on white blood cells (WBCs), platelets, vascular endothelium and in a soluble form in the plasma. Moreover, ABH antigens have also been identified on most epithelial tissues and their secretions (of ABH secretors), but not in connective tissues (Ravn and Dabelsteen, 2000). The expression of ABH antigens varies with age and development. A, B and H antigens can be detected on the RBCs of very young foetuses, but there are less antigen sites and their reactions are weaker than those of adult RBCs. The expression in RBCs increases after birth until about 3 years of age and then remains stable through life. Conversely, the survival of the antigens in other tissues beyond the 12th week after conception is restricted to endothelium and stratified epithelia (integument, oesophagus, lower urinary tract and vagina); while epithelia proceeding towards further morphological and functional differentiation (gastrointestinal tract, lung, thyroid and pituitary) undergo antigenic loss. The secretion-borne antigens first appear at the eighth week (ovulation age) in the salivary glands and in the stomach, to be followed by the rest of the gastrointestinal tract, respiratory system and pancreas. The secretion of these antigens persists throughout life (Szulman, 1964). These tissue-specific oligosaccharides are dependent on cell type-specific expression of a unique glycosyltransferase.

On the surface of erythrocytes, ABH antigens are expressed on glycosphingolipids; on the N-linked glycans of a number of different glycoproteins such as band 3 (anion transporter) and band 4.5 (glucose transporter) and on the O-linked glycan structures of glycophorin A (Podbielska *et al.*, 2004). Similarly in other cell types, and in secretions, ABH determinants may be expressed on both glycolipids and/or glycoproteins. In the plasma, they are carried by glycolipids that may be passively adsorbed onto the RBCs. The life span of RBCs is normal in rare cases lacking carbohydrate ABO antigens (Bombay phenotype) or in cells lacking glycophorin A or B.

The clinical importance of the ABO blood group antigens was initially recognised in relation to blood transfusion practise, where incompatible RBC transfusions resulted in massive intravascular haemolysis. These effects result from circulating anti-A and anti-B antibodies, respectively. Naturally occurring (non-RBC-induced) anti-A and anti-B antibodies appear in serum at about 3–6 months of age, probably in response to bacterial and viral antigens, reach peak prevalence in young adults and decline in old age. In group A and B persons, these antibodies will be almost entirely IgM, but group O individuals will usually have some IgG component.

As previously discussed, a specific biological function for ABH antigenic determinants has not been described. However, it is noteworthy that these structures have been conserved through evolution. Substances with similar immunodominant structures have been found in a variety of living organisms including plants and bacteria (Yamamoto, 2004). Furthermore, Kominato *et al.* (1992) examined the presence of a gene sequence for human A transferase and observed hybridisation with all the mammal species examined. Expression of ABH histo-blood group antigens varies significantly across different vertebrate species. For example, in amphibians and reptiles, the expression of ABH histo-blood group antigens is positive only on epithelial cells of endodermal tissue, while in lower mammals it is also positive on some ectodermal tissues (e.g., epidermis and nervous receptors). Furthermore, vascular endothelial expression has been reported in baboons, but erythrocyte expression is observed only in higher anthropoid primates (chimpanzee and orangutan) and humans (Oriol *et al.*, 1992). The secretor trait is constant in primates (Socha *et al.*, 1995), and although they express the ABO histo-blood group system not all types are found in every species (Moor-Jankowski, 1993; Moor-Jankowski and Wiener, 1969). The genomic sequences of A and B alleles in primates show that crucial amino acid sequences of the glycosyltransferases are conserved during evolution, which reinforces the hypothesis of trans-species inheritance of the ABO polymorphism from a common ancestor (Martinko *et al.*, 1993). However, O alleles are species-specific and result from independent silencing mutations, which may reflect convergent evolution rather than trans-species inheritance of the ABO genes (Kermarrec *et al.*, 1999). The studies based on the accumulation of mutations and the phylogenetic networks of human and non-human primate ABO alleles suggest that some kind of balancing selection may have been operating at the ABO locus, and confirm that in the human lineage group O appeared later than the other ABO phenotypes (Saitou and Yamamoto, 1997). In addition, the fact that blood group genes are less variable in chimpanzees than in humans may reflect selection through infectious diseases in the human lineage (Sumiyama *et al.*, 2000).

Allelic frequency at the ABO blood group locus demonstrates marked geographical variation (Figs. 1.3–1.5). In contrast, the frequency of allelomorphs for other erythrocyte blood group systems (M, N and P) remains remarkably constant in different areas of the world (Mourant *et al.*, 1976). Specially revealing is the high frequency of the non-functional O allele (about 63% in humans) because on the standard mutation-drift balance model null alleles are expected to remain rare. The high frequency of B antigen among Asians, which affects both the Caucasians and the Mongoloids, also suggests an external non-racial influence (Fig. 1.3). To determine whether these geographic variations in ABH antigen expression may have resulted from positive selection pressures, a plethora of studies have examined possible associations between individual ABO blood group phenotypes and specific clinical diseases. Such selection is plausible particularly as these carbohydrate structures can function as receptors and ligands for microbes, and for immunologically important proteins (Weir, 1989). There is evidence that ABO antigens have a role in infections or cancer (i.e., tissue-specific changes with prognostic value), and in multifactorial conditions such as heart and autoimmune diseases. Several studies suggest that group A individuals are at an increased risk of malignancy, thrombosis, high serum cholesterol levels and myocardial infarction, when compared with the O phenotype that has conversely an increased bleeding tendency (Garratty, 2000). Interaction with infectious agents may account for the presence and geographical distribution of ABH polymorphism (Seymour *et al.*, 2004). Many reports have associated some infections with particular ABO phenotypes including *Helicobacter pylori* (O), *Salmonella typhi* (B), *Pasteurella pestis* (O), *Pseudomonas aeruginosa* (A), *Leishmania donovani* (ABO) or Smallpox (A) (Berger *et al.*, 1989). Increased susceptibility to different bacterial agents is observed more frequently for B or O phenotypes, and many times has been related to non-secretor status (Blackwell, 1989). The major differences in ABO frequencies in different parts of the world also have been attributed to epidemics that occurred in the past (Mourant *et al.*, 1976, 1978). Naturally occurring anti-A and -B antibodies are raised in response to bacterial infection (e.g., to *Escherichia coli*). The presence of blood group-like substances in various infecting organisms is one of the factors that could affect host susceptibility through inducing immuno-tolerance. The presence of the ABO and H blood groups in epithelial cells in contact with the environment also suggests another role in the interaction with microorganisms, since many pathogens are known to use cell surface carbohydrates as primary receptors (Karlsson, 1998) and blood groups can also provide a food source for bacteria and protozoa in the gut and the urogenital tract (Greenwell, 1997).

In summary, glycans have numerous and overlapping roles. They are mainly involved in structural functions and in the recognition by endogenous (self) and exogenous (non-self) receptors. Multicellular organisms with long life cycles must constantly change in order to survive lethal pathogens with shorter life cycles. It is clear that glycosidic linkages are much more flexible than peptide bounds, and they allow enough variation to survive lethal pathogens that evolve faster. Glycan-negative individuals in a population may slow or change the spread and profile of a disease. Receptor variation provided by glycan diversity between species, individuals within a species or even organs within an individual provides the opportunity for some of the individuals or cells to resist infections, especially those produced by enveloped viruses that will carry host antigens with them to the next host. In this way, ABO polymorphism may protect groups of individuals from particular infections. The final mapping of these polymorphisms in a particular area represents a balanced equilibrium reached for the associated susceptibilities to the endemic infections (Gagneux and Varki, 1999).

4. ABO PHENOTYPES AND MALARIA RISK

4.1. Population genetics

Athreya and Coriell (1967) reported a high incidence of blood group B in areas where *falciparum* malaria was endemic, compared to similar racial stocks living in non-malarial areas. Despite this association, it should be noted that the highest frequency of blood group B actually occurs in areas without endemic malaria (Figs. 1.2 (which is plate 1.2 in the separate Colour Plate Section) and 1.3). In contrast, if the presence of blood group A has negative survival value in the presence of malaria, as suggested by some studies (see the later description), O frequency should be higher in areas endemic for malaria. In sub-Saharan Africa, there is a higher frequency of the O allele, which could support this hypothesis (Figs. 1.4 and 1.5), and the frequencies of both the A and B genes tend to fall between 10 and 20% although this is not as apparent when ethnic

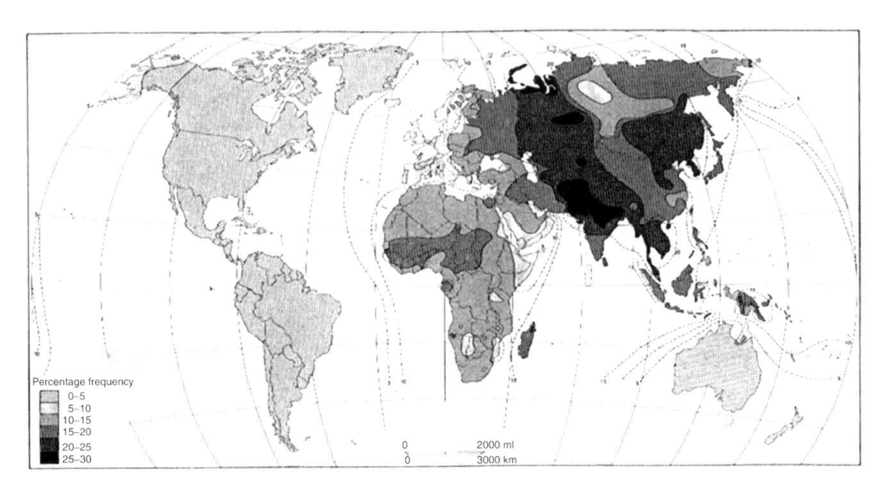

FIGURE 1.3 B allele distribution. Stippled lines represent contours for B allele prevalence estimates by region. This map showing the distribution of the B type allele in native populations of the world is reproduced from Mourant *et al.* (1976) with permission from Oxford University Press.

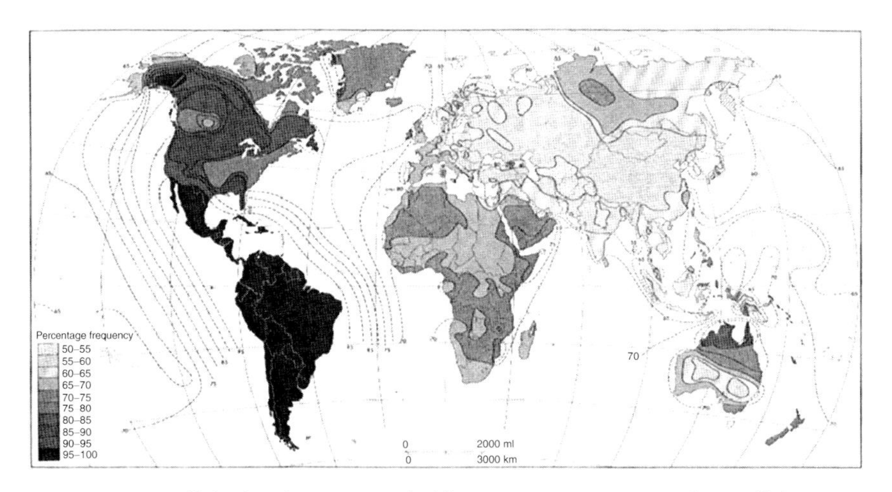

FIGURE 1.4 O allele distribution. Stippled lines represent contours for O allele prevalence estimates by region. This map showing the distribution of the O type allele in native populations of the world is reproduced from Mourant *et al.* (1976) with permission from Oxford University Press.

or geographic groups are taken into account when comparing group O frequency among people living within areas of differing malaria endemicity (e.g., holoendemic, hyperendemic, mesoendemic and hypoendemic). In addition to this higher frequency of O allele found in sub-Saharan Africa, a higher percentage of the A allele belongs to the A_2 subtype (around

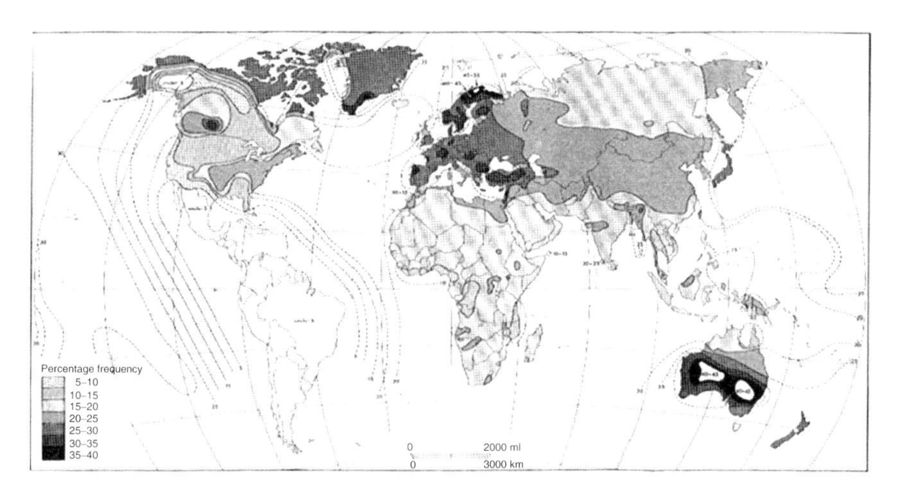

FIGURE 1.5 A allele distribution. Stippled lines represent contours for A allele prevalence estimates by region. This map showing the distribution of the A type allele in native populations of the world is reproduced from Mourant *et al.* (1976), with permission from Oxford University Press.

50% compared with 25% elsewhere), which shows lower enzymatic activity and therefore leaves more unaltered H antigen on the RBCs. Other weaker variants of blood group A (A_{int} and A_{Bantu}) are not uncommon in Africans, and they are also associated with more H antigen present on the RBCs. A study in South Africa revealed that the Bantu have in general more H antigen on their RBCs than do Europeans, and that this was true for all groups O, A_1 and A_2 (Mourant *et al.*, 1976). Since susceptibility to bacterial infections seems to be associated more often with B or O phenotypes (Berger *et al.*, 1989) in contrast to malaria, the geographical distribution of the different phenotypes may depend on the balance between reduced susceptibility to malaria and increased risk of other infections, although falciparum malaria may have been the strongest selection pressure to have shaped ABO blood group distribution in humans (Cserti and Dzik, 2007).

4.2. Infection risk

Studies reporting the association between risk of infection and ABO blood group type are summarised in Tables 1.1 and 1.2 (A, B and AB phenotypes vs O) and in Figs. 1.6 and 1.7 (O vs non-O). Ten were undertaken in Africa, three in Europe, four in America and ten in India. Among 12 cross-sectional studies (Table 1.1; Fig. 1.6), only three detected significant differences in febrile patients: Pant *et al.* (1992a) observed a significant association with parasitaemia prevalence which was higher in subjects with the AB blood group, and the same authors in a separate study (Pant *et al.*, 1993) reported parasitaemia prevalence which was

TABLE 1.1 Descriptive studies of *P. falciparum* prevalence by ABO phenotype: Risk of infection

								A			B			AB			O
Reference	Country	Age range (years)	Urban/ rural	Place	Inclusion criteria		*N*	%pos	Relative risk	95%CI	%pos	Relative risk	95%CI	%pos	Relative risk	95%CI	%pos
Akinboye and Ogunrinade, 1987[a]	Nigeria	18–55	U	HC	Donors	Pf	115	6.1	0.59	0.13–2.74	5.0	0.48	0.06–3.77	0.0	–	–	10.3
Ibhanesebhor *et al.*, 1996[b]	Nigeria	NS	U	HC	Donors	Pf	50	33.3	0.77	0.23–2.53	33.3	0.77	0.23–2.53	0.0	0.00	–	43.2
Kassim and Ejezie, 1982[c]	Nigeria	7–14	R	Comm	School children	Pf	681	37.3	0.90	0.70–1.16	36.8	0.89	0.70–1.13	35.3	0.85	0.44–1.64	41.5
Domarle *et al.*, 1999[d]	Gabon	NS	R	Comm	Children	Pf	59	–	–	–	–	–	–	–	–	–	68.2
Pant *et al.*, 1992a[e]	India (Gujarat)	NS	R	Comm + HC	Fever	Pf (Pv)	783	5.6	2.33	0.81–6.68	2.8	1.14	0.39–3.36	9.3	**3.85**	**1.26–11.75**[*]	2.4
Pant *et al.*, 1992b[f]	India (Gujarat)	All	R	Comm + HC	Fever	Pf (Pv)	769	6.0	0.62	0.61–1.22	8.3	0.85	0.48–1.51	6.9	0.71	0.29–1.70	9.8
Thakur and Verma, 1992[g]	India	5–65	R	Comm	Households	Pf (Pv)	258	24.7	**2.30**	**0.98–5.39**[*]	22.1	2.06	0.88–4.84	10.3	0.96	0.29–3.17	10.7
Pant *et al.*, 1993[h]	India (Gujarat)	All	R	Comm	NS	Pf (Pv)	1781	8.6	**1.71**	**1.04–2.81**[*]	7.9	**1.57**	**0.97–2.53**[*]	8.6	1.70	0.90–3.21	5.1
Singh *et al.*, 1995[i]	India (Madhya Pradesh)	NS	R?	HC	Fever	Pf[*] (Pv)	2095	14.4	0.96	0.73–1.27	18.8	1.26	0.98–1.62	11.4	0.77	0.51–1.14	14.9
Pant and Srivastava, 1997[j]	India (Gujarat)	All	R	Comm + HC	Fever	Pf (Pv)	1287	11.9	1.34	0.86–2.11	6.9	0.78	0.49–1.24	10.2	1.15	0.62–3.12	8.8
Pant *et al.*, 1998[k]	India (Maharashtra)	2–60	U/R?	Comm	Fever	Pf	121	17.2	0.80	0.28–2.34	18.6	0.87	0.34–2.23	9.5	0.44	0.10–1.99	21.4

Pf(Pv), *P. falciparum* (*P. vivax*); HC, health centre; Comm., community survey; NS, not stated; numbers in bold, statistically significant relative risk and confidence interval (CI); $^{*}p < 0.05$.

[a] One hundred and fifteen randomly selected blood samples from voluntary blood donors (apparently healthy adults, six females). Blood Bank, University College Hospital, Ibadan. No significant association was found between ABO blood group and malaria parasitaemia ($\chi^2 = 1.6$, $p > 0.05$).

[b] Fifty blood donor samples for all consecutive neonates requiring a blood transfusion. University of Benin Teaching Hospital. Age of donors not specified. No statistically significant association was found with ABO blood groups ($\chi^2 = 1.02$, d.f. = 3, $p = 0.05$).

[c] Blood smears from 681 school children from Epe township (southwestern Nigeria) were examined for malaria parasites.

[d] Sixty-one children from 10 families in the village of Dienga (southeastern Gabon) were enrolled. The mean age was 10.08 ± 4.99.

[e] In 1992a, blood samples were collected from 783 persons with malaria symptoms or history of fever in the previous week, in different villages of Kheda district and a malaria clinic in Nadiad (Gujarat, India) during 1990–1991. There is a significant difference in the distribution of ABO system among *P. falciparum* cases versus normal individuals ($\chi^2 = 3$, d.f. = 9.96, $p < 0.05$).

[f] In 1992b, blood samples were collected from 769 people with fever at the time of the survey or a history of fever within a week, at Kheda district (Gujarat, India) between June 1989 and April 1990.

[g] A multi-stage stratified random sampling procedure was used to select 100 households in Bastar District, central India, an hyperendemic area for malaria. Two hundred and fifty-eight subjects agreed to participate.

[h] One thousand six hundred and fifty-nine individuals from upper caste Hindus from Kheda district were screened. χ^2 for ABO distribution among malaria cases (*P. vivax* and *P. falciparum*) and non-malarious healthy subjects = 7.893, $p < 0.05$.

[i] Blood was taken for thick and thin smears from patients with fever who came to MRC clinic in Bizandi. In the table, we show the number of *P. falciparum* parasitaemia and mixed infections (both *P. falciparum* and *P. vivax* parasitaemia). *P. vivax* infections are omitted.

[j] Blood samples from 1287 persons of all ages presenting with malaria symptoms or history of fever in the previous week were collected from designated villages of Kheda district and a malaria clinic in Nadiad (Gujarat, India).

[k] Rapid fever survey among the population of JNPT, a seaport in Raigad (population 3500). Finger-prick blood was obtained in all 121 persons with active fever. χ^2 showed no relation between ABO and malaria ($p > 0.05$). Only data on *P. falciparum* has been included in the table.

Review: Blood groups and malaria
Comparison: 01 O versus non-O blood groups in prevalence studies
Outcome: 01 Malaria or no malaria

Study or sub-category	O in malaria n/N	O in no malaria n/N	RR (fixed) 95% CI	Weight %	RR (fixed) 95% CI
01 Africa (all Nigeria)					
Akimboye 1987	6/9	52/106		3.83	1.36 [0.82, 2.24]
Dormale 1999	15/39	7/20		4.35	1.10 [0.54, 2.25]
Ibhanesebhor 1996	16/20	21/30		7.90	1.14 [0.83, 1.58]
Kassim 1982	160/269	226/412		83.93	1.08 [0.95, 1.24]
Subtotal (95% CI)	337	568		100.00	1.10 [0.98, 1.24]
Total events: 197 (O in malaria), 306 (O in no malaria)					
Test for heterogeneity: Chi² = 0.78, df = 3 (P = 0.85), I² = 0%					
Test for overall effect: Z = 1.57 (P = 0.12)					
02 Asia (all India)					
Pant 1992a	5/31	201/752		4.64	0.60 [0.27, 1.36]
Pant 1992b	20/61	185/708		8.56	1.25 [0.86, 1.83]
Pant 1993a	24/122	451/1537		19.34	0.67 [0.46, 0.97]
Pant 1997	32/115	330/1172		17.19	0.99 [0.73, 1.35]
Pant 1998	6/21	22/100		2.23	1.30 [0.60, 2.81]
Singh 1995	82/328	467/1767		42.63	0.95 [0.77, 1.16]
Thakur and Verma 1992	6/48	50/210		5.42	0.53 [0.24, 1.15]
Subtotal (95% CI)	726	6246		100.00	0.90 [0.78, 1.03]
Total events: 175 (O in malaria), 1706 (O in no malaria)					
Test for heterogeneity: Chi² = 9.66, df = 6 (P = 0.14), I² = 37.9%					
Test for overall effect: Z = 1.58 (P = 0.11)					

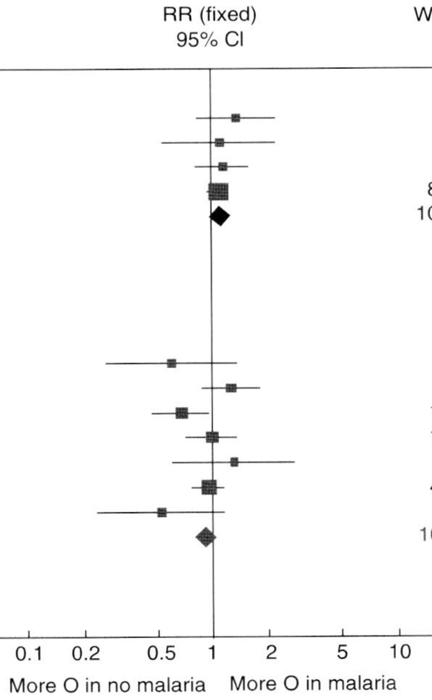

0.1 0.2 0.5 1 2 5 10
More O in no malaria — More O in malaria

FIGURE 1.6 Risk estimates for malaria infection versus no malaria in blood group O versus non-O phenotypes (cross-sectional studies).

higher in A and B patients. Data from Thakur and Verma (1992) also showed higher malaria prevalence in group A individuals. The remaining studies in India showed no blood group associations (Pant and Srivastava, 1997; Pant *et al.*, 1992b, 1998; Singh *et al.*, 1995). Among African studies, four showed no significant differences in infection risk (Akinboye and Ogunrinade, 1987; Domarle *et al.*, 1999; Ibhanesebhor *et al.*, 1996; Kassim and Ejezie, 1982); although a fifth which examined *P. falciparum* malaria in Gabonese school children showed a higher prevalence of asymptomatic malaria in blood group O (44.8%) versus non-O (29.6%) ($p = 0.05$), suggesting a protective effect of the O antigen against clinical forms of malaria (Mombo *et al.*, 2003). In seven case-control studies (Table 1.2; Fig. 1.7), only the study of Joshi *et al.* (1987) found a significant difference with AB phenotype protective against malaria. The other six case-control studies showed no significant differences (Bayoumi *et al.*, 1986; Cavasini *et al.*, 2006; Facer and Brown, 1979; Montoya *et al.*, 1994; Osisanya, 1983; Santos *et al.*, 1983). Early data by Parr (1930) reported increased risk in group A versus non-A, although the *Plasmodium* species studied is not stated. Among the three cohort studies, two showed significant differences. Migot-Nabias *et al.* (2000) reported that blood group O protected against higher parasitaemia in asymptomatic Gabonese school children (Student *t*-test, $p = 0.043$), and a cohort study undertaken in Brazil (Beiguelman *et al.*, 2003) found a significant association between individuals with A and/or B antigens and the number of malaria episodes (Kruskal–Wallis 4.054, $p = 0.044$). Conversely, Molineaux and Gramiccia (1980) in Nigeria found no differences in malaria episodes for the different phenotypes. Hill (1992) found a protective effect associated with O phenotype in Gambians. Among five studies with no detailed data, three of them showed no significant differences (Farr, 1960; Pant *et al.*, 1993; Vasantha *et al.*, 1982). A large study undertaken in Russia by Rubaschkin and Leisermann in 1929 and reported by Parr (1930), showed a very significant association between malaria and the ABO blood groups, with AB having a greater frequency in malaria patients. Hill (1992) also reported differences in malaria prevalence for the different ABO phenotypes (Table 1.4). Mourant *et al.* (1978) presented a review of early case-control data, with two studies from Italy showing increased malaria risk in blood group A, one of them associated to a reduced risk for B phenotype. This data is not included in the tables of this review because the *Plasmodium* species was not specified, and most studies were carried out in areas not endemic for *P. falciparum*. The summary of these early data is also reviewed by Singh *et al.* (1986).

The evidence is not conclusive about group A individuals and susceptibility to *P. falciparum* infection and disease. The meta-analysis shows a low heterogeneity for the groups of studies looking at risk of infection (Fig. 1.6 and 1.7). The overall effects of O blood group versus non-O in

TABLE 1.2 Odds ratio for ABO phenotype by malaria category in case-control studies

Reference	Country	Age range	Urban/ rural	Place	Inclusion criteria	Total Case / Control	N	A Odds ratio (A vs O) (95% CI)	N	B Odds ratio (B vs O) (95% CI)	N	AB Odds ratio (AB vs O) (95% CI)	O N
Bayoumi *et al.*, 1986[a]	Sudan	10–54	R	HC	Malaria	93	31	1.24	21	1.04	2	1.28	39
		13–70	R	Comm	Resistance	75	16	0.53–2.94	13	0.41–2.66	1	0.08–37.83	25
Facer and Brown, 1979[b]	The Gambia	Children	R?	HC	Malaria	80	20	0.93	16	0.72	4	0.74	40
		Children	R?	School	Non-malarious	165	40	0.45–1.88	41	0.34–1.52	10	0.18–2.80	74
Joshi *et al.*, 1987[c]	India (Delhi)	10–60	U	HC	Malaria	85	23	0.87	31	0.99	3	**0.25***	28
			U	Staff/HC	Non-malarious[†]	244	66	0.43–1.75	78	0.52–1.90	30	**0.06–0.96**	70
Osisanya, 1983[d]	UK	NS	U	HC	Malaria	73	20	0.6	10	0.77	1	0.35	42
		NS	U	Reference	Expected ABO distribution	38901	23	0.26–1.38	9	0.25–2.38	2	0.01–5.22	29
Parr, 1930[e]	NS	NS	NS	NS	Malaria	279	130	1.29	31	0.89	29	1.08	80
		NS	NS	Reference	Expected ABO distribution	6696	116	0.86–1.94	40	0.49–1.61	31	0.57–2.02	92
Cavasini *et al.*, 2006[f]	Brazil	>18	R	NS	Malaria	121	25	0.83	16	1.02	2	0.47	78
		>18	R	Donor	Non-malarious	417	98	0.48–1.42	51	0.53–1.97	14	0.07–2.21	254
Santos *et al.*, 1983[g]	Brazil	NS	U + R	HC	Malaria	495	158	1.08	54	1.42	18	1.84	265
		NS	U	Comm	Non-malarious	423	135	0.80–1.45	35	0.88–2.31	9	0.77–4.52	244

NS, not stated; HC, health centre; Comm, community sample; numbers in bold, statistically significant odds ratio (OR) and confidence interval (CI); [*] $p < 0.05$; [†] only non-malarious subjects considered as controls, *P. vivax* parasitaemia excluded.

[a] Ninety-three patients were selected from OPD clinics with confirmed malaria and history of repeated episodes. Seventy-five cases claiming resistance to malaria were found in a community survey.

[b] Blood from 250 children from Western division of The Gambia was ABO typed. Eighty presented with *P. falciparum* malaria at the OPD clinic. The rest were primary school children.

[c] Eighty-five patients with fever attending the malaria clinic in Delhi had *P. falciparum* parasitaemia, and a group of 171 febrile patients and 73 members of MRC staff are the control group. *Note:* 253 *P. vivax* cases were excluded from the analysis.

[d] One hundred and eight blood/serum samples were obtained from patients attending the London Hospital for Tropical Diseases and the malaria reference laboratory of the London School of Hygiene and Tropical Medicine. The patients were divided according to their ethnic background. Data from Asian patient is not included in the table since most cases were *P. vivax* infections. Most infections in the table were by *P. falciparum*, except 12 *P. vivax*, 2 *P. ovale* and 1 *P. malariae*. The controls were obtained from previous studies and a reference population.

[e] Early data. Subject selection and malaria type is not specified. Controls are expected cases according to blood group distribution in the general population. Group A versus non-A nearly significant 1.38 (95% CI 0.98–1.96) p 0.05.

[f] Four hundred and seventeen blood-smear negative donors and 409 patients with confirmed clinical malaria, all older than 18 years, were selected from four areas of Brazilian Amazon. One hundred and twenty-one of the patients had *P. falciparum* malaria. For the 288 cases with *P. vivax* infection, no association was found with ABO phenotypes (data not included in the table).

[g] Two hundred and sixty-five persons affected by malaria were detected by public health personnel or presented at the Hospital of Tropical Diseases in Manaos (Brazil). Controls were 244 apparently healthy persons who were matched by sex, age and morphological racial characteristics with the cases.

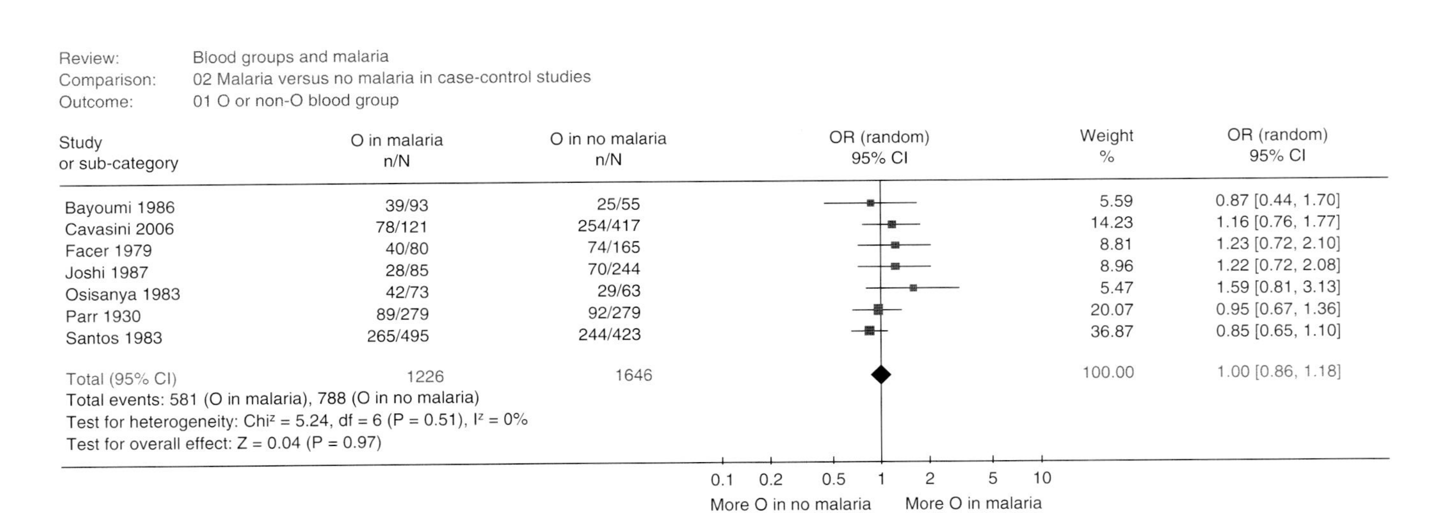

FIGURE 1.7 Risk estimates for malaria infection versus no malaria in blood group O versus non-O phenotypes (case-control studies).

malaria risk are not significant. For prevalence studies, the overall relative risk shows an opposite tendency in African compared with Asian studies. Although many studies showed small differences, the most relevant finding is that there was consistency in the findings of A, B or AB being associated with increased malaria risk and the O phenotype appearing as protective factor in some studies (Table 1.4). Increased susceptibility associated with the A phenotype could also influence *P. vivax* infection, as reported by Gupta and Chowdhuri (1980).

4.3. Severe malaria

We identified eight studies examining the association of ABO phenotypes with severe malaria (Table 1.3; Fig. 1.8). Five of them were undertaken in Africa and three in the Asian-Pacific region. A ninth study was not included in any of the tables because missing information on the blood group type distribution did not allow a clear interpretation (Waiz and Chakraborty, 1990).

4.3.1. Africa

The earliest study was undertaken by Martin *et al.* (1979) in Ibadan (Nigeria). They reported that the prevalence of ABO groups in 91 children presenting with severe malaria (defined as fever plus high parasitaemia, mean 152,000 $\pm$ 17,000 parasites/μl) was similar to healthy adults from the same Nigerian population. In a study from Gabon (Lell *et al.*, 1999), blood group (among other RBC polymorphisms) was determined in 100 children with severe malaria matched with 100 mild malaria cases. Blood group A was significantly associated with severe malaria compared with blood group O (OR 2.91, 95% CI 1.24–6.92), as well as with a trend towards higher parasitaemia, but not with haemoglobin level. Blood group O was seen more often in the mild group. Among patients in this study, a higher frequency of rosette-forming isolates was found in the severe malaria group. In a Zimbabwean study of 489 patients (Fischer and Boone, 1998), those with malaria and blood group A had lower haemoglobin levels and a higher risk of coma than infected patients with other blood group types (OR A vs non-A 3.22, 95% CI 1.19–8.66). Groups AB and O had significantly higher levels of Hb. Prevalence of the clinical features of seizures, confusion, jaundice, hospitalisation and death did not differ significantly among blood group types. Heddini *et al.* (2001) reported a significant correlation between malaria severity and blood group A ($p = 0.031$), cerebral malaria ($p = 0.014$) but not severe anaemia ($p = 0.145$), in a study which examined the parasitised erythrocytes of 111 fresh isolates of children with malaria from Kilifi in Kenya. In 1012 Ghanaian children with severe malaria, it was observed that blood group O was associated with reduced levels of blood lactate, raised levels

TABLE 1.3 Odds ratio for ABO phenotype by severe malaria category

				Total		A		B		AB	O
				Case	*N*	Odds ratio (A vs O)	*N*	Odds ratio (B vs O)	*N*	Odds Rario (AB vs O)	*N*
Reference	Country	Age range	Inclusion criteria	Control	*N*	95% CI	*N*	95% CI	*N*	95% CI	*N*
Martin *et al.*, 1979[a]	Nigeria	Children NS	Malaria high count	91	20	1.05	21	1.00	3	0.75	47
		Adults	*Reference population*	91	19	0.47–2.37	21	0.45–2.20	4	0.12–4.27	47
Lell *et al.*, 1999[b]	Gabon	Children	Severe	100	27	**2.91***	–	–	–	–	54
		Children	Uncomplicated	100	11	**1.24–6.92**	–	–	–	–	64
Fischer and Boone, 1998[c]	Zimbabwe	NS	Confusion or coma	57	14	1.37	13	1.23	2	1.31	28
		NS	No coma	428	87	0.65–2.85	90	0.57–2.60	13	0.0–6.59	238
Donkor *et al.*, 2004[d]	Ghana	6 months–6 years	Lactic acidosis (lactate > 5 mmol/l)	255	58	1.17	93	**1.75**	18	1.61	86
			Malaria without lactic acidosis	753	187	0.79–1.74	200	**1.23–2.50***	42	0.85–3.06	324
Singh, 1985[e]	India	Children NS	Admitted malaria	170	63	**1.93***	59	1.23	9	1.08	39
		Children NS	Healthy	200	51	**1.08–3.46**	75	0.70–2.16	13	0.38–3.04	61
Pathirana *et al.*, 2005[f]	Sri Lanka	All	Severe malaria	80	26	**2.67***	22	**2.44***	13	**6.67‡**	19
		All	Uncomplicated	163	40	**1.25–5.74**	37	**1.11–5.39**	8	**2.18–20.87**	78
Al-Yaman *et al.*, 1995[g]	PNG	Children NS	Cerebral	103	29	0.64	10	0.43	14	0.85	44
		Children NS	Uncomplicated	158	55	0.33–1.21	28	0.17–1.05	20	0.35–2.00	53

NS, not stated; numbers in bold, statistically significant OR and CI; $^{*}p < 0.05$; $^{\ddagger}p < 0.01$.

[a] Ninety-one children presenting to the OPD between June and September 1976, with fever and parasitaemia 152,000 + 17,000. No data on ABO is available from controls. Figures are percentages from a reference adult population for the area.

[b] Cases were 100 children with severe malaria (hyperparasitaemia > 250,000 ul^{-1}, Hb < 50 g/litre and other signs). Controls were age, sex and area matched with cases (mean age 44 months). Uncomplicated malaria defined by: parasitaemia 1,000–50,000, Hb > 8.0 g/litre, platelets > 50 nl^{-1} and schizontaemia zero.

[c] Four hundred and eighty-nine patients with malaria were selected for the study. Hb level was significantly different: lower in A and B. Risk of coma was higher for group A patients compared with non-A group. The data presented in the Table 3 is unpublished and is an analysis of the original data provided by the author.

[d] Unpublished data. One thousand and eight children from 6 months to 6 years admitted in Komfo Anokye Teaching Hospital in Kumasi (Ghana) with WHO criteria of severe malaria.

[e] Cases: 170 children admitted to the hospital with malaria without other associated problems. Controls: 200 normal children from the same area, age and sex matched.

[f] Cases: 103 children with CM (WHO) admitted at Madang General Hospital. Controls: 158 children with uncomplicated malaria.

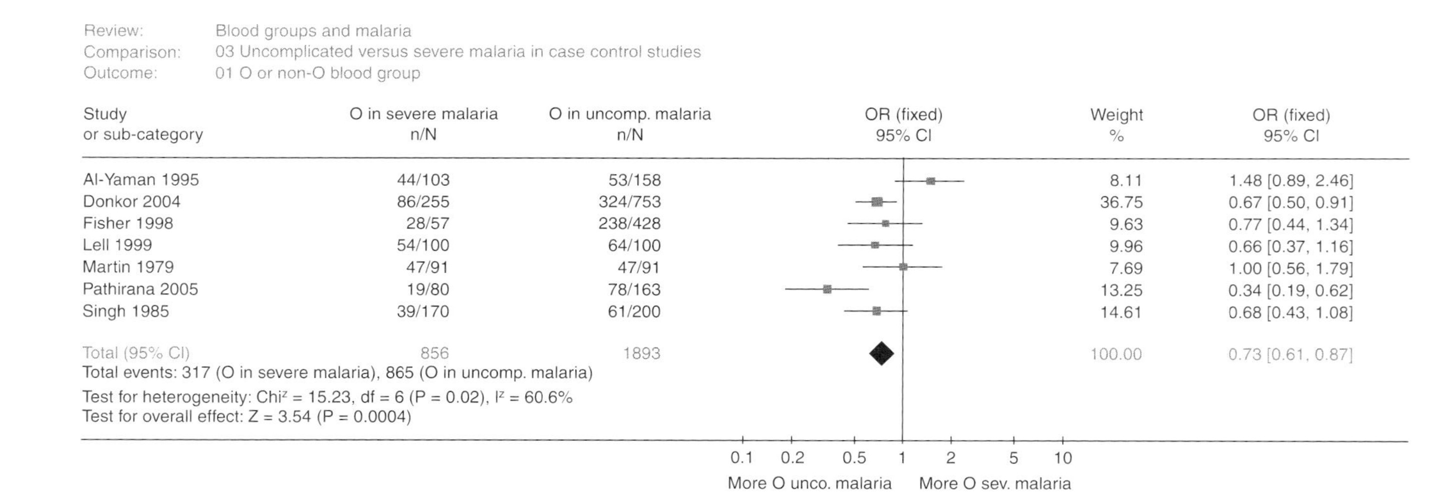

FIGURE 1.8 Risk estimates for uncomplicated versus severe malaria severity in blood group O versus non-O phenotypes.

TABLE 1.4 Summary ABO phenotypes identified as risk factors

Reference		Age	*N*	Outcome	Risk factor	Protective phenotype
Risk of infection: Cross-sectional studies						
Akinboye and Ogunrinade, 1987	Nigeria	Adults	115	Parasitaemia	–	–
Ibhanesebhor *et al.*, 1996	Nigeria	Adults	50	Parasitaemia	–	–
Kassim and Ejezie, 1982	Nigeria	Children	681	Parasitaemia	–	–
Domarle *et al.*, 1999	Gabon	Children	61	Parasitaemia	–	–
Mombo *et al.*, 2003	Gabon	Children	271	Parasitaemia	–	O
Pant *et al.*, 1992a	India	NS	783	Malaria (Pf, Pv)	AB	–
Pant *et al.*, 1992b	India	All	769	Malaria (Pf, Pv)	–	–
Pant *et al.*, 1993	India	All	1659	Parasitaemia	A and B	–
Thakur and Verma, 1992	India	All	258	Parasitaemia	A	–
Singh *et al.*, 1995	India	NS	2095	Malaria	–	–
Pant and Srivastava, 1997	India	All	1287	Malaria (Pf, Pv)	–	–
Pant *et al.*, 1998	India	All	121	Malaria (Pf, Pv)	–	–
Risk of infection: Case-control studies						
Bayoumi *et al.*, 1986	Sudan	All	93/75	Malaria susceptibility	–	–
Facer and Brown, 1979	The Gambia	Children	80/165	Malaria	–	–
Joshi *et al.*, 1987	India	All	85/244	Malaria	–	AB

(*continued*)

TABLE 1.4 *(continued)*

Reference		Age	*N*	Outcome	Risk factor	Protective phenotype
Osisanya, 1983	UK	NS	73/63	Malaria	–	–
Parr, 1930	Lebanon	NS	279/6696	Malaria	A	–
Cavasini *et al.*, 2006	Brazil	Adults	121/417	Malaria	–	–
Santos *et al.*, 1983	Brazil	NS	495/423	Malaria	–	–
Montoya *et al.*, 1994	Colombia	NS	412	Malaria	–	–
Risk of infection: Cohort studies						
Molineaux and Gramiccia, 1980	Nigeria	All	3534	Malaria episodes	–	–
Migot-Nabias *et al.*, 2000	Gabon	Children	300	Parasitaemia	–	O
Beiguelman *et al.*, 2003	Brazil	All	182	Malaria episodes	A and B	–
Risk of infection: No detailed data						
Hill, 1992	The Gambia	NA	NA	Malaria	–	O
Vasantha *et al.*, 1982	India	NA	NA	Malaria	–	–
Farr, 1960	UK	Adults	112/8257	History of malaria		
Rubaschkin and Leisermann, 1929 quoted by Parr, 1930	Russia	NS	4397/13484	Malaria	AB	O
Mourant *et al.*, 1976	–	NS	Review	NS	A, AB	O
Malaria severity studies						
Martin *et al.*,1979	Nigeria	Children	91/100	High parasite count	–	–
Lell *et al.*, 1999	Gabon	Children	100/100	Severe malaria	A	–

Fischer and Boone, 1998	Zimbabwe	NS	20/469	Malaria (coma)	A	O
Heddini *et al.*, 2001	Kenya	Children	55/56	Cerebral malaria	A	–
Donkor *et al.*, 2003*	Ghana	Children	978	Lactic acidosis	B	O
Singh, 1985	India	Children	170/200	Admitted malaria	A	–
Pathirana *et al.*, 2005	Sri Lanka	All	80/163	Severe malaria	–	O
Al-Yaman *et al.*, 1995	PNG	Children	103/158	Cerebral malaria	–	–

NA, not available; NS, not stated; dash signifies no ABO phenotypic association identified.
*Unpublished data.

of which are a marker for disease severity (Donkor, C., Loscertales, M.P., Osei-Akoto, A., and Bunn, J., 2004, unpublished data).

4.3.2. Asia and Pacific

Singh (1985) compared the frequencies of ABO blood group types among 170 children admitted with malaria (*P. falciparum*, 54.1%), and 200 age- and sex-matched healthy controls from the same area. The relative risk of malaria (both *P. vivax* and *P. falciparum*) in group A versus O was 1.93 (95% CI 1.08–3.46). Pathirana *et al.* (2005), in a study from Sri Lanka found that cases of severe malaria were significantly less likely to have blood group O ($p = 0.0003$), and significantly more likely have group AB ($p < 0.0001$) than patients with non-severe malaria. Al-Yaman *et al.* (1995) studied the association between rosetting and malaria severity in Papua New Guinean children. A total of 103 cerebral malaria cases (established using World Health Organization criteria) were recruited and 158 patients with uncomplicated malaria. ABO blood group (together with *P. falciparum* parasitaemia and mean temperature) did not differ significantly between the two groups ($p = 0.17$) (Table 1.3). Blood group and parasite density were associated with rosetting, as group AB had a significantly higher geometric mean rosetting rate than other groups ($p = 0.007$). No relationship was observed between rosetting and cerebral malaria.

Despite the heterogeneity in the indicators used to assess severity, most studies found significant differences that were consistent with the association of the A phenotype with a higher risk and O phenotype with reduced risk for malaria severity (Table 1.4). Among the two studies that showed no significant differences, one of them (Martin *et al.*, 1979) used a high parasite count as the criteria for severity and not a clinical definition of severe malaria, which makes questionable its inclusion in this group. The overall effect of blood group on malaria severity is significant ($p = 0.0004$) with reduced risk for O blood group (Fig. 1.8). The test for data heterogeneity in the meta-analysis was significant limiting the strength of this conclusion.

4.4. Placental malaria

In the only published analysis of ABO phenotypes in relation to placental malaria from the Gambia, blood group O was associated with an increased prevalence of active placental infection in primiparae and with reduced risk of placental malaria in multiparae (Loscertales and Brabin, 2006). Placental *P. falciparum* parasitaemia occurred twice as frequently in primiparae, but only among blood group O women. This effect of parity, which is one of the cardinal features of placental malaria was not observed in non-O phenotypes. In multiparae and specifically in mothers with past or non-infected placentae, blood group O was also associated with higher foeto-placental weight ratios compared with non-O individuals. Similar findings have been observed in a study from Malawi which showed blood group O was a risk factor for increased malaria risk in first pregnancies and reduced risk in later pregnancies (Senga, E.L., Loscertales, M.P., Makwakwa, D.E.B., Liomba, G.N., Dzamalala, C., Kazembe, P., and Brabin, B.J., 2007, unpublished observation).

5. BIOLOGICAL BASIS FOR ABO PHENOTYPE-RELATED SUSCEPTIBILITY TO MALARIA

Host factor susceptibility, both acquired and inherited, plays a role in malaria infection and severity. The first evidence of the importance of inherited host factors was the limited host range of the various Plasmodia species. A growing number of genes have been described that are associated with *P. falciparum* malaria, but their contribution relative to the environmental factors that also influence malarial risk has rarely been estimated. Using a pedigree-based genetic variance component analysis, Mackinnon *et al.* (2005) found that genetic and unidentified household factors each accounted for around one quarter of the total variability in malaria incidence in their study population. This genetic effect was beyond that attributable to the haemoglobinopathies alone, suggesting the existence of many protective genes each causing small population effects.

5.1. Affinity for *Anopheles gambiae*

Anopheles gambiae can effectively and consistently express host-selection behaviour that results in non-random biting. An association between ABO group and attractiveness for *A. gambiae* has been described with preferential feeding for blood group O subjects (Wood, 1975; Wood *et al.*, 1972), although evidence for this was not found by other investigators (Dore *et al.*, 1975). At high levels of transmission and immunity, even sizable differences between persons in attractiveness to vectors may have no easily detectable parasitological or serological expression. There has been little research on this topic: for example, the relationship between the ABO blood groups and anti-sporozoite antibodies has not been studied (Molineaux, 1988).

5.2. Shared antigens

Some antigenic determinants of the ABO system are shared by malaria parasites as well as by other infectious agents (Athreya and Coriell, 1967). It has been shown that malaria parasites share group A antigens (Oliver-Gonzalez and Torregrossa, 1944) and may be better tolerated by hosts that have the A blood group antigen. However, the evidence for the presence of blood group antigens in malaria parasites is mostly indirect. There is an increase of anti-A and anti-B titre in group O individuals exposed to malaria (Kano *et al.*, 1968). These titres are higher in subjects living in a malaria endemic area or who have suffered several attacks than in recently infected patients (Druilhe *et al.*, 1983; Oliver-Gonzalez and Torregrossa, 1944). It has been suggested that individuals with the O phenotype may be selectively advantaged because they have both anti-A and anti-B antibodies shared by microorganisms, which could explain the persistence of the O gene (Saitou and Yamamoto, 1997). An early study by Raper (1968) found no differences in the level of attained parasitaemia (high vs normal) in 3500 patients after successful inoculation of malarious blood for the treatment of neuro-syphilis, in relation to their ability to produce anti-A or anti-B activity. Several early studies showed a significantly longer incubation period in blood-induced *P. vivax* malaria (as a treatment for syphilis) when blood groups of donor and recipient were incompatible, although this evidence was controversial (Myatt *et al.*, 1954).

The *in vivo* observation of phagocytes engulfing normal RBCs (Russel *et al.*, 1963), probably related to the frequent finding of positive Coombs antiglobulin reactions in malaria patients (Facer *et al.*, 1979), suggests that immunological factors are important in the development of malaria-related anaemia. Various mechanisms have been suggested to explain the positive direct Coombs reaction including fixation of immune complexes to the RBC, adsorption of malaria antigens to the surface of normal RBC and sensitisation by anti-RBC antibodies (Casals-Pascual and Roberts, 2006; Facer, 1980). It is proposed that during malaria infection, infected and non-infected RBC become sensitised to anti-RBC antibodies, for example, anti-D (Lee *et al.*, 1989). Anti-RBC antibodies may be addressed to cryptic antigens of the RBCs that have been exposed following changes in the sialic acid of the membrane caused by the parasite (Zouali *et al.*, 1982). It has also been proposed that cold agglutinins (Penalba *et al.*, 1984) of the IgM class may lead to fixation of C3d on RBC and hence to haemolytic anaemia yet in childhood malaria only insignificant titres of cold agglutinins have been detected (Facer *et al.*, 1979). The cause of direct Coombs (DAT) positivity in West African (Gambian) children was investigated and the results obtained provided evidence that the antibodies were directed against *P. falciparum* and not to the RBC, suggesting adsorption of malaria antigens as the main hypothesis (Facer, 1980).

5.3. Inflammatory response

Protection against asexual blood stages is mediated by pro-inflammatory responses, mainly related to tumour-necrosis factor-alpha (TNF-α), interferon-γ (IFN-γ) and nitric oxide (NO). In parallel with these responses, different clinical presentations of severe malaria are associated with specific patterns of inflammatory mediators. In particular, children with respiratory distress have significantly higher plasma levels of TNF-α, IL-10 and neopterin and a significantly higher TNF-α:IL-10 ratio than those without respiratory distress (Awandare *et al.*, 2006). This may explain the pathogenesis of the underlying metabolic acidosis. Circulating levels of neopterin have been inversely correlated with haemoglobin. Neopterin is produced by human monocyte-derived macrophages on stimulation with IFN-γ and is a sensitive indicator for cell immune activation. Neopterin levels are significantly higher in donors with blood group O than other phenotypes (Murr *et al.*, 2005). No associations have been reported between ABO blood group and other pro-inflammatory cytokines related to malaria pathogenesis.

5.4. Invasion of RBCs

Invasion of RBCs is mediated by specific receptors (Miller *et al.*, 1973). The invasion process is a sequence of steps that start with merozoite reversible attachment to the RBC surface, followed by apical reorientation, formation of an irreversible junction and entry into a parasitophorous vacuole (Fig. 1.9). Different blood groups could mediate or act as receptors for invasion, as has been clearly shown for the Duffy blood group in *P. vivax* infection. However, *P. vivax* and *P. falciparum* have different receptors on the RBC membrane (Gaur *et al.*, 2004) as suggested by early epidemiological studies (Spencer *et al.*, 1978; Welch *et al.*, 1977). The fact that there is no single RBC polymorphism which confers complete resistance to *P. falciparum* merozoite invasion suggests that the parasite has ligands for an essential protein or for an abundant membrane substance (e.g., sialic acid) which could not be suppressed without compromising the RBC structure or function. Part of the difficulties of describing *P. falciparum* invasion pathways are attributable to major strain differences in ligands as described by Hadley *et al.* (1987) and Duraisingh *et al.* (2003).

Some of the molecules involved in the different phases of RBC invasion are associated in some extent with ABO phenotypes. The first step of invasion is the attachment of the merozoite to the RBC surface that is reversible and host cell specific. There is significant preference for *P. falciparum* invasion into blood group A_1 erythrocytes compared with A_2, B or O erythrocytes. This preference for sub-group A_1 erythrocytes was observed in parasites utilising both sialic acid-dependent and sialic acid-independent invasion pathways and could be due to differences in initial parasite attachment to the RBCs (Chung *et al.*, 2005). A similar study failed to find any invasion preference by ABO phenotype (Osisanya, 1983). Merozoite surface proteins (MSPs), especially MSP-1, have been related to initial attachment during the invasion process

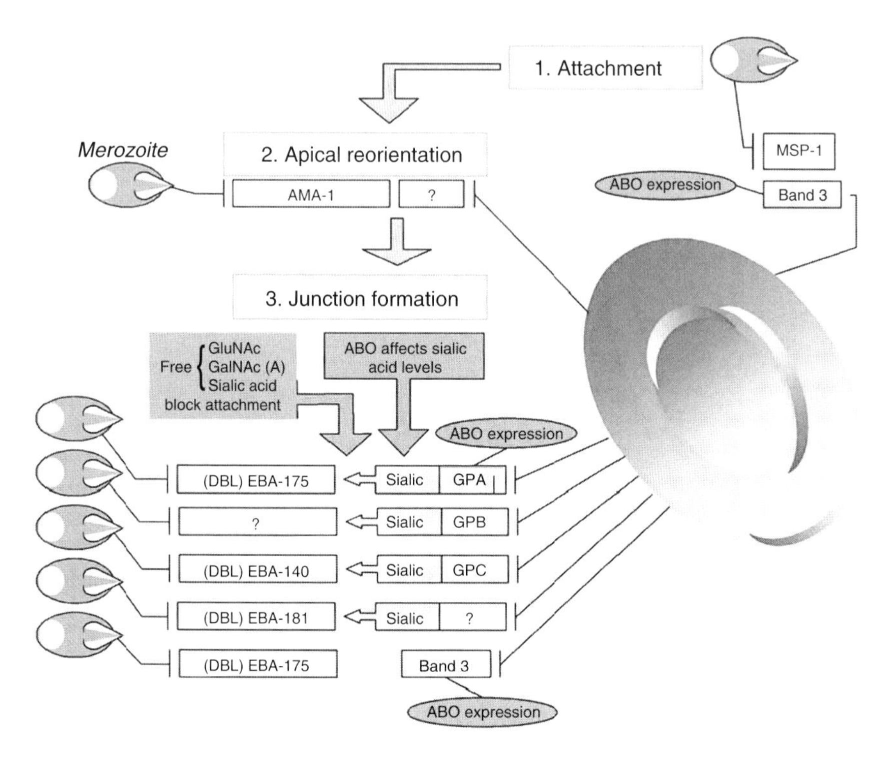

FIGURE 1.9 ABO associations with RBC invasion pathways. This shows ABO expression sites for three stages of RBC invasion by *P. falciparum.* MSP-1, merozoite surface protein 1; AMA-1, apical membrane antigen 1; EBA: erythrocyte-binding antigen; DBL, Duffy-binding-like; GPA, glycophorin A; GPB, glycophorin B; GPC/D, glycophorin C/D; GluNAc, *N*-acetyl-glucosamine; GalNAc, *N*-acetyl-galactosamine.

(Gaur *et al.*, 2004). MSP-1 binds RBCs in a sialic acid-dependent manner and if associated with MSP-9, they form a co-ligand complex that binds to band 3 (Li *et al.*, 2004). Band 3, a membrane protein that acts as an anion carrier, may be a receptor for *Plasmodium* invasion of RBCs (Okoye and Bennett, 1985) and conjointly with GPA expresses the Diego blood group antigens (Telen and Chasis, 1990) that have also been proposed as parasite receptors (Ridgwell *et al.*, 1983). Most importantly, band 3 is known to carry the ABH antigens. Deletion in band 3 gene results in southeast Asian ovalocytosis, that confers some resistance to *P. falciparum* invasion (Mgone *et al.*, 1996), and may protect against cerebral malaria (Allen *et al.*, 1999) (Table 1.5).

Apical reorientation is the second step during merozoite invasion. Several studies suggest that the *P. falciparum* apical membrane antigen 1 (AMA-1), a protein localised in the apical organelles of the merozoite, is involved in the invasion of human RBCs (Gaur *et al.*, 2004) and its

TABLE 1.5 ABO associations with molecules involved in RBC invasion pathways

P. falciparum ligands	Sialic acid dependent	RBC receptor	Protective RBC receptor polymorphisms	Population affected by RBC polymorphism	Effect of polymorphism		Known ABO relationship with pathway molecules
					In vitro	*In vivo*	
MSP-1	No	Band 3	Southeast Asian ovalocytosis	Papua New Guinea	Reduced invasion	Protects against cerebral malaria	Band 3 expresses ABO
AMA-1	No	?					None
EBA-175	Yes	GPA	En(a−): [absent GPA]		Reduced invasion		GPA expresses ABO determinants
			Wrb− [absent Wrb antigen]		Reduced invasion?		Sialic acid level differs between ABO phenotypes
?	Yes	GPB	Dantu [hybrid GPA–GPB]	Southern African	Reduced invasion		Sialic acid level and composition differ between ABO phenotypes

(continued)

TABLE 1.5 (*continued*)

P. falciparum ligands	Sialic acid dependent	RBC receptor	Protective RBC receptor polymorphisms	Population affected by RBC polymorphism	Effect of polymorphism: *In vitro*	Effect of polymorphism: *In vivo*	Known ABO relationship with pathway molecules
			S-s-U- [absent GPB]	Africa	Reduced invasion		
EBA-140 (BAEBL)	Yes	GPC/D	Gerbich [absent GPD, truncated GPC] (elliptocytosis)	Papua-New Guinea	Reduced invasion	Reduced infection?	Sialic acid level differs between ABO phenotypes
			Leach [absent GPC/D]		Reduced invasion		
EBA-181 (JESEBL)	Yes	Unknown					Sialic acid level differs between ABO phenotypes
PfRh1	Yes	Unknown					Sialic acid level differs between ABO phenotypes
PfRh2b	No	Unknown					

MSP-1, merozoite surface protein 1; AMA-1, apical membrane antigen 1; EBA, erythrocyte-binding antigen; PfRh, *P. falciparum* reticulocyte-binding protein homologue; GPA, glycophorin A; GPB, glycophorin B; GPC/D, glycophorin C, D.

erythrocyte-binding sites are not sialic acid dependent (Urquiza *et al.*, 2000). AMA-1 may be directly associated with apical reorientation (Mitchell *et al.*, 2004). No relation has been found in the literature between AMA-1 and ABO blood groups.

After apical re-orientation of the merozoite, parasite ligands and erythrocyte receptor molecules lead to irreversible junction formation and mediate erythrocyte invasion. The merozoite ligands are located in the apical organelles and they belong to two protein superfamilies: the Duffy-binding-like (DBL) proteins of *P. falciparum* (Gaur *et al.*, 2004) and the reticulocyte-binding-like (RBL) protein family. Glycophorins and sialic acid are common receptor elements of those invasion pathways. In the first pathway described, receptor-binding residues of parasite ligands (as the DBL parasite protein EBA-175) bind sialic acid on glycophorin A for invasion (Orlandi *et al.*, 1992), indicating that parasites using EBA-175 need both glycophorin A and the attached carbohydrates (Sim *et al.*, 1994) and explaining the resistance to invasion described for En(a) erythrocytes (Pasvol *et al.*, 1982a) which lack glycophorin A and possibly for Wrb− cells, a polymorphism of Diego blood group expressed on the GPA molecule (Facer and Mitchell, 1984; Pasvol *et al.*, 1982b). Other invasion pathways are also sialic acid dependent and use glycophorin B (GPB) (Dolan *et al.*, 1994) or glycophorin C (GPC) and glycophorin D (GPD) (Mayer *et al.*, 2001). Accordingly, some blood group variants related to these glycophorins show a protective effect on malaria invasion as the Dantu and S-s-U- phenotypes (MNSs variants related to GPB) (Facer, 1983; Field *et al.*, 1994; Vos *et al.*, 1971); the GPC variant Gerbich phenotype (Ge-2,-3) (Booth and McLoughlin, 1972; Pasvol *et al.*, 1984; Patel *et al.*, 2001; Serjeantson, 1989) and the Leach type RBCs (Ge-2,-3) null for GPC/D proteins that is also resistant to merozoite invasion (Chishti *et al.*, 1996). Another invasion pathway associated to sialic acid is related to the DBL protein EBA-181/JESEBL that can recognise multiple receptors on the erythrocyte (Mayer *et al.*, 2004). The protein PfRh1 that belongs to the RBL protein family, also binds to erythrocytes in a sialic acid-dependent manner but using a receptor distinct from glycophorins and known as receptor Y (Rayner *et al.*, 2001). Other RBP homologues (as PfRh4) have been implicated in erythrocyte binding, but this activity has not been demonstrated (Duraisingh *et al.*, 2003).

There are some points of associations between ABO phenotypes and invasion pathways involving glycophorins and/or sialic acid. Podbielska *et al.* (2004) have demonstrated the presence of minor amounts of ABH determinants in the O-glycans of glycophorin A. However, there is no evidence in the literature of ABH determinant expression in GPB or GPC. Sugar components present on glycophorin, in particular *N*-acetyl-glucosamine (GluNAc) and *N*-acetyl-galactosamine (GalNAc) appear to act as important determinants for attachment to the erythrocyte (Facer, 1983), and together with sialic acid, they can specifically block parasite invasion *in vitro* as free monosaccharides, suggesting that merozoites bind to the RBCs in a lectin-like fashion (Jungery, 1985). This may be very relevant since GalNAc is the terminal sugar of the A antigen, and could partly explain the increased susceptibility for A phenotype. The association between particular sugars attached to GPA or GPB and merozoite invasion is complex, since Cad erythrocytes which have normal sialic content but present additional GalNAc residues attached to each O-linked glycan in GPA and GPB resist invasion by *P. falciparum* (Cartron *et al.*, 1983). Sialic acid that is common to some pathways that use glycophorins as ligands has also been related to ABO phenotypes. Sialic acid is one of the essential molecules for parasite attachment and it has even been suggested that the human–chimpanzee differences in malaria susceptibility are related to the genetic changes in sialic acid molecules present in glycophorins (Martin *et al.*, 2005). The erythrocyte membrane in β-thalassemia has lower sialic acid levels in glycophorin (Kahane and Rachmilewitz, 1976) and Tn cells deficient in sialic acid and galactose are particularly resistant to *P. falciparum* invasion (Pasvol *et al.*, 1982b). Membrane sialic acid content is reported as higher in group O; with a higher percentage bound to glycolipids in this phenotype but a lower percentage associated to sialoglycoproteins. These differences were independent of the Rhesus grouping. The sialic acid densities in sialoglycoproteins and the particular classes of sialic molecules also differ from one erythrocyte type to another (Bulai *et al.*, 2003; Udoh, 1991). GluNAc, a component of glycophorins A and C but not B is an effective inhibitor of RBC invasion. Sialic acid is almost as effective as GluNAc, presumably because of the structural similarities between these sugars. The inhibitory ability of sialic acid is considerably enhanced when presented to the parasite in a clustered form, as in an oligosaccharide (Vanderberg *et al.*, 1985). We can speculate that the binding of EBA-175 and other ligands may be optimal if particular 'clustered saccharide patches' are present in GPA rather than isolated oligosacharidae or proteins, as suggested by these studies. This will also explain why those specific oligosacharidae do not inhibit binding of *Plasmodium* to RBCs when they are purified or attached to another Glycophorin (Mayer *et al.*, 2006). In that case, a minimum change in the overall saccharide cluster composition, as happens for the different ABO phenotypes, may impair invasion. The high prevalence among Africans of ABO phenotype variants associated to higher expression of the H antigen (see the earlier description) could also contribute to a change in the saccharide profile of the cell, with reduced A antigen expression and changes in sialic acid amount and composition that may have a protective effect on invasion.

In summary, there is a high diversity of invasion pathways and ligands for *P. falciparum* invasion. It has been suggested that the diversity of surface glycan molecules represents a mechanism of pathogen evasion by long-lived vertebrates such as humans (Gagneux and Varki, 1999). Consequently, *P. falciparum* may have evolved polymorphisms generated by single-point mutations in their receptors to use the diversity in carbohydrate ligands (Mayer *et al.*, 2006). Table 1.5 summarises ABO associations with molecules involved in RBC invasion pathways.

5.5. Rosetting

Cerebral malaria is characterised histologically by deep vascular sequestration of mature parasites in the brain (Aikawa *et al.*, 1980), although it is still discussed to which extent this phenomenon contributes to the pathogenesis of cerebral malaria (Miller *et al.*, 2002; Seydel *et al.*, 2006). Rosetting is defined by the agglutination of non-parasitised erythrocytes around RBCs containing mature forms of the parasite and probably contributes to intravascular erythrocyte sequestration by further clogging the microvasculature (Handunnetti *et al.*, 1989). This phenomenon occurs *in vitro*, and isolates from patients with cerebral malaria appear to have increased rosetting properties (Carlson *et al.*, 1990; Treutiger *et al.*, 1992). Rosetting was established as a *P. falciparum* virulence factor (Carlson *et al.*, 1994; Rowe *et al.*, 1995), the expression of which is modified by a variety of host factors, such as host immunity, ABO blood group and haemoglobin phenotype. The molecules involved in rosetting seem to be distinct from those involved in endothelial cytoadherence, although they are often co-expressed on the same parasitised RBCs, and they share the common ligand *P. falciparum* erythrocyte membrane protein 1 (PfEMP1), a protein expressed by the parasite and exported to the RBC membrane. The rosetting receptor in RBCs is the complement receptor type 1 (CR1). CR1 appears to carry the Knops blood group antigens with the S1(a) phenotype forming fewer rosettes (Rowe *et al.*, 1997). Differences in rosetting ability were also observed between RBCs of different ABO blood groups, with a diminished rosetting potential in blood group O RBCs (Carlson, 1993; Kun *et al.*, 1998). The ability of RBCs from healthy donors to form rosettes appeared to be greater in cells from group A or B than those of group O patients (Barragan *et al.*, 2000; Chotivanich *et al.*, 1998). An *in vitro* study in fresh clinical isolates from the Gambia confirmed that parasitised RBCs form rosettes more readily if belonging to group A or B than group O isolates (Udomsangpetch *et al.*, 1993). While the rosettes formed by blood group O RBCs may be inhibited by several different saccharides, the binding of A or B RBC to *P. falciparum*-infected RBC could only be blocked with constituents of the terminal trisaccharides of the A or B blood group antigens. These data thus suggest that the A and B blood group antigens present on RBC glycoproteins and glycolipids constitute specific receptors for rosette formation that seem to be mediated by lectin-like interactions to several carbohydrate moieties, some that are constituents of heparin or ABO blood group antigens (Carlson and Wahlgren, 1992). Impaired rosette formation may thus contribute to innate resistance to severe *P. falciparum* malaria that occurs in certain RBC disorders and in individuals of blood group O phenotype. The lack of correlation between rosetting and cytoadherence (Iqbal *et al.*, 1993; Reeder *et al.*, 1994) suggests that the linkage of mechanical blockage to pathogenesis remains hypothetical (Al-Yaman *et al.*, 1995).

5.6. Cytoadherence

Sequestration, a mechanism related to severity in *falciparum* malaria, occurs by means of cytoadherence to placental syncytiotrophoblast or to endothelial cells. Adherence facilitates sequestration and reduces splenic destruction, and as a consequence only ring forms are observed in peripheral blood. The knobs that appear in the surface of infected RBCs are the contact point with host cells. The sequence of events is not precisely known. Several host receptors in endothelial cells and syncytiotrophoblast have been identified which may mediate this interaction, although only CD36 and chondroitin sulphate A (CSA), respectively are associated to stable adherence (Miller *et al.*, 2002). Many of the molecules that mediate adherence in the different tissues are associated with ABO phenotypes (Fig. 1.10), and these associations are reviewed in this chapter.

The *P. falciparum* erythrocyte membrane protein 1 (PfEMP1) is a parasite polypeptide expressed in infected RBCs which mediates the attachment to host receptors. PfEMP1 contains several extracellular DBL domains (DBL 1–5), with one to two cysteine-rich inter-domain regions (CIDRs). PfEMP1 is encoded by the multiple *var* gene family and produces multiple adhesion phenotypes, so different parasites can bind to variable host receptors, contributing to *P. falciparum* pathogenesis (Chen *et al.*, 2000a).

One example of these differences in adhesion phenotypes is that the parasites which adhere to the placenta are different from those found in non-pregnant individuals (Beeson *et al.*, 1999). In the placenta, the parasites bind to the proteoglycans hyaluronic acid and CSA through a DBL-γ domain of the PfEMP1 (Beeson *et al.*, 2000; Fried and Duffy, 1996). There are no reports of associations between DBL-γ and ABO antigen expression, although it has been described that DBL-α has affinity for the A blood group (Chen *et al.*, 2000b). Both hyaluronic acid and chondroitin sulphate are closely related to A antigen, since both consist of repeating disaccharide units of glucuronic acid with GluNAc or GalNAc, respectively. As outlined previously, GalNAc is the terminal sugar of A antigen. GalNAc and GluNAc are important determinants for the attachment of

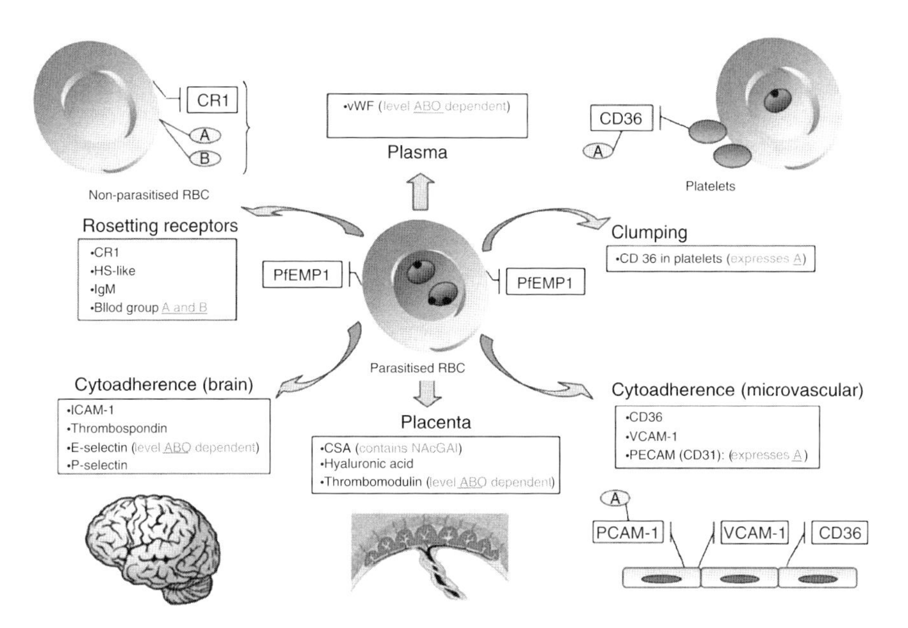

FIGURE 1.10 Associations ABO-rosetting and cytoadherence. The Figure shows ABO blood group interactions and expression in relation to RBC interactions and organs where cytoadherence occurs. PfRh, *P. falciparum* reticulocyte-binding protein homologue; vWF, von Willebrand factor; CR1, complement receptor type 1; HS-like, heparin sulphate-like; ICAM-1, intercellular adhesion molecule 1; CSA, chondroitin sulphate A; VCAM-1, vascular cell adhesion molecule-1; PECAM (CD31), platelet–endothelial cell adhesion molecule 1; PfEMP1, *P. falciparum* erythrocyte membrane protein 1.

the erythrocyte (Facer, 1983) and their structure is very similar. The binding of infected RBCs with CSA may be mediated by the proteoglycan thrombomodulin, that is present in placenta (Rogerson *et al.*, 1997). It has also been described that the levels of thrombomodulin and other soluble adhesion molecules in serum are affected by ABO phenotype (Blann *et al.*, 1996). These factors may contribute to the differences in susceptibility and parity-specific associations of ABO phenotypes with placental malaria.

Conversely, CD36-binding parasites are isolated in non-pregnant individuals, and they express a PfEMP1 with a CD36-binding CIDR1. The NH (2)-terminal of the PfEMP1 (which includes DBL1α and CIDR1α) mediates adherence to multiple host receptors including platelet–endothelial cell adhesion molecule 1 (PECAM-1)/CD31, blood group A antigen and CD36 among others (Chen *et al.*, 2000b). PECAM-1/CD31 is an integral membrane glycoprotein found on the surface of human platelets, leukocytes and the intercellular junction of endothelial cells. PECAM-1 expresses blood group A determinant (Curtis *et al.*, 2000). It has also been described that CD36 in platelets expresses the A antigen (Stockelberg *et al.*, 1996). This may be relevant to pathogenesis because platelet-mediated clumping is strongly associated with severe malaria: Some malaria-infected erythrocytes can auto-agglutinate, and this is mediated by platelets that express surface glycoprotein CD36 (Pain *et al.*, 2001).

Cytoadherence in the brain may be associated with the intercellular adhesion molecule 1 (ICAM-1) and to the level of expression of ICAM-1 in brain endothelium (Ockenhouse *et al.*, 1991). No association has been found between ICAM-1 level (in brain or in serum) and ABO phenotype. Other receptors involved in brain sequestration are thrombospondin, E-selectin and P-selectin. Soluble levels of E-selectin but not P-selectin differ according to ABO phenotype (Blann *et al.*, 1996). No association has been described between ABO and thrombospondin, although this molecule is a repeat within the ADAMTS metalloprotease which produces proteolysis of von Willebrand factor, a substance related to ABO phenotypes.

The von Willebrand factor (vWF) has gained considerable interest in recent years as a marker of endothelial cell activation or insult and by virtue of its interactions with platelets and vessel walls. However, little is known about its association and other markers with malaria. ABO is associated with different levels of vWF, and it accounts for 19% of the total variance in vWF:Ag ($p < 0.0001$) and race for 7% ($p < 0.0001$), for a total of 26%. O subjects had significantly lower levels (Miller *et al.*, 2003). In addition, the susceptibility of vWF of blood group O, A, B and AB to proteolysis by the ADAMTS13 metalloprotease was greater for group O vWF than for non-O vWF in the rank order O $\geq$ B $>$ A $\geq$ AB. The loss of collagen-binding activity of vWF following proteolysis was greater for group O compared with non-O vWF, in the same rank order (Bowen, 2003). Since there is an influence of the ABO blood group on von Willebrand factor and Factor VIII, and Von Willebrand factor interacts with at least two platelet membrane receptors, ABO could also have an effect on platelet function. Sweeney *et al.* (1989) found that the aggregation response to ristocetin was better in group O than in group A despite the lower vWf:Ag levels, suggesting an influence of blood group antigens on the interaction between von Willebrand's factor and platelets. Hollestelle *et al.* (2006) found a significant rise in vWF and vWF propeptide, which were highest in severe malaria, indicating that acute endothelial cell activation is a hallmark of malaria in children. Hence, ABO could also affect susceptibility to severe malaria through its association with vWF levels.

All three members of the family of adhesion molecules (L-, P- and E-selectin) have amino terminal C-type lectin domains that bind carbohydrates. The role of sialic acids in ligand recognition by the selectins was also evident from early studies. Lewis is a blood group closely related to ABO in terms of its synthetic pathway that shows some sialylated forms. Several groups reported that the selectins recognise sialylated structures of the Lewis blood group (Le) as the sialyl Lea, and other Lewis-related sequences (Berg *et al.*, 1991; Fukuda *et al.*, 1999; Grinnell *et al.*, 1983; Lowe *et al.*, 1990; Phillips *et al.*, 1990; Walz *et al.*, 1990). In relation to malaria, it has been shown that naturally occurring glycoproteins of malaria-infected RBCs had an enhanced lectin-binding ability with wheat-germ agglutinin (WGA) lectin among others (Kumar *et al.*, 2006). This particular WGA precipitated up to 50% when put in contact with a cyst glycoprotein from a human blood group Le(a +) containing unusually high amount (18%) of sialic acid (Wu *et al.*, 1996). There are some differences in Lewis antigen level according to ABO phenotype since A, B and AB cells carry less Le antigen than O cells, because their respective transferases use the same precursors. A recent hypothesis has been presented that falciparum malaria may account for the three-fold higher prevalence of the Le(a−b−) phenotype among people of African ancestry (Cserti and Dzik, 2007).

Until recently, the sequestration of erythrocytes infected with *P. falciparum* was thought to be due to one of a number of protein–protein interactions. There is now evidence that sacharidae are relevant in cytoadherence, and that infected erythrocytes can also adhere to proteoglycans. Binding of infected erythrocytes to CSA and other proteoglycans could be crucial to the development of malarial infection of the placenta, and possibly to sequestration in the lung and brain (Rogerson and Brown, 1997), since chondroitin sulphate proteoglycans constitute the major population of proteoglycans in the central nervous system (Schwartz and Domowicz, 2004). ABO phenotype interacts in an indirect or direct form in many aspects and organs where cytoadherence occurs (Fig. 1.10).

6. CONCLUSIONS AND RESEARCH IMPLICATIONS

There are clear associations between group A (and also B but to a lesser extent) and clinical severity, and group O and milder disease in children. The meta-analysis conclusion should be interpreted cautiously because of the heterogeneity of study outcomes; but the consistency of the results for A and O phenotypes supports the association. A limitation in drawing this conclusion concerns publication bias towards the reporting of positive results. A further limitation is the lack of genotype information on ABO groups, since all studies were based on phenotypic data. Genotypes (i.e., AO vs AA, A_1 vs A_2) show differences at a molecular level that could be relevant to these findings. The geographic distribution of the O phenotype in Africa is consistent with selection pressure related to malaria transmission in sub-Saharan Africa. The higher prevalence in Africans of phenotypes associated with higher expression of H antigen in cells could reinforce the hypothesis for a protective effect in malaria severity. Among pregnant subjects, two studies reported an association of similar magnitude between blood group O and induction of parity-specific immunity to placental malaria in multigravidae living under conditions of high malaria transmission.

Several mechanisms relate to these associations including affinity for *A. gambiae*; shared ABO antigens with *P. falciparum*; impairment of merozoite penetration of RBCs; and cytoadherence, endothelial activation and rosetting. In particular, ABO blood group associations with molecules involved with RBC invasion pathways suggest that ABO characteristics are central to the integrity of the RBC invasion process. Sugar components on glycophorin, in particular GluNAc and GalNAc, are critical for parasite attachment. As GalNAc is the terminal sugar of the A antigen, this could relate to the observed malaria susceptibility associated with the A phenotype. ABO phenotypes could also influence cytoadherence through effects on platelet function or endothelial adhesion mediators.

We identified no information on association with therapeutic response to antimalarials, to the emergence of drug resistant parasites or to their potential influence on efficacy of candidate malaria vaccines. However, the genetic influence of ABO phenotypes in relation to *falciparum* malaria susceptibility indicates their central importance for pathogenesis. Future clinical studies of malarial disease should include characterisation of these phenotypes, as well as genotype determination in order to establish more precise associations.

In summary, ABO antigens and oligosaccharide chains attached to cell surfaces and extracellular proteins or lipids provide new insights into the different steps of *P. falciparum* pathogenesis. The picture could be more complex than previously thought because the same glycans may participate in variable functions independent of malaria, including structural roles and specific recognition by endogenous or exogenous lectin receptors. From an evolutionary point of view, when a lethal pathogen recognises a glycan sequence, that also serves as a ligand for one or more unrelated endogenous lectins (as happens with sialic acid), then the elimination of this sequence would be detrimental, independent of selection pressure of malaria. Sialic acid bound to RBCs may act as a decoy for viruses that require nucleated cells for replication and at the same time act as a receptor for malaria (Gagneux and Varki, 1999). These essential functions of the glycan residues (in the form of sialic or ABO) would lead to varied selection pressure and contribute further to the reduced ability to evade malaria infection through molecule suppression.

The role of glycans in the pathogenesis of malaria may not relate solely to a single molecule, but to a cluster of saccharidae patches, each of which could be associated with higher or lower susceptibility within a whole spectrum.

Research into the relationship between the ABO blood group phenotypes and malaria pathogenesis may open a new insight into the glycobiology of infectious diseases and contribute to a wider understanding of malaria pathogenesis. Hopefully, this will lead to the development of improved strategies to reduce the magnitude of the burden of malaria experienced by so many populations.

ACKNOWLEDGEMENTS

We would like to express our appreciation to the following colleagues for providing unpublished data: Dr. Cynthia Donkor, Dr. Phil Fischer and Dr. Edward Senga.

REFERENCES

Aikawa, M., Suzuki, M., and Gutierrez, Y. (1980). Pathology of malaria. *In* "Malaria" (J. P. Kreier, ed.), Vol. 2, pp. 47–102. Academic Press, New York.

Akinboye, D. O., and Ogunrinade, A. F. (1987). Malaria and loaisis among blood donors at Ibadan, Nigeria. *Trans. R. Soc. Trop. Med. Hyg.* **81,** 398–399.

Allen, S. J., O'Donnell, A., Alexander, N. D., Mgone, C. S., Peto, T. E., Clegg, J. B., Alpers, M. P., and Weatherall, D. J. (1999). Prevention of cerebral malaria in children in Papua New Guinea by southeast Asian ovalocytosis band 3. *Am. J. Trop. Med. Hyg.* **60,** 1056–1060.

Al-Yaman, F., Genton, B., Mokela, D., Raiko, A., Kati, S., Rogerson, S., Reeder, J., and Alpers, M. (1995). Human cerebral malaria: Lack of significant association between erythrocyte rosetting and disease severity. *Trans. R. Soc. Trop. Med. Hyg.* **89,** 55–58.

Athreya, B. H., and Coriell, L. L. (1967). Relation to blood groups to infection. I. A survey and review of data suggesting possible relationship between malaria and blood groups. *Am. J. Epidemiol.* **86,** 292–304.

Awandare, G. A., Goka, B., Boeuf, P., Tetteh, J. K., Kurtzhals, J. A., Behr, C., and Akanmori, B. D. (2006). Increased levels of inflammatory mediators in children with severe *Plasmodium falciparum* malaria with respiratory distress. *J. Infect. Dis.* **194,** 1438–1446.

Barragan, A., Kremsner, P. G., Wahlgren, M., and Carlson, J. (2000). Blood group A antigen is a coreceptor in *Plasmodium falciparum* rosetting. *Infect. Immun.* **68,** 2971–2975.

Bayoumi, R. A., Bashir, A. H., and Abdulhadi, N. H. (1986). Resistance to *falciparum* malaria among adults in central Sudan. *Am. J. Trop. Med. Hyg.* **35,** 45–55.

Beeson, J. G., Brown, G. V., Molyneux, M. E., Mhango, C., Dzinjalamala, F., and Rogerson, S. J. (1999). *Plasmodium falciparum* isolates from infected pregnant women and children are associated with distinct adhesive and antigenic properties. *J. Infect. Dis.* **180,** 464–472.

Beeson, J. G., Rogerson, S. J., Cooke, B. M., Reeder, J. C., Chai, W., Lawson, A. M., Molyneux, M. E., and Brown, G. V. (2000). Adhesion of *Plasmodium falciparum*-infected erythrocytes to hyaluronic acid in placental malaria. *Nat. Med.* **6,** 86–90.

Beiguelman, B., Alves, F. P., Moura, M. M., Engracia, V., Nunes, A. C., Heckmann, M. I., Ferreira, R. G., da Silva, L. H., Camargo, E. P., and Krieger, H. (2003). The association of genetic markers and malaria infection in the Brazilian Western Amazonian region. *Mem. Inst. Oswaldo Cruz* **98,** 455–460.

Berg, E. L., Robinson, M. K., Mansson, O., Butcher, E. C., and Magnani, J. L. (1991). A carbohydrate domain common to both sialyl Le(a) and sialyl Le(X) is recognized by the endothelial cell leukocyte adhesion molecule ELAM-1. *J. Biol. Chem.* **266,** 14869–14872.

Berger, S. A., Young, N. A., and Edberg, S. C. (1989). Relationship between infectious diseases and human blood type. *Eur. J. Clin. Microbiol. Infect. Dis.* **8,** 681–689.

Blackwell, C. C. (1989). The role of ABO blood groups and secretor status in host defences. *FEMS Microbiol. Immunol.* **1,** 341–349.
Blann, A. D., Daly, R. J., and Amiral, J. (1996). The influence of age, gender and ABO blood group on soluble endothelial cell markers and adhesion molecules. *Br. J. Haematol.* **92,** 498–500.
Booth, P. B., and McLoughlin, K. (1972). The Gerbich blood group system, especially in Melanesians. *Vox Sang.* **22,** 73–84.
Bowen, D. J. (2003). An influence of ABO blood group on the rate of proteolysis of von Willebrand factor by ADAMTS13. *J. Thromb. Haemost.* **1,** 33–40.
Bulai, T., Bratosin, D., Pons, A., Montreuil, J., and Zanetta, J. P. (2003). Diversity of the human erythrocyte membrane sialic acids in relation with blood groups. *FEBS Lett.* **534,** 185–189.
Carlson, J. (1993). Erythrocyte rosetting in *Plasmodium falciparum* malaria—with special reference to the pathogenesis of cerebral malaria. *Scand. J. Infect. Dis.* **86**(Suppl), 1–79.
Carlson, J., and Wahlgren, M. (1992). *Plasmodium falciparum* erythrocyte rosetting is mediated by promiscuous lectin-like interactions. *J. Exp. Med.* **176,** 1311–1317.
Carlson, J., Helmby, H., Hill, A. V., Brewster, D., Greenwood, B. M., and Wahlgren, M. (1990). Human cerebral malaria: Association with erythrocyte rosetting and lack of anti-rosetting antibodies. *Lancet* **336,** 1457–1460.
Carlson, J., Nash, G. B., Gabutti, V., Al-Yaman, F., and Wahlgren, M. (1994). Natural protection against severe *Plasmodium falciparum* malaria due to impaired rosette formation. *Blood* **84,** 3909–3914.
Cartron, J. P., Prou, O., Luilier, M., and Soulier, J. P. (1983). Susceptibility to invasion by *Plasmodium falciparum* of some human erythrocytes carrying rare blood group antigens. *Br. J. Haematol.* **55,** 639–647.
Casals-Pascual, C., and Roberts, D. J. (2006). Severe malarial anaemia. *Curr. Mol. Med.* **6,** 155–168.
Cavasini, C. E., De Mattos, L. C., Alves, R. T., Couto, A. A., Calvosa, V. S., Domingos, C. R., Castilho, L., Rossit, A. R., and Machado, R. L. (2006). Frequencies of ABO, MNSs, and Duffy phenotypes among blood donors and malaria patients from four Brazilian Amazon areas. *Hum. Biol.* **78,** 215–219.
Chen, Q., Heddini, A., Barragan, A., Fernandez, V., Pearce, S. F., and Wahlgren, M. (2000a). The semiconserved head structure of *Plasmodium falciparum* erythrocyte membrane protein 1 mediates binding to multiple independent host receptors. *J. Exp. Med.* **192,** 1–10.
Chen, Q., Schlichtherle, M., and Wahlgren, M. (2000b). Molecular aspects of severe malaria. *Clin. Microbiol. Rev.* **13,** 439–450.
Chester, M. A., and Olsson, M. L. (2001). The ABO blood group gene: A locus of considerable genetic diversity. *Transfus. Med. Rev.* **15,** 177–200.
Chishti, A. H., Palek, J., Fisher, D., Maalouf, G. J., and Liu, S. C. (1996). Reduced invasion and growth of *Plasmodium falciparum* into elliptocytic red blood cells with a combined deficiency of protein 4.1, glycophorin C, and p55. *Blood* **87,** 3462–3469.
Chotivanich, K. T., Udomsangpetch, R., Pipitaporn, B., Angus, B., Suputtamongkol, Y., Pukrittayakamee, S., and White, N. J. (1998). Rosetting characteristics of uninfected erythrocytes from healthy individuals and malaria patients. *Ann. Trop. Med. Parasitol.* **92,** 45–56.
Chung, W. Y., Gardiner, D. L., Hyland, C., Gatton, M., Kemp, D. J., and Trenholme, K. R. (2005). Enhanced invasion of blood group A1 erythrocytes by *Plasmodium falciparum*. *Mol. Biochem. Parasitol.* **144,** 128–130.
Clausen, H., and Hakomori, S. (1989). ABH and related histo-blood group antigens; immunochemical differences in carrier isotypes and their distribution. *Vox Sang.* **56,** 1–20.
Contreras, M., and Lubenko, A. (1999). Antigens in human blood. *In* "Postgraduate Haematology" (A. V. Hoffbrand, S. M. Lewis, and E. G. D. Tuddenham, eds.), 4th edn., pp. 182–215. Butterworth Heinemann, Oxford.

Cserti, C. M., and Dzik, W. H. (2007). The ABO blood group system and *Plasmodium falciparum* malaria. *Blood,* in press.

Curtis, B. R., Edwards, J. T., Hessner, M. J., Klein, J. P., and Aster, R. H. (2000). Blood group A and B antigens are strongly expressed on platelets of some individuals. *Blood* **96,** 1574–1581.

Dolan, S. A., Proctor, J. L., Alling, D. W., Okubo, Y., Wellems, T. E., and Miller, L. H. (1994). Glycophorin B as an EBA-175 independent *Plasmodium falciparum* receptor of human erythrocytes. *Mol. Biochem. Parasitol.* **64,** 55–63.

Domarle, O., Migot-Nabias, F., Mvoukani, J. L., Lu, C. Y., Nabias, R., Mayombo, J., Tiga, H., and Deloron, P. (1999). Factors influencing resistance to reinfection with *Plasmodium falciparum. Am. J. Trop. Med. Hyg.* **61,** 926–931.

Dore, C., Hubbard, J. L., Thornton, C., and Weiner, J. S. (1975). On *Anopheles* behaviour and blood groups. *Curr. Anthropol.* **16,** 470–472.

Druilhe, P., Zouali, M., Gentilini, M., and Eyquem, A. (1983). Demonstration of an abnormal increase of anti-T hemagglutinin titers in malaria infected patients. *Comptes Rendus des Séances de l'Académie des Sciences. Série III, Sciences de la Vie* **296,** 339–344.

Duraisingh, M. T., Triglia, T., Ralph, S. A., Rayner, J. C., Barnwell, J. W., McFadden, G. I., and Cowman, A. F. (2003). Phenotypic variation of *Plasmodium falciparum* merozoite proteins directs receptor targeting for invasion of human erythrocytes. *EMBO J.* **22,** 1047–1057.

Facer, C. A. (1980). Direct Coombs antiglobulin reactions in Gambian children with *Plasmodium falciparum* malaria. II. Specificity of erythrocyte-bound IgG. *Clin. Exp. Immunol.* **39,** 279–288.

Facer, C. A. (1983). Merozoites of *P. falciparum* require glycophorin for invasion into red cells. *Bull. Soc. Pathol. Exot.* **76,** 463–469.

Facer, C. A., and Brown, J. (1979). ABO blood groups and *falciparum* malaria. *Trans. R. Soc. Trop. Med. Hyg.* **73,** 599–600.

Facer, C. A., and Mitchell, G. H. (1984). Wrb negative erythrocytes are susceptible to invasion by malaria parasites. *Lancet* **2**(8405), 758–759.

Facer, C. A., Bray, R. S., and Brown, J. (1979). Direct Coombs antiglobulin reactions in Gambian children with *Plasmodium falciparum* malaria. I. Incidence and class specificity. *Clin. Exp. Immunol.* **35,** 119–127.

Farr, A. D. (1960). Blood group frequencies in some allergic conditions and in malaria. *J. Inst. Sci. Technol.* **6,** 32–33.

Field, S. P., Hempelmann, E., Mendelow, B. V., and Fleming, A. F. (1994). Glycophorin variants and *Plasmodium falciparum*: Protective effect of the Dantu phenotype *in vitro. Hum. Genet.* **93,** 148–150.

Fischer, P. R., and Boone, P. (1998). Short report: Severe malaria associated with blood group. *Am. J. Trop. Med. Hyg.* **58,** 122–123.

Fried, M., and Duffy, P. E. (1996). Adherence of *Plasmodium falciparum* to chondroitin sulfate A in the human placenta. *Science* **272,** 1502–1504.

Fukuda, M., Hiraoka, N., and Yeh, J. C. (1999). C-type lectins and sialyl Lewis X oligosaccharides. Versatile roles in cell-cell interaction. *J. Cell Biol.* **147,** 467–470.

Gagneux, P., and Varki, A. (1999). Evolutionary considerations in relating oligosaccharide diversity to biological function. *Glycobiology* **9,** 747–755.

Garratty, G. (2000). Blood groups and disease: A historical perspective. *Transfus. Med. Rev.* **14,** 291–301.

Gaur, D., Mayer, D. C., and Miller, L. H. (2004). Parasite ligand-host receptor interactions during invasion of erythrocytes by *Plasmodium merozoites. Int. J. Parasitol.* **34,** 1413–1429.

Greenwell, P. (1997). Blood group antigens: Molecules seeking a function? *Glycoconj. J.* **14,** 159–173.

Grinnell, B. W., Hermann, R. B., and Yan, S. B. (1983). Human protein C inhibits selectin-mediated cell adhesion: Role of unique fucosylated oligosaccharide. *Glycobiology* **4,** 221–225.

Guerra, C. A., Snow, R. W., and Hay, S. I. (2006a). Defining the global spatial limits of malaria transmission in 2005. *Adv. Parasitol.* **62,** 157–179.

Guerra, C. A., Snow, R. W., and Hay, S. I. (2006b). Mapping the global extent of malaria in 2005. *Trends Parasitol.* **22,** 353–358.

Gupta, M., and Chowdhuri, A. N. (1980). Relationship between ABO blood groups and malaria. *Bull. World Health Organ.* **56,** 913–915.

Hadley, T. J., Klotz, F. W., Pasvol, G., Haynes, J. D., McGinniss, M. H., Okubo, Y., and Miller, L. H. (1987). *Falciparum* malaria parasites invade erythrocytes that lack glycophorin A and B (MkMk). Strain differences indicate receptor heterogeneity and two pathways for invasion. *J. Clin. Invest.* **80,** 1190–1193.

Handunnetti, S. M., David, P. H., Perera, K. L., and Mendis, K. N. (1989). Uninfected erythrocytes form "rosettes" around *Plasmodium falciparum* infected erythrocytes. *Am. J. Trop. Med. Hyg.* **40,** 115–118.

Hartmann, G. (1941). Thesis "Group Antigens in Human Organs." US Army Medical Research Laboratory, Ford Knox (republished 1970), Copenhagen.

Heddini, A., Pettersson, F., Kai, O., Shafi, J., Obiero, J., Chen, Q., Barragan, A., Wahlgren, M., and Marsh, K. (2001). Fresh isolates from children with severe *Plasmodium falciparum* malaria bind to multiple receptors. *Infect. Immun.* **69,** 5849–5856.

Hill, A. V. (1992). Malaria resistance genes: A natural selection. *Trans. R. Soc. Trop. Med. Hyg.* **86,** 225–226.

Hollestelle, M. J., Donkor, C., Mantey, E. A., Chakravorty, S. J., Craig, A., Akoto, A. O., O'Donnell, J., van Mourik, J. A., and Bunn, J. (2006). Von Willebrand factor propeptide in malaria: Evidence of acute endothelial cell activation. *Br. J. Haematol.* **133,** 562–569.

Ibhanesebhor, S. E., Otobo, E. S., and Ladipo, O. A. (1996). Prevalence of malaria parasitaemia in transfused donor blood in Benin City, Nigeria. *Ann. Trop. Paediatr.* **16,** 93–95.

Iqbal, J., Perlmann, P., and Berzins, K. (1993). Serological diversity of antigens expressed on the surface of erythrocytes infected with *Plasmodium falciparum*. *Trans. R. Soc. Trop. Med. Hyg.* **87,** 583–588.

Joshi, H., Raghavendra, K., Subbarao, S. K., and Sharma, V. P. (1987). Genetic markers in malaria patients of Delhi. *Indian J. Malariol.* **24,** 33–38.

Jungery, M. (1985). Studies on the biochemical basis of the interaction of the merozoites of *Plasmodium falciparum* and the human red cell. *Trans. R. Soc. Trop. Med. Hyg.* **79,** 591–597.

Kahane, I., and Rachmilewitz, E. A. (1976). Alterations in the red blood cell membrane and the effect of vitamin E on osmotic fragility in beta-thalassemia major. *Isr. J. Med. Sci.* **12,** 11–15.

Kano, K., McGregor, I. A., and Milgrom, F. (1968). Hemagglutinins in sera of Africans of Gambia. *Proc. Soc. Exp. Biol. Med.* **129,** 849–853.

Karlsson, K. A. (1998). Meaning and therapeutic potential of microbial recognition of host glycoconjugates. *Mol. Microbiol.* **29,** 1–11.

Kassim, O. O., and Ejezie, G. C. (1982). ABO blood groups in malaria and schistosomiasis haematobium. *Acta Trop.* **39,** 179–184.

Kermarrec, N., Roubinet, F., Apoil, P. A., and Blancher, A. (1999). Comparison of allele O sequences of the human and non-human primate ABO system. *Immunogenetics* **49,** 517–526.

Kominato, Y., McNeill, P. D., Yamamoto, M., Russell, M., Hakomori, S., and Yamamoto, F. (1992). Animal histo-blood group ABO genes. *Biochem. Biophys. Res. Commun.* **189,** 154–164.

Kumar, K. A., Singh, S., and Babu, P. P. (2006). Studies on the glycoprotein modification in erythrocyte membrane during experimental cerebral malaria. *Exp. Parasitol.* **114,** 173–179.

Kun, J. F., Schmidt-Ott, R. J., Lehman, L. G., Lell, B., Luckner, D., Greve, B., Matousek, P., and Kremsner, P. G. (1998). Merozoite surface antigen 1 and 2 genotypes and rosetting of *Plasmodium falciparum* in severe and mild malaria in Lambarene, Gabon. *Trans. R. Soc. Trop. Med. Hyg.* **92,** 110–114.

Landsteiner, K. (1900). Zur Kenntnis der antifermentativen, lytischen und agglutinierenden Wirkungen des Blutserums und der Lymphe. *Zentralblatt für Bakteriologie, Parasitenkunde, Infektionskrankheiten und Hygiene. 1. Abt. Medizinisch-hygienische Bakteriologie, Virusforschung und Parasitologie. Originale* **27,** 357–362.

Lansdsteiner, K. (1901). Ueber Agglutinationserscheinungen normalen menschlichen Blutes. Wien. *Klinische Wochenschrift* **14,** 1132–1134. Translation (1901): On agglutination phenomena of normal human blood. *In* "Papers on Human Genetics" (S. H. Boyer, ed.), pp. 27–31. Prentice-Hall, New Jersey.

Lee, S. H., Looareesuwan, S., Wattanagoon, Y., Ho, M., Wuthiekanun, V., Vilaiwanna, N., Weatherall, D. J., and White, N. J. (1989). Antibody-dependent red cell removal during *P. falciparum* malaria: The clearance of red cells sensitized with an IgG anti-D. *Br. J. Haematol.* **73,** 396–402.

Lell, B., May, J., Schmidt-Ott, R. J., Lehman, L. G., Luckner, D., Greve, B., Matousek, P., Schmid, D., Herbich, K., Mockenhaupt, F. P., Meyer, C. G., Bienzle, U., *et al.* (1999). The role of red blood cell polymorphism in resistance and susceptibility to malaria. *Clin. Infect. Dis.* **28,** 794–799.

Li, X., Chen, H., Oo, T. H., Daly, T. M., Bergman, L. W., Liu, S. C., Chishti, A. H., and Oh, S. S. (2004). A co-ligand complex anchors *Plasmodium falciparum* merozoites to the erythrocyte invasion receptor band 3. *J. Biol. Chem.* **279,** 5765–5771.

Loscertales, M. P., and Brabin, B. J. (2006). ABO phenotypes and malaria related outcomes in mothers and babies in The Gambia: A role for histo-blood groups in placental malaria? *Malar. J.* **17,** 72.

Lowe, J. B., Stoolman, L. M., Nair, R. P., Larsen, R. D., Berhend, T. L., and Marks, R. M. (1990). ELAM-1-dependent cell adhesion to vascular endothelium determined by a transfected human fucosyltransferase cDNA. *Cell* **63,** 475–484.

Mackinnon, M. J., Mwangi, T. W., Snow, R. W., Marsh, K., and Williams, T. N. (2005). Heritability of malaria in Africa. *Public Libr. Sci. Med.* **2,** e340.

Martin, S. K., Miller, L. H., Hicks, C. U., David-West, A., Ugbode, C., and Deane, M. (1979). Frequency of blood group antigens in Nigerian children with *falciparum* malaria. *Trans. R. Soc. Trop. Med. Hyg.* **73,** 216–218.

Martin, M. J., Rayner, J. C., Gagneux, P., Barnwell, J. W., and Varki, A. (2005). Evolution of human-chimpanzee differences in malaria susceptibility: Relationship to human genetic loss of N-glycolylneuraminic acid. *Proc. Natl. Acad. Sci. USA* **102,** 12819–12824. Erratum in: *Proc. Natl. Acad. Sci.* USA **103,** 9745.

Martinko, J. M., Vincek, V., Klein, D., and Klein, J. (1993). Primate ABO glycosyltransferases: Evidence for trans-species evolution. *Immunogenetics* **37,** 274–278.

Mayer, D. C., Kaneko, O., Hudson-Taylor, D. E., Reid, M. E., and Millar, L. H. (2001). Characterization of a *Plasmodium falciparum* erythrocyte-binding protein paralogous to EBA-175. *Proc. Natl. Acad. Sci. USA* **98,** 5222–5227.

Mayer, D. C., Mu, J. B., Kaneko, O., Duan, J., Su, X. Z., and Millar, L. H. (2004). Polymorphism in the *Plasmodium falciparum* erythrocyte-binding ligand JESEBL/EBA-181 alters its receptor specificity. *Proc. Natl. Acad. Sci. USA* **101,** 2518–2523.

Mayer, D. C., Jiang, L., Achur, R. N., Kakizaki, I., Gowda, D. C., and Miller, L. H. (2006). The glycophorin C N-linked glycan is a critical component of the ligand for the *Plasmodium falciparum* erythrocyte receptor BAEBL. *Proc. Natl. Acad. Sci. USA* **103,** 2358–2562.

Mgone, C. S., Koki, G., Paniu, M. M., Kono, J., Bhatia, K. K., Genton, B., Alexander, N. D., and Alpers, M. P. (1996). Occurrence of the erythrocyte band 3 (AE1) gene deletion in relation to malaria endemicity in Papua New Guinea. *Trans. R. Soc. Trop. Med. Hyg.* **90,** 228–231.

Migot-Nabias, F., Mombo, L. E., Luty, A. J., Dubois, B., Nabias, R., Bisseye, C., Millet, P., Lu, C. Y., and Deloron, P. (2000). Human genetic factors related to susceptibility to mild malaria in Gabon. *Genes Immunol.* **1,** 435–441.

Miller, L. H., Dvorak, J. A., Shiroishi, T., and Durocher, J. R. (1973). Influence of erythrocyte membrane components on malaria merozoite invasion. *J. Exp. Med.* **138,** 1597–1601.

Miller, L. H., Baruch, D. I., Marsh, K., and Doumbo, O. K. (2002). The pathogenic basis of malaria. *Nature* **415,** 673–679.

Miller, C. H., Haff, E., Platt, S. J., Rawlins, P., Drews, C. D., Dilley, A. B., and Evatt, B. (2003). Measurement of von Willebrand factor activity: Relative effects of ABO blood type and race. *J. Thromb. Haemost.* **1,** 2191–2197.

Min-Oo, G., and Gros, P. (2005). Erythrocyte variants and the nature of their malaria protective effect. *Cell. Microbiol.* **7,** 753–763.

Mitchell, G. H., Thomas, A. W., Margos, G., Dluzewski, A. R., and Bannister, L. H. (2004). Apical membrane antigen 1, a major malaria vaccine candidate, mediates the close attachment of invasive merozoites to host red blood cells. *Infect. Immun.* **72,** 154–158.

Molineaux, L. (1988). The epidemiology of human malaria as an explanation of its distribution, including some implications for its control. *In* "Malaria: Principles and Practice of Malariology" (W. H. Wernsdorfer, and I. McGregor, eds.), p. 936. Churchill Livingstone, Edinburgh.

Molineaux, L., and Gramiccia, G. (1980). "The Garki Project: Research on the Epidemiology and Control of Malaria in the Sudan Savanna of West Africa." World Health Organization, Geneva, Switzerland.

Mombo, L. E., Ntoumi, F., Bisseye, C., Ossari, S., Lu, C. Y., Nagel, R. L., and Krishnamoorthy, R. (2003). Human genetic polymorphisms and asymptomatic *Plasmodium falciparum* malaria in Gabonese schoolchildren. *Am. J. Trop. Med. Hyg.* **68,** 186–190.

Montoya, F., Restrepo, M., Montoya, A. E., and Rojas, W. (1994). Blood groups and malaria. *Rev. Inst. Med. Exot. Trop. São Paulo.* **36,** 33–38.

Moor-Jankowski, J. (1993). Blood groups of primates: Historical perspective and synopsis. *J. Med. Primatol.* **22,** 1–2.

Moor-Jankowski, J., and Wiener, A. S. (1969). Blood group antigens in primate animals and their relation to human blood groups. *Primates Med.* **3,** 64–77.

Mourant, A. E., Kopec, A. C., and Domaniewska-Sobczak, K. (1976). "The Distribution of the Human Blood Groups and Other Polymorphisms," 2nd edn. Oxford University Press, London.

Mourant, A. E., Kopec, A. C., and Domaniewska-Sobczak, K. (1978). "Blood Groups and Diseases." Oxford University Press, Oxford.

Murr, C., Schroecksnadel, K., Schonitzer, D., Fuchs, D., and Schennach, H. (2005). Neopterin concentrations in blood donors differ between ABO blood group phenotypes. *Clin. Biochem.* **38,** 916–919.

Myatt, A. V., Coatney, G. R., Hernandez, T., and Guinn, E. (1954). Effect of blood group on the prepatent period of inoculated vivax malaria. *Am. J. Trop. Med. Hyg.* **3,** 981–984.

Ockenhouse, C. F., Ho, M., Tandon, N. N., Van Seventer, G. A., Shaw, S., White, N. J., Jamieson, G. A., Chulay, J. D., and Webster, H. K. (1991). Molecular basis of sequestration in severe and uncomplicated *Plasmodium falciparum* malaria: Differential adhesion of infected erythrocytes to CD36 and ICAM-1. *J. Infect. Dis.* **164,** 163–169.

Okoye, V. C., and Bennett, V. (1985). *Plasmodium falciparum* malaria: Band 3 as a possible receptor during invasion of human erythrocytes. *Science* **227,** 169–171.

Oliver-Gonzalez, J., and Torregrossa, M. R. (1944). A substance in animal parasites related to the human isoagglutinogens. *J. Infect. Dis.* **74,** 173–177.

Oriol, R., Mollicone, R., Coullin, P., Dalix, A. M., and Candelier, J. J. (1992). Genetic regulation of the expression of ABH and Lewis antigens in tissues. *APMIS Suppl.* **27,** 28–38.

Orlandi, P. A., Klotz, F. W., and Haynes, J. D. (1992). A malaria invasion receptor, the 175-kilodalton erythrocyte binding antigen of *Plasmodium falciparum* recognizes the terminal Neu5Ac(alpha 2–3)Gal sequences of glycophorin A. *J. Cell Biol.* **116,** 901–909.

Osisanya, J. O. (1983). ABO blood groups and infections with human malarial parasites *in vivo* and *in vitro*. *East Afr. Med. J.* **60,** 616–621.
Owen, R. (2000). Karl Landsteiner and the first human marker locus. *Genetics* **155,** 995–998.
Pain, A., Ferguson, D. J., Kai, O., Urban, B. C., Lowe, B., Marsh, K., and Roberts, D. J. (2001). Platelet-mediated clumping of *Plasmodium falciparum*-infected erythrocytes is a common adhesive phenotype and is associated with severe malaria. *Proc. Natl. Acad. Sci. USA* **98,** 1805–1810.
Pant, C. S., and Srivastava, H. C. (1997). Distribution of three genetic markers and malaria in other backward castes of Kheda district, Gujarat. *Indian J. Malariol.* **34,** 42–46.
Pant, C. S., Gupta, D. K., Bhatt, R. M., Gautam, A. S., and Sharma, R. C. (1992a). An epidemiological study of G-6-PD deficiency, sickle cell haemoglobin, and ABO blood groups in relation to malaria incidence in Muslim and Christian communities of Kheda, Gujarat (India). *J. Commun. Dis.* **24,** 199–205.
Pant, C. S., Gupta, D. K., Sharma, R. C., Gautam, A. S., and Bhatt, R. M. (1992b). Frequency of ABO blood groups, sickle-cell haemoglobin, G-6-PD deficiency and their relation with malaria in scheduled castes and scheduled tribes of Kheda District, Gujarat. *Indian J. Malariol.* **24,** 235–239.
Pant, C. S., Gupta, D. K., Bhatt, R. M., Gautam, A. S., and Sharma, R. C. (1993). Three genetic markers and malaria in upper caste Hindus of Kheda District of Gujarat State. *Indian J. Malariol.* **30,** 229–233.
Pant, C. S., Srivastava, H. C., and Yadav, R. S. (1998). Prevalence of malaria and ABO blood groups in a seaport area in Raigad, Maharashtra. *Indian J. Malariol.* **38,** 225–228.
Parr, L. W. (1930). On isohemagglutination, the haemolytic index and heterohemagglutination. *J. Infect. Dis.* **46,** 173–185.
Pasvol, G. (2003). How many pathways for invasion of the red blood cell by the malaria parasite? *Trends Parasitol.* **19,** 430–432.
Pasvol, G., Jungery, M., Weatherall, D. J., Parsons, S. F., Anstee, D. J., and Tanner, M. J. (1982a). Glycophorin as a possible receptor for *Plasmodium falciparum*. *Lancet* **2,** 947–950.
Pasvol, G., Wainscoat, J. S., and Weatherall, D. J. (1982b). Erythrocytes deficient in glycophorin resist invasion by the malarial parasite, *Plasmodium falciparum*. *Nature* **297,** 64–66.
Pasvol, G., Anstee, D., and Tanner, M. J. (1984). Glycophorin C and the invasion of red cells by *Plasmodium falciparum*. *Lancet* **1,** 907–908.
Patel, S. S., Mehlotra, R. K., Kastens, W., Mgone, C. S., Kazura, J. W., and Zimmerman, P. A. (2001). The association of the glycophorin C exon 3 deletion with ovalocytosis and malaria susceptibility in the Wosera, Papua New Guinea. *Blood* **98,** 3489–3491.
Pathirana, S. L., Alles, H. K., Bandara, S., Phone-Kyaw, M., Perera, M. K., Wickremasinghe, A. R., Mendis, K. N., and Handunnetti, S. M. (2005). ABO-blood-group types and protection against severe, *Plasmodium falciparum* malaria. *Ann. Trop. Med. Parasitol.* **99,** 119–124.
Penalba, C., Simonneau, M., Autran, B., Bouvet Koskas, E., Vroclans, M., Vachon, F., Coulaud, J. P., and Saimot, A. G. (1984). Anti-I anti-erythrocyte auto-immunization in malaria. *Bull. Soc. Pathol. Exot.* **77,** 469–480.
Phillips, M. L., Nudelman, E., Gaeta, F. C., Perez, M., Singhal, A. K., Hakomori, S., and Paulson, J. C. (1990). ELAM-1 mediates cell adhesion by recognition of a carbohydrate ligand, sialyl-Lex. *Science* **250,** 1130–1132.
Podbielska, M., Fredriksson, S. A., Nilsson, B., Lisowska, E., and Krotkiewski, H. (2004). ABH blood group antigens in O-glycans of human glycophorin A. *Arch. Biochem. Biophys.* **429,** 45–53.
Raper, A. B. (1968). ABO blood groups and malaria. *Trans. R. Soc. Trop. Med. Hyg.* **62,** 158–159.
Ravn, V., and Dabelsteen, E. (2000). Tissue distribution of histo-blood group antigens. *APMIS Acta Pathol. Microbiol. Immunol. Scand.* **108,** 1–28.
Rayner, J. C., Vargas-Serrato, E., Huber, C. S., Galinski, M. R., and Barnwell, J. W. (2001). A *Plasmodium falciparum* homologue of *Plasmodium vivax* reticulocyte binding protein

(PvRBP1) defines a trypsin-resistant erythrocyte invasion pathway. *J. Exp. Med.* **194,** 1571–1581.

Reeder, J. C., Rogerson, S. J., Al-Yaman, F., Anders, R. F., Coppel, R. L., Novakovic, S., Alpers, M. P., and Brown, G. V. (1994). Diversity of agglutinating phenotype, cytoadherence, and rosette-forming characteristics of *Plasmodium falciparum* isolates from Papua New Guinean children. *Am. J. Trop. Med. Hyg.* **51,** 45–55.

Ridgwell, K., Tanner, M. J., and Anstee, D. J. (1983). The Wrb antigen, a receptor for *Plasmodium falciparum* malaria, is located on a helical region of the major membrane sialoglycoprotein of human red blood cells. *Biochem. J.* **209,** 273–276.

Rogerson, S. J., and Brown, G. V. (1997). Chondroitin sulphate A as an adherence receptor for *Plasmodium falciparum*-infected erythrocytes. *Parasitol. Today* **13,** 76–79.

Rogerson, S. J., Novakovic, S., Cooke, B. M., and Brown, G. V. (1997). *Plasmodium falciparum*-infected erythrocytes adhere to the proteoglycan thrombomodulin in static and flow-based systems. *Exp. Parasitol.* **86,** 8–18.

Rowe, A., Obeiro, J., Newbold, C. I., and Marsh, K. (1995). *Plasmodium falciparum* rosetting is associated with malaria severity in Kenya. *Infect. Immun.* **63,** 2323–2326.

Rowe, J. A., Moulds, J. M., Newbold, C. I., and Miller, L. H. (1997). *P. falciparum* rosetting mediated by a parasite-variant erythrocyte membrane protein and complement-receptor 1. *Nature* **388,** 292–295.

Russel, P. F., West, L. S., Macdonald, G., and Mauwell, R. D. (1963). Pathology. *In* "Practical Malariology," 2nd edn., pp. 365–367. Oxford Medical Publications, London.

Saitou, N., and Yamamoto, F. (1997). Evolution of primate ABO blood group genes and their homologous genes. *Mol. Biol. Evol.* **14,** 399–411.

Santos, S. E. B., Salzano, F. M., Franco, M. H., and de Melo Freitas, M. J. (1983). Mobility, genetic markers, susceptibility to malaria and race mixture in Manaus, Brazil. *J. Human Evol.* **12,** 373–381.

Schwartz, N. B., and Domowicz, M. (2004). Proteoglycans in brain development. *Glycoconj. J.* **21,** 329–341.

Serjeantson, S. W. (1989). A selective advantage for the Gerbich-negative phenotype in malarious areas of Papua New Guinea. *Papua New Guinea Med. J.* **32,** 5–9.

Seydel, K. B., Milner, D. A., Jr., Kamiza, S. B., Molyneux, M. E., and Taylor, T. E. (2006). The distribution and intensity of parasite sequestration in comatose Malawian children. *J. Infect. Dis.* **194,** 208–215.

Seymour, R. M., Allan, M. J., Pomiankowski, A., and Gustafsson, K. (2004). Evolution of the human ABO polymorphism by two complementary selective pressures. *Proc. R. Soc. Lond. Ser. B* **271,** 1065–1072.

Sim, B. K., Chitnis, C. E., Wasniowska, K., Hadley, T. J., and Miller, L. H. (1994). Receptor and ligand domains for invasion of erythrocytes by *Plasmodium falciparum. Science* **264,** 1941–1944.

Singh, T. (1985). Malaria and ABO blood groups. *Indian Pediatr.* **22,** 857–858.

Singh, I. P., Walter, H., Bhasin, M. K., Bhardwaj, V., and Sudhakar, K. (1986). Genetic markers and malaria. Observations in Gujarat, India. *Human Hered.* **36,** 31–36.

Singh, N., Shukla, M. M., Uniyal, V. P., and Sharma, V. P. (1995). ABO blood groups among malaria cases from district Mandla, Madhya Pradesh. *Indian J. Malariol.* **32,** 59–63.

Snow, R. W., Guerra, C. A., Noor, A. M., Myint, H. Y., and Hay, S. I. (2005). The global distribution of clinical episodes of *Plasmodium falciparum* malaria. *Nature* **434,** 214–217.

Socha, W. W., Blancher, A., and Moor-Jankowski, J. (1995). Red cell polymorphisms in nonhuman primates: A review. *J. Med. Primatol.* **24,** 282–305.

Spencer, H. C., Miller, L. H., Collins, W. E., Knud-Hansen, C., McGinnis, M. H., Shiroishi, T., Lobos, R. A., and Feldman, R. A. (1978). The Duffy blood group and resistance to *Plasmodium vivax* in Honduras. *Am. J. Trop. Med. Hyg.* **27,** 664–670.

Stockelberg, D., Hou, M., Rydberg, L., Kutti, J., and Wadenvik, H. (1996). Evidence for an expression of blood group A antigen on platelet glycoproteins IV and V. *Transfus. Med. Rev.* **6,** 243–248.
Sumiyama, K., Kitano, T., Noda, R., Ferrell, R. E., and Saitou, N. (2000). Gene diversity of chimpanzee ABO blood group genes elucidated from exon 7 sequences. *Gene* **259,** 75–79.
Sweeney, J. D., Labuzetta, J. W., Hoernig, L. A., and Fitzpatrick, J. E. (1989). Platelet function and ABO blood group. *Am. J. Clin. Pathol.* **91,** 79–81.
Szulman, A. E. (1964). The histological distribution of the blood group substances in man as disclosed by immunofluorescence. III. The A, B and H antigens in embryos and fetuses from 18 mm in length. *J. Exp. Med.* **119,** 503–516.
Telen, M. J., and Chasis, J. A. (1990). Relationship of the human erythrocyte Wrb antigen to an interaction between glycophorin A and band 3. *Blood* **15,** 842–848.
Thakur, A., and Verma, I. C. (1992). Malaria and ABO blood groups. *Indian J. Malariol.* **29,** 241–244.
Treutiger, C. J., Hedlund, I., Helmby, H., Carlson, J., Jepson, A., Twumasi, P., Kwiatkowski, D., Greenwood, B. M., and Wahlgren, M. (1992). Rosette formation in *Plasmodium falciparum* isolates and antirosette activity of sera from Gambians with cerebral or uncomplicated malaria. *Am. J. Trop. Med. Hyg.* **46,** 503–510.
Udoh, A. E. (1991). Distribution of sialic acid between sialoglycoproteins and other membrane components of different erythrocyte phenotypes. *Acta Physiol. Hung.* **78,** 265–273.
Udomsangpetch, R., Todd, J., Carlson, J., and Greenwood, B. M. (1993). The effects of hemoglobin genotype and ABO blood group on the formation of rosettes by *Plasmodium falciparum*-infected red blood cells. *Am. J. Trop. Med. Hyg.* **48,** 149–153.
Uneke, C. J. (2007). *Plasmodium falciparum* malaria and ABO blood group: Is there any relationship? *Parasitol. Res.* **100,** 759–765.
Urquiza, M., Suarez, J. E., Cardenas, C., Lopez, R., Puentes, A., Chavez, F., Calvo, J. C., and Patarroyo, M. E. (2000). *Plasmodium falciparum* AMA-1 erythrocyte binding peptides implicate AMA-1 as erythrocyte binding protein. *Vaccine* **19,** 508–513.
Vanderberg, J. P., Gupta, S. K., Schulman, S., Oppenheim, J. D., and Furthmayr, H. (1985). Role of the carbohydrate domains of glycophorins as erythrocyte receptors for invasion by *Plasmodium falciparum* merozoites. *Infect. Immun.* **47,** 201–210.
Vasantha, K., Gorakhshakar, A. C., Pavri, R. S., and Bhatia, H. M. (1982). Genetic markers in malaria endemic tribal populations of Dadra and Nagar Haveli. *In* "Proceedings of Recent Trends in Immunohaematology", pp. 162–168. Institute of Haematology, ICMR, Bombay.
Vos, G. H., Moores, P., and Lowe, R. F. (1971). A comparative study between the S–s–U–phenotype found in Central African Negroes and the S–s + U–phenotype in Rhnull individuals. *S. Afr. J. Med. Sci.* **36,** 1–6.
Waiz, A., and Chakraborty, B. (1990). Cerebral malaria—An analysis of 55 cases. *Bangladesh Med. Res. Counc. Bull.* **16,** 46–51.
Walz, G., Aruffo, A., Kolanus, W., Bevilacqua, M., and Seed, B. (1990). Recognition by ELAM-1 of the sialyl-Lex determinant on myeloid and tumor cells. *Science* **250,** 1132–1135.
Watkins, W. M., and Morgan, W. T. (1959). Possible genetical pathways for the biosynthesis of blood group mucopolysaccharides. *Vox Sang.* **4,** 97–119.
Weir, D. M. (1989). Carbohydrates as recognition molecules in infection and immunity. *FEMS Microbiol. Immunol.* **1,** 331–340.
Welch, S. G., McGregor, I. A., and Williams, K. (1977). The Duffy blood group and malaria prevalence in Gambian West Africans. *Trans. R. Soc. Trop. Med. Hyg.* **71,** 295–296.
Wood, C. S. (1975). New evidence for a late introduction of malaria into the New World. *Curr. Anthropol.* **16,** 93–104.
Wood, C. S., Harrison, G. A., Dore, C., and Weiner, J. S. (1972). Selective feeding of *Anopheles gambiae* according to ABO blood group status. *Nature* **239,** 165.

Wu, A. M., Wu, J. H., Watkins, W. M., Chen, C. P., and Tsai, M. C. (1996). Binding properties of a blood group Le(a+) active sialoglycoprotein, purified from human ovarian cyst, with applied lectins. *Biochim. Biophys. Acta* **1316,** 139–144.
Yamamoto, F. (2004). Review: ABO blood group system-ABH oligosaccharide antigens, anti-A and anti-B, A and B glycosyltransferases, and ABO genes. *Immunohematology* **20,** 3–22.
Zouali, M., Druilhe, P., Gentilini, M., and Eyquem, A. (1982). High titres of anti-T antibodies and other haemagglutinins in human malaria. *Clin. Exp. Immunol.* **50,** 83–91.

CHAPTER 2

Structure and Content of the *Entamoeba histolytica* Genome

C. G. Clark,[*] U. C. M. Alsmark,[†] M. Tazreiter,[‡] Y. Saito-Nakano,[§] V. Ali,[¶] S. Marion,[‖,1] C. Weber,[‖] C. Mukherjee,[#] I. Bruchhaus,[**] E. Tannich,[**] M. Leippe,[††] T. Sicheritz-Ponten,[‡‡] P. G. Foster,[§§] J. Samuelson,[¶¶] C. J. Noël,[†] R. P. Hirt,[†] T. M. Embley,[†] C. A. Gilchrist,[‖‖] B. J. Mann,[‖‖] U. Singh,[##] J. P. Ackers,[*] S. Bhattacharya,[a] A. Bhattacharya,[b] A. Lohia,[#] N. Guillén,[‖] M. Duchêne,[‡] T. Nozaki,[¶] and **N. Hall[c,2]**

[*] Department of Infectious and Tropical Diseases, London School of Hygiene and Tropical Medicine, London WC1E 7HT, UK
[†] Division of Biology, Newcastle University, Newcastle NE1 7RU, UK
[‡] Department of Specific Prophylaxis and Tropical Medicine, Center for Physiology and Pathophysiology, Medical University of Vienna, A-1090 Vienna, Austria
[§] Department of Parasitology, National Institute of Infectious Diseases, Tokyo, Japan
[¶] Department of Parasitology, Gunma University Graduate School of Medicine, Maebashi, Japan
[‖] Institut Pasteur, Unité Biologie Cellulaire du Parasitisme and INSERM U786, F-75015 Paris, France
[#] Department of Biochemistry, Bose Institute, Kolkata 700054, India
[**] Bernhard Nocht Institute for Tropical Medicine, D-20359 Hamburg, Germany
[††] Zoologisches Institut der Universität Kiel, D-24098 Kiel, Germany
[‡‡] Center for Biological Sequence Analysis, BioCentrum-DTU, Technical University of Denmark, DK-2800 Lyngby, Denmark
[§§] Department of Zoology, Natural History Museum, London, SW7 5BD, UK
[¶¶] Department of Molecular and Cell Biology, Boston University Goldman School of Dental Medicine, Boston, Massachusetts 02118
[‖‖] Department of Medicine, Division of Infectious Diseases, University of Virginia Health Sciences Center, Charlottesville, Virginia 22908
[##] Departments of Internal Medicine, Microbiology, and Immunology, Stanford University School of Medicine, Stanford, California 94305
[a] School of Environmental Sciences, Jawaharlal Nehru University, New Delhi 110067, India
[b] School of Life Sciences and Information Technology, Jawaharlal Nehru University, New Delhi 110067, India
[c] The Institute for Genomic Research, Rockville, Maryland 20850
[1] Present address: Cell Biology and Biophysics Program, European Molecular Biology Laboratory, 69117 Heidelberg, Germany
[2] Present address: School of Biological Sciences, University of Liverpool, Liverpool L69 7ZB, United Kingdom

Advances in Parasitology, Volume 65
ISSN 0065-308X, DOI: 10.1016/S0065-308X(07)65002-7

Contents

Abstract

The intestinal parasite *Entamoeba histolytica* is one of the first protists for which a draft genome sequence has been published. Although the genome is still incomplete, it is unlikely that many genes are missing from the list of those already identified. In this chapter we summarise the features of the genome as they are currently understood and provide previously unpublished analyses of many of the genes.

1. INTRODUCTION

Entamoeba histolytica is one of the most widespread and clinically important parasites, causing both serious intestinal (amoebic colitis) and extraintestinal (amoebic liver abscess) diseases throughout the world. A recent World Health Organization estimate (WHO, 1998) places *E. histolytica* second after *Plasmodium falciparum* as causing the most deaths annually (70,000) among protistan parasites.

Recently a draft of the complete genome of *E. histolytica* was published (Loftus *et al.*, 2005) making it one of the first protist genomes to be sequenced. The *E. histolytica* genome project was initiated in 2000 with funding from the Wellcome Trust and the National Institute of Allergy and Infectious Diseases to the Wellcome Trust Sanger Institute and The Institute for Genomic Research (TIGR) in the UK and the USA, respectively. The publication describing the draft sequence concentrated on the expanded gene families, metabolism and the role of horizontal gene transfer in the evolution of *E. histolytica*. In this chapter we summarise the structure and content of the *E. histolytica* genome in comparison to other sequenced parasitic eukaryotes, provide a description of the current assembly and annotation, place the inferred gene content in the context of what is known about the biology of the organism and discuss plans for completing the *E. histolytica* genome project and extending genome sequencing to other species of *Entamoeba.*

The fact that the genome sequence is still a draft has several important consequences. The first is that a few genes may be missing from the sequence data we have at present, although the number is likely to be small. For example, at least one gene (*amoeba pore* B) is not present in the genome data despite it having been cloned, sequenced and the protein extensively characterised well before the start of the genome project. The second consequence is that the assembly contains a number of large duplicated regions that may be assembly artefacts, meaning that the number of gene copies is overestimated in several cases. These problems cannot as yet be resolved but should be eventually as more data becomes available. Nevertheless, it is important to remember these issues when reading the rest of this chapter.

As the number of genes in *E. histolytica* runs into several thousands, it is not possible to discuss all of them. However, we have generated a number of tables that identify many genes and link them to their entries in GenBank using the relevant protein identifier. Only a few tables are included in the text of this chapter, but the others are available online as supplementary material, http://pathema.tigr.org/pathema/entamoeba_resources.shtml. The *E. histolytica* genome project data are being 'curated' at the J. Craig Venter Institute (JCVI, formerly TIGR), and it is on that site that the most current version of the assembled genome will be found. The 'Pathema' database will hold the data and the annotation (http://pathema.tigr.org/). The gene tables are also linked to the appropriate entry in the Pathema database, and the links will be maintained as the genome structure is refined over time.

Reference is made throughout the text to other species of *Entamoeba* where data are available. *Entamoeba dispar* is the sister species to *E. histolytica* and infects humans without causing symptoms. *Entamoeba invadens* is a reptilian parasite that causes invasive disease, primarily in snakes and lizards, and is widely used as a model for *E. histolytica* in the study of encystation, although the two species are not very closely related (Clark *et al.*, 2006b). Genome projects for both these species are under way at TIGR, and it is anticipated that high-quality draft sequences will be produced for both in the near future. It is hoped that the *E. dispar* sequence will prove useful in identifying genomic differences linked to disease causation while that of *E. invadens* will be used to study patterns of gene expression during encystation. Small-scale genome surveys have been performed for two other species: *Entamoeba moshkovskii*, which is primarily a free-living species although it occasionally infects humans, and *Entamoeba terrapinae*, a reptilian commensal species, http://www.sanger.ac.uk/Projects/Comp_Entamoeba/

2. GENOME STRUCTURE

2.1. The *E. histolytica* genome sequencing, assembly and annotation process

The first choice to be made in the genome project was perhaps the easiest—the identity of the strain to be used for sequencing. A significant majority of the existing sequence data prior to the genome project was derived from one strain: HM-1:IMSS. This culture was established in 1967 from a rectal biopsy of a Mexican man with amoebic dysentery and axenised shortly thereafter. It has been used widely for virulence, immunology, cell biology and biochemistry in addition to genetic studies. In an attempt to minimise the effects of long-term culture cryopreserved cells that had been frozen in the early 1970s were revived and this uncloned culture used to generate the DNA for sequencing.

Before undertaking a genome scale analysis, it is important to understand the quality and provenance of the underlying data. The *E. histolytica* genome was sequenced by whole genome shotgun approach with each centre generating roughly half of the reads. Several different DNA libraries containing inserts of different sizes were produced using DNA that had been randomly sheared and sequences were obtained from both ends of each cloned fragment. The Phusion assembler (Mullikin and Ning, 2003) was used to assemble the 450,000 short reads into larger contigs (contiguous sequences), resulting in 1819 genome fragments that were ~12× deep, which means that each base has been sequenced 12 times, on average. While the genome shotgun sequence provides high coverage of each base, it is inevitable that there will be misassemblies and sequencing errors in the final consensus particularly towards each end of the contigs. Another problem with draft sequence is that it contains gaps, and while most of these will be small and will mostly contain repetitive non-coding 'junk' sequence, some of the gaps will probably contain genes. This makes it impossible to be absolutely certain of the absence of particular genes in *E. histolytica* and, in some cases, the presence or absence of particular biological pathways. Due to the high repeat content and low GC content (24.1%) of the *E. histolytica* genome, closure of the remaining gaps is likely to be a lengthy process. Therefore, it was decided to undertake and publish an analysis of the genome draft following assembly of the shotgun reads.

Annotation of the protein coding regions of the genome was initially carried out using two genefinders [GlimmerHMM (Majoros *et al.*, 2004) and Phat (Cawley *et al.*, 2001)] previously used successfully on another low G + C genome, that of *P. falciparum*. The software was re-trained specifically for analysis of the *E. histolytica* genome. The training process involved preparing a set of 600 manually edited genes to be used as models with the subsequent genefinding then being carried out on all of the assembled contigs to generate a 'complete' gene set. Predicted gene functions were generated automatically by homology searches using public protein and protein-domain databases, with subsequent refinement of identifications being carried out by manual inspection. For particular genes and gene families of special interest, members of the *Entamoeba* scientific community were involved throughout this process as expert curators with each individual assisting in the analysis and annotation of their genes of interest. Therefore although the manual curation of the genome has not been systematic, those areas of biology that are of primary interest to the *Entamoeba* community have been annotated most thoroughly. The publication of the genome by Loftus *et al.* therefore represents a 'first draft' of the complete genome sequence and the level of annotation is similar to the initial publications of other genomes such as *Drosophila* (Adams *et al.*, 2000; Myers *et al.*, 2000) and human (Lander *et al.*, 2001).

2.2. Karyotype and chromosome structure

The current *E. histolytica* genome assembly is ~23.7 million basepairs (Mbp) in size (Table 2.1). This figure is not likely to be a very accurate measure. In part this is due to misassembly of repetitive regions, which will cause the genome to appear smaller and in part because of the possibility of aneuploidy in some regions of the genome, which would cause them to appear more than once in the assembly. Overall, however, this size is not inconsistent with data from pulse-field gels (Willhoeft and Tannich, 1999) and kinetic experiments (Gelderman *et al.*, 1971a,b) making the *E. histolytica* genome comparable in size (24 Mbp) to that of *P. falciparum* (23 Mbp) (Gardner *et al.*, 2002), *Trypanosoma brucei* (26 Mbp) (Berriman *et al.*, 2005) and the free-living amoeba *Dictyostelium discoideum* (34 Mbp) (Eichinger *et al.*, 2005).

The current assembly does not represent complete chromosomes. Analysis of pulse-field gels predicts 14 chromosomes ranging in sizes from 0.3 to 2.2 Mbp and possibly a ploidy of 4 (Willhoeft and Tannich, 1999). There is no current information regarding the size and nature of the centromeres, and there are no contigs that appear to contain likely centromeric regions based on comparisons with other organisms. A search for signature telomeric repeats within the data indicates that these are either not present in the genome, not present in our contigs, or are diverged enough to be unidentifiable. However, there is circumstantial evidence

TABLE 2.1 Genome summary statistics for selected single celled organisms with sequenced genomes

Statistic[a]	*Entamoeba histolytica*	*Plasmodium falciparum*	*Dictyostelium discoideum*	*Saccharomyces cerevisiae*	*Encephalitozoon cuniculi*
Genome Size (Mbp)	23.7	22.8	33.8	12.5	2.5
G + C content (%)	24.1	19.4	22.5	38	45.5
Gene number	9938	5268	12,500	5538	1997
Av. gene size (bp)	1167	2534	1756	1428	1077
% coding DNA	49.2	52.6	ND	70.5	ND
Av. protein size (aa)	389	761	518	475	359
Av. intergenic dist. (kb)	0.8	1.7	0.8	0.6	0.1
Gene density (kb per gene)	1.9	4.3	2.5	2.2 kb	1.1
% Genes with introns	25.2	54	69	5	<1
Av. intron size (bp)	102.1	179	146	ND	–
Av. number of introns/gene	1.5	2.6	1.9	1	1

[a] Abbreviations: Mbp: million basepairs; kb: kilobasepairs; bp: basepairs; aa: amino acids, ND: not determined.

that the chromosome ends may contain arrays of transfer RNA (tRNA) genes (see Section 2.4).

2.3. Ribosomal RNA genes

The organisation of the structural RNA genes in *E. histolytica* is unusual with the ribosomal RNA (rRNA) genes carried exclusively on 24 kilobasepair (kb) circular episomes (Bhattacharya *et al.*, 1998) that have two transcription units in an inverted repeat. These episomes are believed to make up about 20% of the total cellular DNA; indeed, roughly 15% of all of the sequencing reads generated in the genome project were derived from this molecule with the exception of certain libraries where attempts were made to exclude it. There are thought to be numerous other circular DNA molecules of varying sizes present with unknown functions (Dhar *et al.*, 1995; Lioutas *et al.*, 1995), but unfortunately they have not yet been identified in the genome shotgun sequence data. The exact reasons for this are unknown, but the small size of the DNA may have prevented proper shearing during the library construction process. These molecules represent an intriguing unsolved aspect of the *E. histolytica* genome.

2.4. tRNA genes

Perhaps the most unusual structural feature identified in the *E. histolytica* genome is the unprecedented number and organisation of its tRNA genes (Clark *et al.*, 2006a). Over 10% of the sequence reads contained tRNA genes, and these are (with a few exceptions) organised in linear arrays. The array organisation of the tRNAs was immediately obvious in some cases from the presence of more than one repeat unit in individual sequence reads and in other cases from their presence in both reads from the two ends of the same clone. However, because of the near complete identity of the array units they were impossible to assemble by the software used and therefore the size of the arrays cannot be estimated accurately.

By manual assembly of tRNA gene-containing reads, 25 distinct arrays with unit sizes ranging from under 500 bp to over 1750 bp were identified (Clark *et al.*, 2006a). The arrayed genes are predicted to be functional because of the 42 acceptor types found in arrays none has been found elsewhere in the genome. These array units encode between one and five tRNAs and a few tRNA genes are found in more than one unit. Three arrays also encode the 5S RNA and one encodes what is thought to be a small nuclear RNA. Experimental quantitative hybridisations suggest a copy number of between about 70 and 250 for various array units. In total it is estimated that there are about 4500 tRNA genes in the genome.

The frequency of a particular tRNA gene appears to be independent of the codon usage in *E. histolytica* protein-coding genes.

Between the genes in the array units are complex, non-coding, short tandem repeats ranging in size from 5 to over 36 bp. Some variation in short tandem repeat number is observed between copies of the same array unit, but this variation is usually minor and not visible when inter-tRNA polymerase chain reaction (PCR) amplification is performed. However, these regions often exhibit substantial variation when different isolates of *E. histolytica* are compared and this is the basis of a recently described genotyping method for this organism (Ali *et al.*, 2005).

There is indirect evidence to suggest that the tRNA arrays are present at the ends of chromosomes. Although allelic *E. histolytica* chromosomes often differ substantially in size in pulse-field gels, a central protein-encoding region appears to be conserved as DNA digested with rare cutting enzymes gives only a single band in Southern blots when most protein-coding genes are used as probes. In contrast, when some tRNA arrays are used as probes on such blots, the same number of bands is seen in digested and undigested DNA. It is therefore tempting to conclude that the tRNA genes are at the ends of the chromosomes and to speculate that these repeat units may perform a structural role. In *D. discoideum* it is thought that rDNA may function as a telomere in some cases (Eichinger *et al.*, 2005) and the tRNA arrays in *E. histolytica* may perform a similar role.

The chromosomal regions flanking the tRNA arrays are generally devoid of protein coding genes but often contain incomplete transposable elements (see next section) and other repetitive sequences (Clark *et al.*, 2006a). This is also consistent with a telomeric location.

2.5. LINEs

The *E. histolytica* genome is littered with transposable elements. There are two major types of autonomous LINEs (long interspersed elements) of which there are three subtypes (EhLINE 1, 2 and 3) and there are two types of SINEs (short interspersed elements) (Eh SINE1 and 2) (Table 2.2a). The classification of these elements and their organisation has been reviewed recently (Bakre *et al.*, 2005). Phylogenetic analysis of the EhLINEs places them in the R4 clade of non-long terminal repeat (LTR) elements, a mixed clade of elements that includes members from nematodes, insects and vertebrates (Van Dellen *et al.*, 2002a). Analysis of the *E. histolytica* genome shows no evidence for the presence of LTR retrotransposons and very few DNA transposons (of the *Mutator* family) (Pritham *et al.*, 2005).

All copies of EhLINEs examined encode non-conservative amino acid changes, frame shifts and/or stop codons and no copy with a continuous

TABLE 2.2 Summary properties of the repeated DNAs

(a) References for data[a]

Repeat type	Size in kb	Estimated copy no. from genome sequence (Ref no.)	Estimated copy no. per haploid genome from hybridisation (Ref no.)	Transcript size in kb (Ref no.)
EhLINE1	4.8	142 (1) 409; 49 full-length (2)	140 (3)	No full-length transcript (4)
EhLINE2	4.72	79 (1) 290; 56 full-length (2)	Not Determined	Not Determined
EhLINE3	4.81	12 (1) 52; 3 full-length (2)	Not Determined	Not Determined
EhSINE1	0.5–0.6	219 (1) 272; 81 full-length (2) 214; >90 full-length (3)	500	0.7 (6)
EhSINE2	0.65	120 (1) 117; 62 full-length (2) 122; ~50 full-length (3)	Not Determined	0.75 (7)
EhSINE3	0.58	1 (1,2)	Not Determined	Not detected (3)
Tr	0.7	1 per rDNA episome (5)	Not Determined	0.7 (5)
BspA-like	0.96	77 (8)	190 (3)	Not detected (3)[b]
Ehssp1	0.9–1.1	Not Determined	306 (9)	1.5 (9)

(b) Consensus sequences of Family 16 and 17 repeats[c]

Family	Sequence
Family 16	GTAATGAATATAYAACTAAGAATTTCATT TAAAATGRATATG
Family 17	CAACAAATAAATRGKTTCAATAAAATA

[a] (1): Van Dellen *et al.* (2002a); (2): Bakre *et al.* (2005); (3): This analysis; (4): A. A. Bakre and S. Bhattacharya (unpublished data); (5): Burch *et al.* (1991); (6): Cruz-Reyes *et al.* (1995); (7): Shire and Ackers (2007); (8): Davis *et al.* (2006); (9): Satish *et al.* (2003).

[b] Although no transcript was detected, the protein has been demonstrated on the cell surface and in Western blots using antibodies (Davis *et al.*, 2006).

[c] Standard abbreviation for degenerate sequence positions are used: R = purine, Y = pyrimidine, K = G or T.

open reading frame (ORF) has yet been found. This suggests that the majority of these elements are inactive. However, a large number of EhLINE1 copies do contain long ORFs without mutations in the conserved protein motifs of the reverse transcriptase (RT) and restriction enzyme-like endonuclease (EN) domains, suggesting that inactivity is quite recent. ESTs corresponding to EhLINEs have been found suggesting that transcription of these elements still occurs. Although most R4 elements insert in a site-specific manner, EhLINEs do not show strict site-specificity and are widely dispersed in the genome. They are quite frequently found close to protein-coding genes and inserted near T-rich stretches (Bakre *et al.*, 2005).

All three EhLINE subtypes are of approximately equal size ranging from 4715 to 4811 bp in length. Individual members within an EhLINE family typically share >85% identity, while between families they are <60% identical. By aligning the available sequences, each EhLINE can be interpreted to encode a single predicted ORF that spans almost the entire element (EhLINE1, 1589 aa; EhLINE2, 1567 aa; EhLINE3, 1587 aa). However, a precise 5-bp duplication at nucleotide position 1442 in about 80% of the copies of EhLINE1 creates a stop codon, dividing the single ORF into two. Similarly in 92% of EhLINE2 copies, the single ORF contains a precise deletion of two nucleotides at position 1272, resulting in two ORFs. Very few intact copies of EhLINE3 are found. The location of the stop codon leading to two ORFs appears to be conserved since in both EhLINE1 and EhLINE2 the size of ORF1 is about half that of ORF2 (Bakre *et al.*, 2005). Among the identifiable domains in the predicted proteins are RT and EN. The putative 5′ and 3′ untranslated regions are very short (3–44 bp).

EhLINEs 1 and 2 appear to be capable of mobilising partner SINEs (see next section) for which abundant transcripts have been detected in *E. histolytica*. Putative LINE/SINE partners can be assigned on the basis of conserved sequences at the 3′-ends of certain pairs, which otherwise showed no sequence similarity. The relevance of this assignment for the EhLINE1/SINE1 pair has recently been demonstrated (Mandal *et al.*, 2004).

2.6. SINEs

The two EhSINEs are clearly related to the EhLINEs, as they have a conserved 3′ sequence. They are nonautonomous, non-LTR retrotransposons (nonautonomous SINEs). The genetic elements encoding the abundant polyadenylated but untranslatable transcripts found in *E. histolytica* cDNA libraries [initially designated interspersed elements (Cruz-Reyes and Ackers, 1992; Cruz-Reyes *et al.*, 1995) or *ehapt2* (Willhoeft *et al.*, 2002)] have now been designated EhSINE1 (Van Dellen *et al.*, 2002a; Willhoeft *et al.*, 2002). BLAST searching of databases with representative examples of the first 44 EhSINE1s detected has identified 90 full-length (≥99% complete) copies and at least a further 120 partial (≥50% of full length) copies in the genome. Length variation is observed among EhSINE1s and is largely due to variable numbers of internal 26–27 bp repeats (J. P. Ackers, unpublished data). The majority contain 2 internal repeats and cluster closely around 546 bp in length.

A second *E. histolytica* SINE (EhSINE2) has recently been described (Van Dellen *et al.*, 2002a; Willhoeft *et al.*, 2002). Examination of the 4 published sequences again suggests the presence of variable numbers of short (20 bp) imperfect repeats. BLAST searching identified a total of 47 full-length (≥99%) and at least 60 partial copies in the genome. The 3′-end of EhSINE2 shows high similarity (76%) to the 3′ end of EhLINE2.

A polyadenylated transcript designated UEE1 found commonly in cDNA libraries from *E. dispar* (Sharma *et al.*, 1999) is also a non-LTR retrotransposon. A single copy of a UEE1-like element has been identified in the *E. histolytica* genome and is here designated EhSINE3. There is no significant sequence identity between EhSINE3 and EhLINE3, but the 3′ end of EhSINE3 is very similar to that of EhLINE1.

Analysis of an *E. histolytica* EST library identified over 500 significant hits to both EhSINE1 and EhSINE2. No convincing transcript from EhSINE3 could be identified, although the nearly identical *E. dispar* UEE elements (EdSINE1; Shire and Ackers, 2007) are abundantly transcribed.

A very abundant polyadenylated transcript, *ehapt1*, was described by Willhoeft *et al.* (1999) in a cDNA library. However, only a small number of partial matches could be found in the current *E. histolytica* assembly and only 10–20 strong hits in the much larger *E. histolytica* EST library now available. *ehapt1* does not appear to be a SINE element, and its nature is currently unclear. The lack of matches in the genome suggests that either it is encoded in regions missing from the current assembly or it contains numerous introns.

2.7. Other repeats

The *E. histolytica* genome contains a number of other repetitive elements whose functions are not always clear. There are over 75 genes encoding leucine-rich tandem repeats (LRR) of the type found in BspA-like proteins of the *Treponema pallidum* LRR (TpLRR) subfamily, which has a consensus sequence of LxxIxIxxVxxIgxxAFxxCxx (Davis *et al.*, 2006). These proteins generally have a surface location and may be involved in cell–cell interaction. Genes encoding such proteins are mainly found in Bacteria and some Archaea; so far they have been identified in only one other eukaryote, *Trichomonas vaginalis* (Hirt *et al.*, 2002). An extensive description of the BspA-like proteins of *E. histolytica* has recently been published (Davis *et al.*, 2006) and one of them has been shown to be surface exposed (Davis *et al.*, 2006).

E. histolytica stress sensitive protein (Ehssp) 1 is a dispersed, polymorphic and multicopy gene family (Satish *et al.*, 2003) and is present in ~300 copies per haploid genome as determined by hybridisation (Table 2.2a). The average Ehssp1 ORF is 1 kb in length with a centrally located acidic-basic region (ABR) that is highly polymorphic. Unlike other such domains no clear repetitive motifs are present. The protein has, on average, 21% acidic (aspartate and glutamate) and 17% basic (arginine and lysine) amino acids, most of which are located in the ABR. The ABR varies in size from 5 to 104 amino acids among the various copies. No size polymorphism is seen outside the central ABR domain. The genes have an unusually long 5′ untranslated region (UTR; 280 nucleotides). Only one or a few copies of the gene are transcribed during normal growth, but many are turned on under stress conditions. Homologues of this gene are present in *E. dispar*, but there is very little size polymorphism in the *E. dispar* gene family.

Eukaryotic genomes usually contain numerous microsatellite loci with repeat sizes of two to three basepairs. With the exception of di- and trinucleotides made up entirely of A+T such sequences are rare in the *E. histolytica* genome. In contrast, two dispersed repeated sequences of unknown function occur far more frequently than would be expected at random. Family 16 has a 42 base consensus sequence and occurs ~38 times in the genome while family 17 has a 27 base consensus sequence and occurs 35 time in the genome (Table 2.2b). The significance of these sequences remains to be determined.

2.8. Gene number

The current assembly predicts that the genome contains around 10,000 genes, almost twice as many as seen in *P. falciparum* (Gardner *et al.*, 2002) or *Saccharomyces cerevisiae* (Goffeau *et al.*, 1996) but closer to that of the free-living protist *D. discoideum* (~12,500; Eichinger *et al.*, 2005). It should be remembered that this number will change as the assembly improves, and is likely to decrease somewhat. Nevertheless, the comparatively large gene number when compared to some other parasitic organisms reflects both the relative complexity of *E. histolytica* and the presence of large gene families, despite the loss of certain genes as a consequence of parasitism. Both gene loss and gain can represent an adaptive response to life in the human host. Gene loss is most evident in the reconstruction of metabolic pathways of *E. histolytica*, which show a consistent pattern of loss of synthetic capacity as a consequence of life in an environment rich in complex nutrient sources. Similarly, analyses of expanded gene families with identifiable functions indicate that many are directly associated with the ability to sense and adapt to the environment within the human host and the ability to ingest and assimilate the nutrients present. One consequence of these gene family expansions being linked to phagocytosis of bacteria and other cells may be an association between many of these gene families and pathogenicity.

2.9. Gene structure

Most *E. histolytica* genes comprise only a single exon; however as many as 25% may be spliced and 6% contain 2 or more introns. Therefore, mRNA splicing is far less common than in the related protist *D. discoideum* or the malaria parasite *P. falciparum*. The genome contains all of the essential machinery for splicing (see Section 2.14) and a comparison of intron positions suggests that *D. discoideum* and *E. histolytica* have both lost introns since their shared common ancestor with *P. falciparum*, although many more have been lost in the *E. histolytica* lineage. A good example of this intron loss is the vacuolar ATP synthase subunit D gene (Fig. 2.1). This protein is highly conserved but the number of introns in each gene varies. *P. falciparum* has five introns, *D. discoideum* has two and *E. histolytica* has one. The positions of three of the five *P. falciparum* introns are conserved in one of the other species, which suggests that these three (at least) were present in the common ancestor and that intron loss has led to the lower number seen in *E. histolytica* today. This loss is consistent with reverse transcriptase mediated 3′ intron loss (Roy and Gilbert, 2005) as the 5′-most introns are retained. It would appear that this process has been more active in the *E. histolytica* and *D. discoideum* lineages than in *P. falciparum*, possibly because *Plasmodium* lacks a reverse transcriptase.

2.10. Gene size

Genes in *E. histolytica* are surprisingly short, not only due to the loss of introns but also in the predicted lengths of the proteins they code for. On average the predicted length of a protein in *E. histolytica* is 389 amino acids (aa) which is 129 aa and 372 aa shorter than in *D. discoideum* and

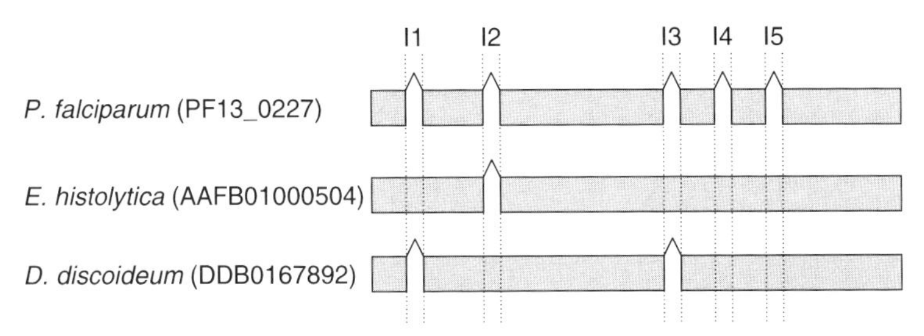

FIGURE 2.1 Positions of introns in the vacuolar ATPase subunit D gene in *P. falciparum*, *D. discoideum* and *E. histolytica*.

P. falciparum respectively. In fact the protein length distribution is most similar to that of the microsporidian *Encephalitozoon cuniculi* (Fig. 2.2) which has a very compact genome of 3 Mbp and <2000 genes. Direct

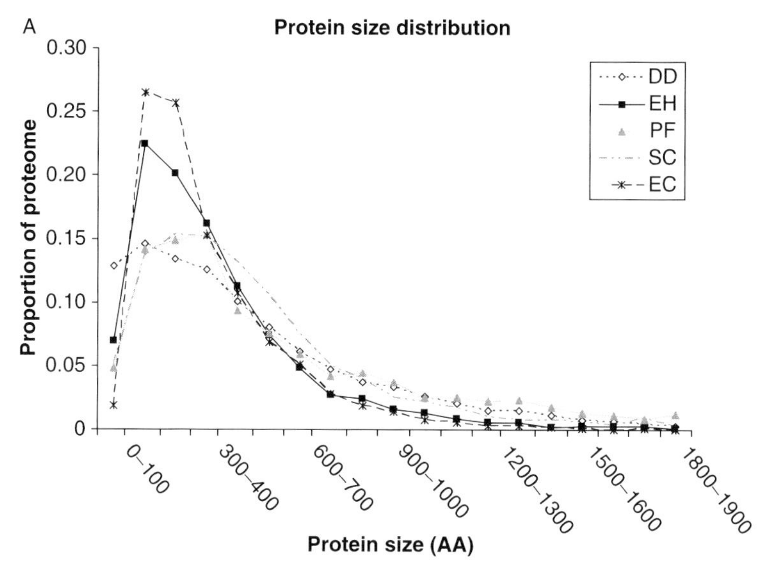

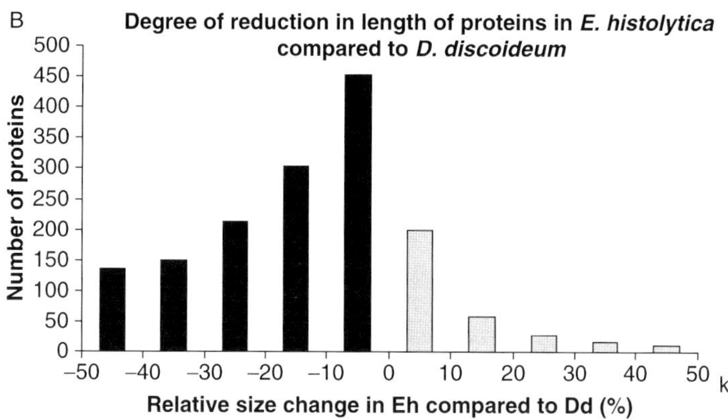

FIGURE 2.2 Comparison of protein sizes in *E. histolytica* and *D. discoideum.* (A) The graph shows the distribution of predicted amino acid length across sequenced genomes from single celled eukaryotes: *D. discoideum* (DD) *Encephalitozoon cuniculi* (EC), *P. falciparum* (PF), *E. histolytica* (EH), and *S. cerevisiae* (SC). *Entamoeba histolytica* and *E. cuniculi* have a distribution that is skewed towards smaller proteins relative to the other species. (B) The histogram displays the degree of size change of genes in *E. histolytica* relative to *D. discoideum* when comparing orthologous genes identified by reciprocal best BLAST hits. The black bars show genes that are smaller in *E. histolytica* where as the gray bars are smaller in *D. discoideum.*

comparison of orthologous genes between *E. histolytica* and its closest sequenced relative *D. discoideum* demonstrates this phenomenon quite well, with the majority of *E. histolytica* proteins being shorter than the *D. discoideum* counterpart (N. Hall, unpublished data). Protein length is normally very well conserved among eukaryotes, so the reason for protein shortening is unclear. It has been postulated that in bacteria reduced protein lengths reflects a reduced capacity for signalling (Zhang, 2000). This would not seem to be the case here as the number of genes identified as having a role in signalling suggests quite the opposite. An alternative theory is that as *E. histolytica* has reduced organelles it is possible that its proteins contain fewer or simpler targeting signals.

2.11. Protein domain content

The most common protein family (Pfam) domains of *E. histolytica* are shown in Table 2.3. The domains that are unusually common in *E. histolytica* reflect some of the more unusual aspects of the biology of this protist. For example, the Rab and Rho families that are involved in signalling and vesicle trafficking are among the most common domains in *E. histolytica* while in other species they are not often among the top 50 families. This could well be due to the fact that *E. histolytica* has a 'predatory' life style, and these domains are intimately involved in environmental sensing, endocytosis and delivery of lysosomes to the phagosome. There are also a number of domains involved in actin dynamics and cytoskeletal rearrangement that are not common in non-phagocytic species, such as the gelsolin and Src-homology 3 (SH3) domains. Myb domains are the most common transcription regulatory domains in *E. histolytica*; this domain is also common in plants where the proteins regulate many plant-specific pathways (Ito, 2005). An important finding from an initial analysis was the presence of unusual multidomain proteins, including five proteins containing both RhoGEF (Rho GTPase guanine nucleotide exchange factor) and Arf-GAP (ADP ribosylation factor GTPase activating protein) domains, suggesting a mechanism for direct communication between the regulators of vesicle budding and cytoskeletal rearrangement. Over 80 receptor kinases were identified (see Section 7.2.2), each containing a kinase domain and a C rich extracellular domain. These kinases fall into distinct classes, depending on the presence of CXC or CXXC repeats. There are also domains that are common in most other sequenced genomes but rare or missing from *E. histolytica*. For example, most mitochondrial carrier domain proteins are not needed in *E. histolytica* as it lacks a normal mitochondrion (Section 8).

TABLE 2.3 Number and ranking of Pfam domains across different genomes

Domain name	Domain detail	EH		EC		PF		SC		AT		CE		DD	
		#	Rank	#	Rank	#	Rank	#	Rank	#	Rank	#	Rank	#	Rank
WD40	WD domain, G-β repeat	249	1	139	1	287	2	414	1	1137	3	694	1	719	2
LRR_1	Leucine-rich repeat	131	2	40	2	55	12	43	17	3793	2	494	5	372	4
Pkinase	Protein kinase domain	95	3	27	5	78	8	116	2	839	4	405	8	225	7
HEAT	HEAT repeat	70	4	13	15	44	17	114	3	220	17	162	26	108	12
efhand	EF hand	58	5	7	28	80	7	29	25	422	8	213	20	153	9
RRM_1	RNA recognition motif	57	6	30	3	95	6	86	6	375	10	223	19	134	10
Ras	Ras family	46	7	9	22	13	44	25	28	78	68	66	76	126	11
TPR_1	Tetratricopeptide repeat	42	8	23	7	48	15	103	4	334	12	180	22	168	8
Ank	Ankyrin repeat	34	9	6	34	55	12	61	9	431	6	629	2	446	3
PUF	Pumilio-family RNA binding repeat	33	10	8	23	15	34	51	13	142	32	75	68	34	62
RhoGAP	RhoGAP domain	27	11	2	118	1	520	11	80	9	559	31	138	45	39
Myb_DNA-binding	Myb-like DNA-binding domain	22	12	15	12	10	62	21	34	424	7	30	141	55	26
RhoGEF	RhoGEF domain	22	12	1	230	0	1215	3	366	0	2581	34	130	47	37

(continued)

TABLE 2.3 *(continued)*

		EH		EC		PF		SC		AT		CE		DD	
Domain name	Domain detail	#	Rank	#	Rank	#	Rank	#	Rank	#	Rank	#	Rank	#	Rank
Helicase_C	Helicase conserved C-terminal domain	20	14	28	4	64	11	74	8	150	31	98	49	84	20
DEAD	DEAD/DEAH box helicase	20	14	22	9	49	14	59	10	103	50	76	67	48	35
PH	PH domain	19	16	1	230	5	123	25	28	22	255	77	63	94	16
Metallophos	Calcineurin-like phosphoesterase	19	16	6	34	16	32	21	34	66	83	78	62	31	67
Gelsolin	Gelsolin repeat	18	18	2	118	2	295	4	255	33	169	12	323	29	68
LIM	LIM domain	17	19	0	703	0	1,215	8	116	16	341	103	47	56	25
CH	Calponin homology (CH) domain	16	20	4	54	1	520	7	137	26	211	57	87	49	33
Filamin	Filamin/ABP280 repeat	16	20	0	703	1	520	0	1842	2	1450	55	91	10	203

Note: Columns labeled '#' give the total number of occurrences of a particular domain. Columns labeled 'Rank' give the ranking of the domain where the most common domain is ranked 1. The organisms shown are *E. histolytica* (EH), *Encephalitozoon cuniculi* (EC), *Plasmodium falciparum* (PF), *Arabidopsis thaliana* (AT), *Saccharomyces cerevisae* (SC), *Dictyostelium discoideum* (DD).

2.12. Translation-related proteins

Two of the predicted tRNAs (Ile^{TAT} and Tyr) need to be spliced because of the presence of an intron. tRNA introns are distinct in structure from those in protein-coding genes and require a distinct splicing machinery. The expected enzymes required for this splicing are present as are a number of tRNA modification enzymes (including those for synthesising queuine and pseudouridine) and rRNA methylases that act on specific bases in their respective RNA molecules. The expected panel of tRNA synthetases necessary for aminoacylating the tRNAs is also present, with one or two gene copies for each type.

The majority of ribosomal protein genes are well conserved in *E. histolytica* and only the gene for large subunit protein L41 could not be identified. The missing protein is only 25 amino acids in length, 17 of which are arginines or lysines, which would make it difficult to identify in this A + T-rich genome, but it is highly conserved, having been reported from Archaea to mammals. However, it also appears to be dispensable, as *S. cerevisiae* can grow relatively normally after deletion of both its copies (Yu and Warner, 2001). Nevertheless, deletion of L41 in *S. cerevisiae* reduces the level of 80S ribosomes, suggesting that it is involved in ribosomal subunit association, reduces peptidyl transferase activity and increases translocation (Dresios *et al.*, 2003). In addition, L41 has been shown to interact with the β subunit of protein kinase CKII and to stimulate phosphorylation of DNA topoisomerase II α by CKII (Lee *et al.*, 1997b). If this gene is truly absent from *E. histolytica*, then it may have important consequences for the cell.

No genes for mitochondrial ribosomal proteins were found. Their absence is not surprising since *E. histolytica* lacks typical mitochondria (see Section 8).

In eukaryotic translation, elongation factor (EF)-1 is activated upon GTP binding and forms a ternary complex with aminoacyl tRNAs and ribosomes. EF-1 β and δ subunits work as GDP-GTP exchange factors to cycle EF-1 α between two forms while EF-1 γ provides structural support for the formation of this multimeric complex. EF2 assists in the translocation of tRNAs on the mRNA by exactly one codon. *E. histolytica* has most of the expected factors except for EF-1 δ, a protein involved in exchanging GDP with GTP. This is also absent from *S. cerevisiae* and *P. falciparum*. It is likely that EF-1 β carries out this activity. It is thought that the EF-1 complex can exist in two forms, EF-1-$\alpha/\beta/\gamma$ and EF-1-$\alpha/\delta/\gamma$. In *E. histolytica*, probably only the former complex exists.

Eukaryotes typically have two polypeptide release factors, eRF1 and eRF3. Both of these factors have been found in *E. histolytica*.

2.13. Analysis of cell cycle genes

Alternation of DNA duplication and chromosome segregation is a hallmark in the cell cycle of most eukaryotes. Carefully orchestrated processes coordinate an ensemble of cell cycle regulating 'checkpoint' proteins that ensure progeny cells receive an exact copy of the parental genetic material (Hartwell and Weinert, 1989). Unlike most eukaryotes, *E. histolytica* cells can reduplicate their genome several times before cell division occurs (Gangopadhyay *et al.*, 1997). Approximately 5–20% of the trophozoites (depending on the growth phase) in axenic cultures are multi-nucleated. Additionally, DNA reduplication may occur without nuclear division so that single nuclei contain 1X-6X or more genome contents (Das and Lohia, 2002). Thus axenically cultured *E. histolytica* trophozoites display heterogeneity in their genome content, suggesting that eukaryotic cell cycle checkpoints are either absent or altered in this organism. Around 200 genes have been identified in yeasts that play a direct role in cell cycle progression.

2.13.1. DNA replication initiation and DNA duplication

The DNA replication licensing system is one of the crucial mechanisms that ensures the alternation of S-phase with mitosis in most cells (Tye, 1999). Initiation of DNA replication involves binding of the replicative helicases to DNA replication origins in late mitosis. Loading of the replicative helicase Mcm2–7 proteins is preceded by formation of the pre-replicative complex (pre-RC) and its subsequent activation. Formation of pre-RC requires the ordered assembly of the origin recognition complex (ORC), cell division cycle 6 (Cdc6), Cdt1 and the Mcm2–7 proteins. The pre-RC is activated by the protein kinase Cdc7p and its regulatory subunit Dbf4 (Masai and Arai, 2002). Other factors that regulate the transition from pre-RC to replication initiation are Mcm10p, Cdc45p, TopBP1, RecQL4 and the GINS complex (Gregan *et al.*, 2003; Machida *et al.*, 2005; Merchant *et al.*, 1997; Wohlschlegel *et al.*, 2002). Two other Mcm (minichromosome maintenance) proteins—Mcm8 and Mcm9—have been identified in metazoan systems and are believed to be part of the replicative helicase (Maiorano *et al.*, 2006). Replication origin licensing is inactivated during S-phase but Mcm2–9p may function as a helicase that unwinds DNA ahead of the replication fork during S-phase (Maiorano *et al.*, 2006). Once S-phase has begun, the formation of new pre-RC is kept in check by high cyclin-dependent kinase (CDK) activity and by the activity of the protein geminin (Bell and Dutta, 2002).

A detailed analysis of the *E. histolytica* genome shows that homologues of several proteins required for DNA replication initiation are absent. These include ORC 2–6, Cdt1, geminin, Cdc7/Dbf4 and Mcm10. A single gene encoding a homologue of the archaeal and human Cdc6/Orc1p

(Capaldi and Berger, 2004) was identified. This suggests that DNA replication initiation in *E. histolytica* is likely similar to archaeal replication initiation where a single Cdc6p/ORC1p replaces the hetero-hexameric ORC complex (Kelman and Kelman, 2004). Several proteins described from metazoa, such as Cdt1, geminin, Mcm8 and Mcm9, have not been found in yeasts. Surprisingly, Mcm8 and Mcm9 were identified in the *E. histolytica* genome.

Of the four known checkpoint genes that regulate DNA replication in *S. cerevisiae* only Mec1 and Mrc1 have homologues in *E. histolytica*. *E. histolytica* homologues of several proteins involved in G1-S transitions are absent, such as Sic1 and Chk1. The S-phase checkpoint genes p21, p27, p53 and retinoblastoma (RB) required for transition from G1 to S-phase in humans were absent in *E. histolytica*. Chk1 and Chk2 genes encode kinases that act downstream from the ATM and ATR kinases (intra-S-phase checkpoint genes). The Chk1 homologue is absent, but a Chk2 homologue has been identified in *E. histolytica* and partially characterised (Iwashita *et al.*, 2005).

2.13.2. Chromosome segregation and cell division

A large number of genes are known to regulate different events during the transition from G2-Mitosis—spindle formation checkpoint, chromosome segregation, mitosis, exit from mitosis and cytokinesis—in *S.cerevisiae*. Many of the proteins required by yeast for kinetochore formation have no obvious homologues in *E. histolytica*, suggesting that amoeba kinetochores may have an altered composition and structure. Proteins of the anaphase promoting complex (APC) regulate transition from metaphase to anaphase. With the exception of APC11, none of the APC proteins could be identified in *E. histolytica*. In contrast two genes encoding CDC 20 homologues, which are known to activate the APC complex, were identified in *E. histolytica* along with ubiquitin and related proteins (Wostmann *et al.*, 1992), indicating that although most APC subunit homologues were absent the pathway of proteasomal degradation for regulation of cell cycle proteins may still be functional in *E. histolytica*. Effectors of the apoptotic pathway and meiosis were also largely absent.

2.13.3. CDKs and cyclins

The CDC28 gene encodes the single CDK in *S. cerevisiae* and regulates cell cycle progression by binding to different cyclins at the G1/S or G2/M boundaries (Reed, 1992; Surana *et al.*, 1991; Wittenberg *et al.*, 1990). Similarly, *Schizosaccharomyces pombe* also encodes a single CDK (*cdc2*) (Simanis and Nurse, 1986). Mammals and plants can encode multiple CDKs and an equally large number of cyclins (Morgan, 1995; Vandepoele *et al.*, 2002). Association of different CDKs with specific cyclins regulates the cell cycle in different developmental stages as well as in specific tissues. CDKs belong to the serine/threonine family of kinases with a conserved PSTAIRE domain where cyclins are believed to bind (Jeffrey *et al.*, 1995; Morgan, 1996), although some mammalian and plant CDKs have been shown to have divergent PSTAIRE motifs. This heterogeneity may or may not affect cyclin binding (Poon *et al.*, 1997). The *E. histolytica* genome encodes at least nine different CDKs among which not even one has the conserved PSTAIRE motif. The closest homologue of the CDC28/cdc2 gene, which shows only conservative substitutions in the PSTAIRE motif (PVSTVRE), was cloned previously (Lohia and Samuelson, 1993). The remaining eight CDK homologues exhibit even greater divergence in this motif. Eleven putative cyclin homologues with a high degree of divergence have been found. Identifying their CDK/cyclin partner along with their roles in the cell cycle is a major task that lies ahead. Some of the CDKs may not function by associating with their functional cyclin partners but may play a role in regulating global gene expression, either by activation from non-cyclin proteins or by other mechanisms (Nebreda, 2006).

E. histolytica presents a novel situation where the eukaryotic paradigm of a strictly alternating S-phase and mitosis is absent. Discrete G1, S and G2 populations of cells are not routinely found in axenic cultures. Instead cells in S-phase show greater than 2× genome contents, suggesting that the G2 phase is extremely short and irregular. This observation together with the absence of a large number of checkpoint genes suggests that regulation of genome partitioning and cell division in *E. histolytica* may be additionally dependant on extracellular signals. *E. histolytica* must, however, contain regulatory mechanisms to ensure that its genome is maintained and transmitted with precision even in the absence of the expected checkpoint controls. The discovery of these mechanisms will be crucial to our understanding of how the *E. histolytica* cell divides.

2.14. Transcription

RNA polymerase II transcription in *E. histolytica* is known to be α-amanitin-resistant (Lioutas and Tannich, 1995). The F homology block of the RNA polymerase II largest subunit (RPB1) has been identified as the putative α-amanitin binding site. This block is highly divergent in the α-amanitin resistant *T. vaginalis* RNA polymerase II (Quon *et al.*, 1996). The *E. histolytica* RPB1 homologue also diverges from the consensus in this region but, interestingly, it is also quite dissimilar to the *T. vaginalis* sequence.

The heptapeptide repeat (TSPTSPS) common to other eukaryotic RNA polymerase II large subunit C-terminal domains (CTD) is not present in the *E. histolytica* protein. Indeed, the *E. histolytica* CTD is not similar to any other RNA polymerase II domain in the current database. However, the CTD of the *E. histolytica* enzyme does remain proline/serine-rich (these amino acids constitute 40% of the CTD sequence). The *E. histolytica* CTD also retains the potential to be highly phosphorylated: of the 24 serines, 6 threonines and 3 tyrosines within the CTD, 9 serines, 3 threonines and 1 tyrosine are predicted to be within potential phosphorylation sites. It is therefore possible that, despite its divergence, modification of the CTD by kinases and phosphatases could modulate protein–protein interactions as is postulated to occur in other RNA polymerases (Yeo *et al.*, 2003). In yeasts, phosphorylation of the CTD regulates association with the mediator protein (Davis *et al.*, 2002; Kang *et al.*, 2001; Kornberg, 2001). The yeast mediator protein complex consists of 20 subunits. However, perhaps due to the divergence of the CTD, only two of these proteins have been identified in *E. histolytica* (Med7 and Med10). Homologues of the Spt4 and Spt5 elongation factors, also thought to interact with the CTD, have been identified.

The RNA polymerase core is composed of 12 putative subunits in *S. cerevisiae* (Young, 1991), while *S. pombe* contains a subset of 10 of these proteins, lacking the equivalents of subunits 4 and 9 (Yasui *et al.*, 1998). In *E. histolytica* only 10 of the RNA polymerase subunits have been identified, identifiable homologues of subunits 4 and 12 being absent. While the homologue of subunit nine was present, it lacks the first of the two characteristic zinc binding motifs of this protein and the DPTLPR motif in the C-terminal region. A similar sequence, DPTYPK, is however present and a homologue of the transcription factor TFIIE large subunit Tfa1, which is proposed to interact with this region of the protein, has been identified (Hemming and Edwards, 2000; Van Mullem *et al.*, 2002). The conserved N-terminal portion (residues 1–52) of Rpb9 is thought to interact with both Rpb1 and Rpb2 in *S. cerevisiae* (Hemming and Edwards, 2000), and homologues of these have been identified.

The core promoter of *E. histolytica* has an unusual tripartite structure consisting of the three conserved elements TATA, GAAC and INR (Purdy *et al.*, 1996; Singh and Rogers, 1998; Singh *et al.*, 1997, 2002). Singh and Rogers (1998) have speculated that the GAAC motif may be the binding site of a second or alternative *E. histolytica* DNA binding protein in the preinitiation complex. It is therefore of interest that, in addition to the *E. histolytica* TATA-binding protein (TBP), two other proteins contain the TATA-binding motif (Hernandez *et al.*, 1997). TBP is a subunit of the TFIID general transcription factor (GTF), which in other organisms is required for the recognition of the core promoter. In the light of the variation in the core promoter previously mentioned, and the divergence in proteins that bind to the core promoter in other parasitic protists, it is not surprising that only 6 of the 14 evolutionary conserved subunits of TFIID, TBP associated factors (TAFs) 1, 5, 6, 10, 12 and 13 were identified. Homologues of some of the global regulatory subunits of the Ccr4/Not complex, which interacts with TBP and TAFs 1 and 13, have also been identified. (A detailed analysis of *E. histolytica* transcription factors can be found at http://www.transcriptionfactor.org).

TAFs 5, 6, 10 and 12 are also components of the histone acetyltransferase (HAT) complexes in other organisms as are SPT6 and SPT16 (Carrozza *et al.*, 2003). While all known components of the HAT complexes have by no means been identified or the role of the previously unknown bromodomain containing proteins encoded in the *E. histolytica* genome understood, histone acetylation complexes are known to be active in *E. histolytica* (Ramakrishnan *et al.*, 2004). Other potential members of chromatin remodelling complexes of *E. histolytica* include the TBP interacting helicase (RVB1 and 2) and the SNF2 subunit of the SWI/SNF complex.

Homologues of some of the other GTFs (TFII E, F and H) but not the large or small subunits were identified. In contrast to the difficulty identifying some of the GTFs, the *E. histolytica* spliceosomes components U1, U2, U4/6, U5 and the Prp19 complex have all been identified. In fact homologues of 10 of the 14 'core' small nuclear ribonucleoproteins (snRNPs), 2 of the U1-specific snRNPs, 7 of the 10 U2-specific snRNPs, 5 of the 6 U5-specific snRNPs, 3 of the U4/6 specific snRNPs and 4 of the 9 subunits of the Prp19 complex have been found. Indeed *E. histolytica* has homologues of ~80% of the *S. cerevisiae* splicing machinery (Jurica and Moore, 2003).

Like *Giardia intestinalis*, *E. histolytica* has short 5′ untranslated regions on its mRNAs. However, unlike those of *G. intestinalis*, *E. histolytica* mRNA has been shown to be capped (Ramos *et al.*, 1997; Vanacova *et al.*, 2003). Identification of homologues of the Ceg1 RNA guanylyltransferase—an enzyme that adds an unmethylated GpppRNA cap to new transcripts—and of Abd1—which methylates the cap to form m7GpppRNA—gives new insight into the probable cap structure in *E. histolytica* (Hausmann *et al.*, 2001; Pillutla *et al.*, 1998). It has been proposed that the capping enzymes interact with the phosphorylated CTD of RNA polymerase (Schroeder *et al.*, 2000). The CTD of *E. histolytica* large subunit is, as discussed earlier, not well conserved but contains several probable phosphorylation sites.

mRNAs in *E. histolytica* are polyadenylated, and the polyadenylation signal is found within the short 3′ untranslated region (Bruchhaus *et al.*, 1993; Li *et al.*, 2001). However, only 8 of the 18 yeast cleavage and polyadenylation specificity factor (CPSF) subunits are identifiable in *E. histolytica*.

3. VIRULENCE FACTORS

3.1. Gal/GalNAc lectin

One of the hallmarks of *E. histolytica* pathogenicity is contact-dependent killing of host cells. *E. histolytica* is capable of killing a variety of cell types, including human intestinal epithelium, erythrocytes, neutrophils and lymphocytes (Burchard and Bilke, 1992; Burchard *et al.*, 1992a,b; Guerrant *et al.*, 1981; Ravdin and Guerrant, 1981). Cytolysis occurs as a stepwise process that begins with adherence to target cells via galactose/*N*-acetyl D-galactosamine-inhibitable (Gal/GalNAc) lectin (Petri *et al.*, 1987; Ravdin and Guerrant, 1982). Adherence via the Gal/GalNAc lectin is a requirement for cell killing because in the presence of galactose or GalNAc target cells are not killed by the amoebae. Target cell death occurs within 5–15 min and is often followed by phagocytosis. Inhibition of the Gal/GalNAc lectin with galactose or specific antibody also blocks phagocytosis (Bailey *et al.*, 1990). Resistance to lysis by the complement system is also mediated in part by the Gal/GalNAc lectin. The lectin contains a CD59-like domain that likely helps protect the trophozoites from complement; CD59 is a surface antigen of many blood cells known to have this property (Braga *et al.*, 1992).

The Gal/GalNAc lectin is a membrane complex that includes heavy (Hgl) 170 kilodalton (kDa), and light (Lgl) 30–35 kDa subunits linked by disulphide bonds, and a non-covalently associated intermediate (Igl) 150 kDa subunit (Cheng *et al.*, 2001; Petri *et al.*, 1989). The structure and function of the Gal/GalNAc lectin has recently been reviewed (Petri *et al.*, 2002). The heavy subunit is a type1 transmembrane protein while the light and intermediate subunits have glycosylphosphatidylinositol (GPI) anchors (Cheng *et al.*, 2001; McCoy *et al.*, 1993). Gal/GalNAc lectin subunits do not share any significant protein identity or similarity to any other known proteins, though Hgl and Igl have some very limited regions of similarity with known classes of proteins that will be discussed below.

3.1.1. The heavy (Hgl) subunit

On the basis of pulse-field gel electrophoresis there are five loci in the genome with similarity to the Hgl subunit. However, the current genome assembly only identifies two complete genes, one of which corresponds to Hgl2 (Tannich *et al.*, 1991b). The predicted proteins encoded by these loci are 92% identical. In initial assemblies there were three other sequences with high similarity to the Hgl subunit that were pseudogenes. These pseudogenes may account for the additional loci detected by pulse-field gel electrophoresis. The large size of these genes means that assembly problems may also be affecting our interpretation.

Hgl subunit sequences can be divided into domains based on amino acid content and distribution (Fig. 2.3). The amino-terminal domain of ~200 amino acids consists of 3.2% cysteine and 2.1% tryptophan residues. The next domain, also ~200 amino acids, is completely devoid of these 2 amino acids. The C-terminal domain of ~930 amino acids is cysteine-rich, comprising 10.8% cysteine. The number and spacing of all predicted tryptophan and cysteine residues are 100% conserved in the 2 complete genes. Although a portion of the C-terminal domain can be said to contain cysteine-rich pseudo-repeats, there is no clear repetitive structure to the protein (Tannich *et al.*, 1991b). The Hgl subunit has a single transmembrane domain and a highly conserved 41 amino acid cytoplasmic domain. In addition to these two *hgl* genes, the genome contains a newly identified divergent member of the Hgl gene family (XP_650534). This ORF shares 43% similarity with the 2 other Hgl isoforms, and is predicted to encode a protein with an almost identical domain structure to that of Hgl described earlier.

3.1.2. The light (Lgl) subunit

The Lgl subunit is encoded by 5 genes (*lgl1–5*) that share 74–85% amino acid identity. A sequence corresponding to Lgl2 is missing from the current genome assembly. The light subunits range from 270 to 294 amino acids in length. Each isoform has a 12 amino acid signal peptide, 5 conserved cysteine residues and a GPI-anchor addition site. Lgl1 has two potential glycosylation sites. Lgl2 has one of these sites, Lgl3 has one different site, and Lgl4 and Lgl5 have none.

3.1.3. The intermediate (Igl) subunit

The Igl subunit was first identified by a monoclonal antibody that blocked amoebic adherence to and cytotoxicity for mammalian cells (Cheng *et al.*, 1998). Co-purification of the Hgl, Lgl and Igl suggests that these three

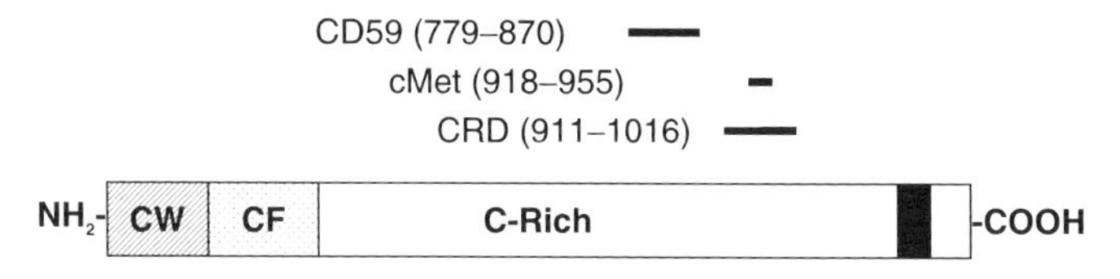

FIGURE 2.3 Domain diagram of the Hgl subunit of the Gal/GalNAc lectin. CW-Cysteine-Tryptophan region; CF-Cysteine free region; C-Rich- Cysteine rich region. The black vertical box near the carboxy terminus of the protein represents the single transmembrane domain. The horizontal black bars above the diagram indicate the location of a carbohydrate recognition domain (CRD), the region with similarity to the hepatic growth factor receptor, c-Met, and the region that has similarity to the CD59, the membrane inhibitor of the complement membrane attack complex. The numbers in parentheses indicate the location of these regions in the Hgl1 isoform (Mann *et al.*, 1991), where the methionine of the immature protein is residue 1.

subunits form a complex (Cheng *et al.*, 1998, 2001). The Igl subunit also has galactose-binding activity (Cheng *et al.*, 1998) and can serve as protective antigen in vaccine trials (Cheng and Tachibana, 2001). There are two loci that encode Igl subunits (Cheng *et al.*, 2001) and the predicted amino acid sequences are 81% identical. The Igl subunit, like the Hgl subunit, does not have any recognisable carbohydrate-binding domain.

3.1.4. Conservation of Gal/GalNAc lectin subunits in other species of *Entamoeba*

There are clearly identifiable orthologues of the Hgl and Lgl subunits among the limited sequences of *E. dispar*, *E. invadens*, *E. moshkovskii* and *E. terrapinae* available at present (Dodson *et al.*, 1997; Pillai *et al.*, 1997; Wang *et al.*, 2003). Because these genomes are incomplete it is possible that as yet unidentified family members will show greater similarity to the *E. histolytica* sequences. Nevertheless, the Lgl subunit is quite conserved among the five *Entamoeba* species. For instance, the *E. terrapinae* gene is 56% identical and 62% similar to *E. histolytica* Lgl1 over a span of 201 amino acids. The Hgl subunits are more diverse. The *E. dispar* Hgl orthologue is highly similar to the *E. histolytica* subunit (86%) but the other species show more diversity, including the region that corresponds to the carbohydrate recognition domain (CRD). However, the number and positions of the cysteine residues are highly conserved, as is the sequence of the cytoplasmic domain, showing only a few changes. It is difficult to put precise numbers to these similarities because the complete sequences of Hgl subunits from the other species are not present in the database. The character of the conservation of the Hgl subunits suggests that the ligand specificity is different for the Hgl subunits of each species but the signalling functions of the cytoplasmic domains are similar, if not perhaps identical. Only *E. dispar* has an identifiable Igl subunit. The other three species clearly have paralogues of the CXXC repeat family to which Igl belongs, but their similarity to Igl is mostly restricted to the CXXC and CXC repeat motifs.

3.2. Cysteine endopeptidases

E. histolytica is characterised by its extraordinary capacity to invade and destroy human tissues. The main lytic activity has been attributed to cysteine endopeptidases. This class of enzymes, which is found in all organisms, plays a major role in the pathogenicity of *E. histolytica* as demonstrated in a large number of *in vitro* and *in vivo* studies (Ankri *et al.*, 1999; Gadasi and Kessler, 1983; Keene *et al.*, 1990; Li et *al.*, 1995; Luaces and Barrett, 1988; Lushbaugh *et al.*, 1985; Reed *et al.*, 1989; Schulte and Scholze, 1989; Stanley *et al.*, 1995). Most striking are results from laboratory animal infections showing that *E. histolytica* trophozoites with reduced cysteine protease activity are greatly impaired in their ability to induce amoebic disease (Ankri *et al.*, 1999; Stanley *et al.*, 1995). In addition, the discovery that *E. histolytica* cysteine proteases possess interleukin-1β convertase activity suggests that these enzymes use a mechanism that is novel in microbial pathogenicity (Zhang *et al.*, 2000).

Thiol-dependent proteolytic activity in *E. histolytica* was first attributed to a neutral sulphydryl protease (McLaughlin and Faubert, 1977) and later to a cytotoxic protease (Lushbaugh *et al.*, 1984). Other terms that have been used to describe closely related or identical enzymes are cathepsin B (Lushbaugh *et al.*, 1985), neutral proteinase (Keene *et al.*, 1990), histolysin (Luaces and Barrett, 1988) (later changed to histolysain; Luaces *et al.*, 1992) and amoebapain (Scholze *et al.*, 1992). *E. histolytica* cysteine endopeptidases were found to be secreted (Leippe *et al.*, 1995) and localised in lysosome-like vesicles or at the surface of the cell (Garcia-Rivera *et al.*, 1999; Jacobs *et al.*, 1998). Molecular cloning has revealed a large number of cysteine endopeptidase genes in the *E. histolytica* genome (Bruchhaus *et al.*, 2003; Garcia-Rivera *et al.*, 1999; Reed *et al.*, 1993; Tannich *et al.*, 1991c., 1992). Interestingly, most of these genes are not expressed during *in vitro* cultivation (Bruchhaus *et al.*, 2003). As our current knowledge of *E. histolytica* biology and pathogenicity is mostly based on analysis of cultured cells, the function of most of the cysteine endopeptidases and their precise role in *E. histolytica* virulence is largely unknown.

Homology searches using conserved active site regions revealed that the *E. histolytica* genome contains at least 44 genes coding for cysteine endopeptidases. Of these, the largest group is structurally related to the C1 papain superfamily (Table 2.4), whereas a few others are more similar to family C2 (calpain-like cysteine proteases), C19 (ubiquitinyl hydrolase), C54 (autophagin) and C65 (otubain), respectively (Table 2.5).

Phylogenetic analyses of the 36 C1-family members revealed that they represent three distinct clades (A, B and C), consisting of 12, 11 and 13 members, respectively. Clades A and B members correspond to the two previously described subfamilies of *E. histolytica* cysteine proteases, designated EhCP-A and EhCP-B (Bruchhaus *et al.*, 2003). In contrast, clade C represents a new group of *E. histolytica* cysteine endopeptidases that has not been described before. EhCP-A and EhCP-B-subfamily members are classical pre-pro enzymes with an overall cathepsin L-like structure (Barrett, 1998) as indicated by the presence of an ERFNIN motif in the pro region of at least 21 of the 23 EhCP-A and EhCP-B enzymes (Fig. 2.4). Interestingly, biochemical studies with purified EhCP-A indicated a cathepsin B-like substrate specificity (Scholze and Schulte, 1988). This is likely due to the substitution of an alanine residue by acidic or charged amino acids in the postulated S2 pocket, corresponding to residue 205 of the papain sequence (Barrett, 1998). As reported previously (Bruchhaus *et al.*, 2003), the EhCP-A and EhCP-B subfamilies differ in the length of the pro regions

TABLE 2.4 Family C1-like cysteine endopeptidases of *E. histolytica*

Protein name	Previous designation	Accession No.	Protein length (aa) Total (pre, pro, mature)	Active site residues	Conserved motifs	Remarks
EhCP-A1	EhCP1	XP_650156	315 (13,80,222)	QCHN	ERFNIN, DWR	
EhCP-A2	EhCP2	XP_650642	315 (13,80,222)	QCHN	ERFNIN, DWR	
EhCP-A3	EhCP3	XP_653254	308 (13,79,216)	QCHN	ERFNIN, DWR	
EhCP-A4	EhCP4	XP_656602	311 (20,73,218)	QCHN	ERFNIN, DWR	
EhCP-A5	EhCP5	XP_650937	318 (20,72,225)	QCHN	ERFNIN, DWR, RGD	Degenerate in *E. dispar*
EhCP-A6	EhCP6	XP_657364	320 (17,79,224)	QCHN	ERFNIN, DWR	
EhCP-A7	EhCP8	XP_648996	315 (13,80,222)	QCHN	ERFNIN, DWR	
EhCP-A8	EhCP9	XP_657446	317 (15,82,220)	QCHN	ERFNIN, DWR	
EhCP-A9	EhCP10	XP_655675	297 (17,90,190)	QCHN	ERFNIN, DWR	
EhCP-A10	EhCP17	XP_651147	420 (18,148,254)	QCHN	ERFNIN, DWR	
EhCP-A11	EhCP19	XP_651690	324 (17,79,228)	QC IN[a]	ERFNIN, DWR	
EhCP-A12	New	XP_653823	317 (14,83,220)	(d)	ERFNIN, DWR	

(*continued*)

TABLE 2.4 *(continued)*

Protein name	Previous designation	Accession No.	Protein length (aa) Total (pre, pro, mature)	Active site residues	Conserved motifs	Remarks
EhCP-B1	EhCP7	XP_651581	426 (15,106,305)	QCHN	ERFNIN, PCNC	Hydrophobic C-terminus
EhCP-B2	EhCP11	AAO03568	431 (15,106,310)	QCHS[a]	ERFNIN, PCNC	GPI cleavage site
EhCP-B3	EhCP12	XP_656747	474 (16,107,351)	QCHN	ERFNIN, PCNC	TMH:444–466 aa
EhCP-B4	EhCP13	XP_648501	379 (16,105,258)	QCHN	ERFNIN, PCNC	TMH or GPI cleavage site
EhCP-B5	EhCP14	XP_652671	434 (12,108,314)	QCHN	ERFNIN, PCNC	GPI cleavage site
EhCP-B6	EhCP15	XP_652465	300 (14,55,231)	QCHN	PCNC	Hydrophobic C-terminus
EhCP-B7	EhCP16	XP_650400	650 (18,144,488)	QCHN	ERFNIN, PCNC	Hydrophobic C-terminus, Cys-rich profile
EhCP-B8	EhCP18	XP_651049	473 (15,105,353)	QCHN	ERFNIN, PCNC, RGD	GPI cleavage site
EhCP-B9	EhCP112	XP_652993	446 (19,112,315)	QCHN	ERFNIN, PCNC, RGD	Hydrophobic C-terminus, Cys-rich profile
EhCP-B10	New	XP_648306	372 (b)	QCHN	ERFNIN, PCNC, RGD	Hydrophobic C-terminus
EhCP-B11	New	XP_648013	133 (b)	Q ? ? ?	PCNC	

EhCP-C1	New	XP_654453	586 (c)	QCIN[a]	HS(X)$_6$ICP	TMH:12–34
EhCP-C2	New	XP_656632	567 (c)	QCHN	HS(X)$_6$ICP	TMH:27–49
EhCP-C3	New	XP_655128	572 (c)	QCHN	HS(X)$_6$LCP	TMH:17–39
EhCP-C4	New	XP_655800	502 (c)	QCHN	LT(X)$_6$LCP	
EhCP-C5	New	XP_654800	557 (c)	QCHN	IS(X)$_6$ICP	TMH:20–42
EhCP-C6	New	XP_651553	557 (c)	QCHD[a]	HS(X)$_6$LCA	TMH:14–36
EhCP-C7	New	XP_657273	595 (c)	QCHN	IS(X)$_6$LCP	TMH:19–41
EhCP-C8	New	XP_655479	627 (c)	QCHN	IS(X)$_6$ICP	TMH:29–51
EhCP-C9	New	XP_655011	518 (c)	(d)	HS(X)$_6$ICP	TMH:12–34
EhCP-C10	New	XP_654829	530 (c)	QCHN	IS(X)$_6$ICP	TMH:15–37
EhCP-C11	New	XP_648083	526 (c)	(d)	HS(X)$_6$ICP	TMH:20–42
EhCP-C12	New	XP_650829	473 (c)	(d)	MS(X)$_6$LCG	TMH:26–48 & 449–471
EhCP-C13	New	XP_656556	564 (c)	QCHN	VS(X)$_6$RCG	TMH:21–43

a: active sites that lack the canonical motif QCHN; b: incomplete sequence; c: cleavage sites to be determined; d: not conserved; ???: incomplete active site sequence.

TABLE 2.5 Family C2-, C19-, C54- and C65-like cysteine endopeptidases of *E. histolytica*

Name	Homology	Family	ProteinID	Protein length	Active site
EhCALP1	Calpain-like	C2	XP_649922	591 aa	Not conserved
EhCALP2	Calpain-like	C2	XP_657312	473 aa	QCHN
EhUBHY	Ubiquitin Hydrolase-like	C19	XP_657356	444 aa	NDTN
EhAUTO1	Autophagin-like	C54	XP_651386	325 aa	YCHS
EhAUTO2	Autophagin-like	C54	XP_653798	364 aa	YCHD
EhAUTO3	Autophagin-like	C54	XP_652043	364 aa	YCHD
EhAUTO4	Autophagin-like	C54	XP_656724	348 aa	YCHD
EhOTU	Otubain-like	C65	XP_654013	259 aa	DCH

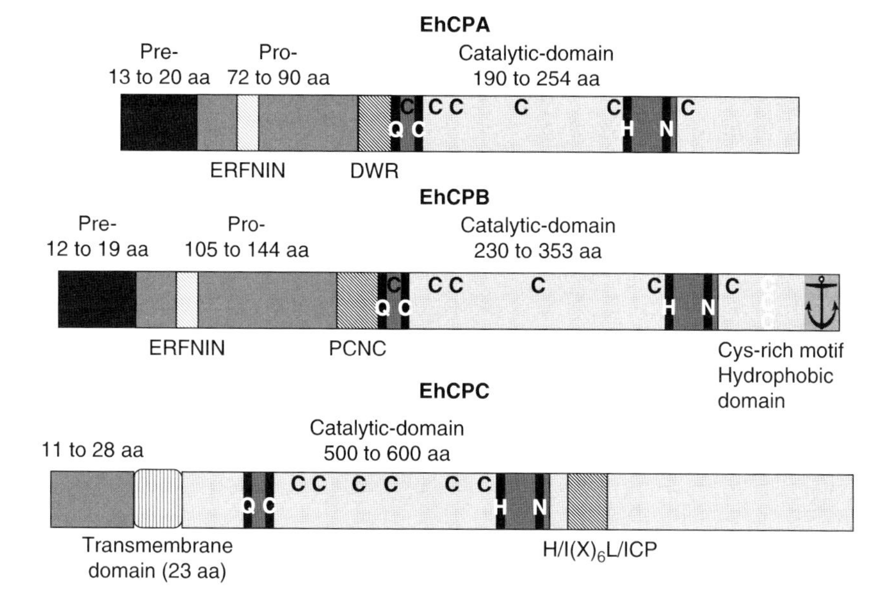

FIGURE 2.4 Structural domains of the three different types of family C1-like cysteine endopeptidases EhCP-A, EhCP-B and EhCP-C. Shown are the location and length of domains specific for each the three types as well as the conserved active site and cysteine residue.

as well as of the catalytic domains, and have distinct sequence motifs in the N-terminal regions of the mature enzymes (DWR vs. PCNC). Moreover, none of the EhCP-A subfamily but 10 of the 11 EhCP-B sequences contain hydrophobic stretches near or at the C-terminus, some of which are predicted to constitute transmembrane helices (TMH) or GPI-attachment moieties. This finding is consistent with previous reports on surface localisation of *E. histolytica* cysteine proteases but, so far, studies on the cellular localisation of the various EhCP-B molecules have not been reported.

In contrast to the EhCP-A and EhCP-B subfamilies, primary structure prediction indicates that EhCP-C members are not pre-pro enzymes, as they lack hydrophobic signal sequences as well as identifiable pro regions. Instead, they contain a hydrophobic region located 11–28 amino acids from the N-terminus, which is predicted to form a TMH (Fig. 2.4). Therefore, this new group of molecules appears to be membrane associated via a signal anchor. All EhCP-C enzymes have a conserved motif of the sequence $H/I(X)_6L/ICP$ in the C-terminal half but they differ substantially in their pI, with values ranging from 4.6 to 8.8. As there is no example of a structurally related cysteine endopeptidase corresponding to the EhCP-C subfamily in other organisms, the specific functions of this group of molecules remain completely unknown.

In addition to the large number of C1 superfamily members, the *E. histolytica* genome contains two genes encoding cysteine endopeptidases homologous to family C2 or calpain-like cysteine proteases (EhCALP1 and EhCALP2). Enzymes of this class contain several calcium-binding domains and have been shown to participate in a variety of cellular processes, including remodelling of the cytoskeleton and membranes, signal transduction pathways and apoptosis.

Another four genes were identified coding for enzymes with homology to the peptidase family C54, also termed autophagins (EhAUTO1–4). The process of autophagy has been studied in human and yeast cells (Kirisako *et al.*, 2000; Marino *et al.*, 2003). Autophagy is a mechanism for the degradation of intracellular proteins and the removal of damaged organelles. During this process the cellular components become enclosed in double membranes and are subsequently degraded by lysosomal peptidases. Autophagins seem to be important for cytoplasm-to-vacuole targeting.

Two other genes encoding putative cysteine endopeptidases of *E. histolytica* show homology to the C19 and C65 families. These two groups of enzymes are known to be involved in ubiquitin degradation. Family C19 are ubiquitinyl hydrolases described as having ubiquitin-specific peptidase activity in humans. C65 or otubains are a group of enzymes with isopeptidase activity, which releases ubiquitin from polyubiquitin.

In summary, the *Entamoeba* genome contains a considerable number of endopeptidase genes. Elucidation of the precise role of each of the various enzymes will be a major challenge but may help us to understand the mechanism(s) of virulence and other unique properties of this protistan parasite.

3.3. *Amoeba pores* and related proteins

In the lysosome-like granular vesicles of *E. histolytica* is found a family of small proteins, *amoeba pores,* that are cytolytic towards human host cells, display potent antibacterial activity and cause ion channel formation in artificial membranes (for a review, see Leippe, 1997). Three *amoeba pore* isoforms have been isolated and biochemically characterised, and their primary structure has been elucidated by molecular cloning of the genes encoding their precursors (Leippe *et al.*, 1991, 1992, 1994b). These membrane-permeabilising polypeptides are discharged by *E. histolytica* into bacteria-containing phagosomes to combat growth of engulfed microorganisms (Andrä *et al.*, 2003). Because of their potent cytolytic activity against human cells *in vitro* (Berninghausen and Leippe, 1997; Leippe *et al.*, 1994a), *amoeba pores* have been viewed as a crucial element of the machinery use by the parasite to kill host cells. Trophozoites of *E. histolytica* lacking the major isoform *amoeba pore* A, whether through antisense inhibition of translation (Bracha *et al.*, 1999) or epigenetic silencing of the gene (Bracha *et al.*, 2003), became avirulent demonstrating that this protein plays a key role in pathogenesis. Relatives of these protistan polypeptides are found in granules of porcine and human cytotoxic lymphocytes where they are termed NK-lysin and granulysin, respectively. All of these polypeptides are 70–80 amino acids in length and are characterised by a compact α-helical, disulphide-bonded structure known as the saposin-like fold. The structures of the amoebic and mammalian polypeptides have been solved and compared (Anderson *et al.*, 2003; Hecht *et al.*, 2004; Leippe *et al.*, 2005; Liepinsh *et al.*, 1997). The biological activities have also been measured in parallel (Bruhn *et al.*, 2003; Gutsmann *et al.*, 2003) to evaluate the similarities and differences of these effector molecules from organisms whose evolutionary paths diverged very early. As they are active against both prokaryotic and eukaryotic target cells, they may be viewed as broad-spectrum effector molecules.

In the genome of *E. histolytica*, 16 genes coding for putative saposin-like proteins (SAPLIPs) were identified. All of these genes are transcribed by cells growing in axenic culture (Winkelmann *et al.*, 2006). Like *amoeba pores*, the predicted proteins all contain one C-terminal SAPLIP domain and (with one exception) a putative signal peptide (Table 2.6). As a

TABLE 2.6 Attributes of the identified SAPLIPs of *E. histolytica*

	Size, aa			SAPLIP domain					
Name	Entire	Signal peptide[a]	Proform/ Mature[c]	Position aa residues	Similarity Name, Acc. no.[e]	Identical to annotated protein	SAPLIP domain found[g]	Similar to (aa sequence identity, %)[f]	Homologous proteins (aa sequence identity, %)
Amoeba pore A	98	21[b]	77	22–98	SAPOSIN B, IPR 008139	*Amoeba pore* A precursor XP_653265		*Amoeba pore* A AAA29111 (100%)	Disparpore A AAA18632 *E. dispar* (94%)
SAPLIP 1	92	15	77	16–92	SAPOSIN B, IPR 008139	Saposin-like protein XP_655836		*Amoeba pore* A AAA29111 (64%)	Disparpore A AAA18632 *E. dispar* (68%)
Amoeba pore B	96	19[b]	77	20–96	SAPOSIN B, IPR 008139	*Amoeba pore* B precursor (EH-APP) Q24824		*Amoeba pore* B CAA54226 (100%)	Disparpore B AAF04195 *E. dispar* (90%)
Amoeba pore C	101	24[b]	77	25–101	SAPOSIN B, IPR 008139	*Amoeba pore* C XP_656029		*Amoeba pore* C CAA54225 (100%)	Disparpore C AAF04196 *E. dispar* (88%)
SAPLIP 2	153	15	138	71–153	SAPOSIN B, IPR 008139	Hypothetical protein XP_656037		—	—
SAPLIP 3	94	16	78	18–94	SAPOSIN B, IPR 008139	Hypothetical protein XP_656682		*Amoeba pore* A AAA29111 (30%)	Invapore X AAP80381 *E. invadens* (67%)
SAPLIP 4	96	17	79	18–96	SAPOSIN B, IPR 008139	Hypothetical proteins XP_652159 and XP_652303		*Amoeba pore* C CAA54225 (27%)	Disparpore C AAF04196 *E.dispar* (30%)
SAPLIP 5	1026	18	1008	946–1026	SAPOSIN B, IPR 008139	Chromosome partition protein XP_655789		—	—
SAPLIP 6	92	15	77	14–92				—	—

(*continued*)

TABLE 2.6 *(continued)*

Name	Size, aa			SAPLIP domain		Identical to annotated protein	SAPLIP domain found[g]	Similar to (aa sequence identity, %)[f]	Homologous proteins (aa sequence identity, %)
	Entire	Signal peptide[a]	Proform/ Mature[c]	Position aa residues	Similarity Name, Acc. no.[e]				
					SAPOSIN B, IPR 008139	Hypothetical protein XP_655820			
SAPLIP 7	926	17	909	855–926	SAPOSIN B, IPR 008139	Conserved hypothetical protein XP_656441		—	—
SAPLIP 8	980	15	965	902–980	SAPOSIN B, IPR 008139	Hypothetical protein XP_656913		—	—
SAPLIP 9	140	15	125	61–140	SAPOSIN B, IPR 008139	Hypothetical protein XP_650376		—	—
SAPLIP 10	657	16	641	577–657	SAPOSIN B, IPR 008139	—	Genomic survey sequence AZ687176	—	—
SAPLIP 11	693	17	676	615–693[d]	—	—	Genomic survey sequence AZ692153	—	—
SAPLIP 12	873	16	857	793–873	SAPOSIN B, IPR 008139	Hypothetical protein XP_652721		—	—

SAPLIP 13	1009	None predicted	1009	931–1005	SAPOSIN B, IPR 008139	Hypothetical protein XP_655089		—	—
SAPLIP 14	915	17	898	834–915	SAPOSIN B, IPR 008139	—	Genomic survey sequence AZ690015	—	—
SAPLIP 15	804	17	787	728–800	SAPOSIN B, IPR 008139	—	Genomic survey sequence BH132588	—	—
SAPLIP 16	921	15	906	842–921[d]	—	—	Genomic survey sequence AZ546519	—	—

Note: SAPLIPs were named according to the similarity of their SAPLIP domain to *amoeba pore* A.

[a] By the programme SignalP and manually corrected if predicted cleavage site is within the SAPLIP domain.
[b] Verified by experimental data.
[c] With the exception of *amoeba pores* it is not possible to decide whether proteins are further processed.
[d] Identified manually.
[e] Extracted from InterPro databases.
[f] If no similarity is reported, there is none outside the SAPLIP domain.
[g] Sequences only found in GSS section of GenBank with given identifier.

transmembrane domain is not apparent in these proteins, it may well be that they are secretory products stored in the cytoplasmic vesicles and act synergistically with the *amoeba pores*. However, only 4 of them have a similar size to amoeba pores, the others being considerably larger (up to 1009 residues). At present, it is not clear whether these larger gene products represent precursor molecules that are processed further. None of the novel SAPLIPs contain the conserved unique histidine residue at the C-terminus that is a key residue for the pore-forming activity of *amoeba pores* (Andrä and Leippe, 1994; Hecht *et al.*, 2004; Leippe *et al.*, 2005). Indeed, it has recently been shown that recombinant SAPLIP3 has no pore-forming or bactericidal activity, although it does cause membrane fusion *in vitro* (Winkelmann *et al.*, 2006). This is in agreement with the experimental evidence for only three pore-forming entities being present in trophozoite extracts. Therefore, it is most likely that the three *amoeba pores* are the sole pore-forming molecules of the parasite. However, the lipid-interacting activity present in all SAPLIP proteins (Munford *et al.*, 1995) and a function that helps to kill bacterial prey may well characterise all members of the *amoeba pore*/SAPLIP superfamily of this voraciously phagocytic cell.

3.4. Antioxidants

E. histolytica trophozoites usually reside and multiply within the human gut, which constitutes an anaerobic or microaerophilic environment. However, during tissue invasion, the amoebae are exposed to an increased oxygen pressure and have to eliminate toxic metabolites such as reactive oxygen or nitrogen species (ROS/RNS) produced by activated phagocytes during the respiratory burst. *E. histolytica* lacks a conventional respiratory electron transport chain that terminates in the reduction of O_2 to H_2O. However, *E. histolytica* does respire and tolerates up to 5% oxygen in the gas phase (Band and Cirrito, 1979; Mehlotra, 1996; Weinbach and Diamond, 1974). Thus, *E. histolytica* trophozoites must use different antioxidant enzymes for the removal of ROS, RNS and oxygen (Fig. 2.5).

Among the enzymes in the first line of oxidative defence are superoxide dismutases (SODs), which are metalloproteins that use copper/zinc (Cu/ZnSOD), manganese (MnSOD) or iron (FeSOD) as metal cofactors. SODs catalyse the dismutation of superoxide radical anions to form H_2O_2 and O_2 (Fridovich, 1995). Analysis of the *E. histolytica* genome revealed only a single gene coding for a FeSOD and no sequences encoding MnSOD or Cu/ZnSOD. This reflects the situation found in most protistan parasites and is consistent with biochemical studies previously performed on *E. histolytica* lysates (Tannich *et al.*, 1991a).

E. histolytica lacks the tripeptide glutathione (Fahey *et al.*, 1984), which constitutes the major low molecular weight thiol found in almost all

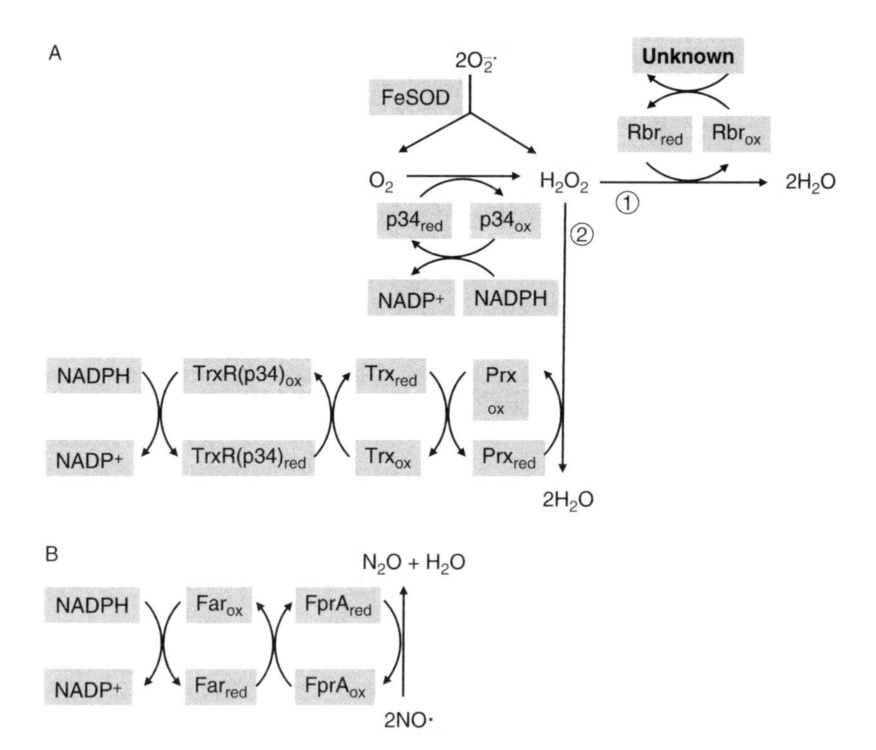

FIGURE 2.5 Predicted antioxidant system of *E. histolytica*. A. Superoxide radical anions are detoxified by an iron-containing superoxide dismutase (FeSOD). Molecular oxygen is reduced to hydrogen peroxide by a NADPH:flavin oxidoreductase (thioredoxin reductase, p34). Hydrogen peroxide is converted to water by rubrerythrin (Rbr). The nature of its redox partner is unknown. Hydrogen peroxide can also be converted to water via a classical thioredoxin redox system consisting of thioredoxin reductase (TrxR, p34), thioredoxin (Trx) and peroxiredoxin (Prx). B. Nitric oxide is reduced by an A-type flavoprotein (FprA) to nitrous oxide and water. For this reaction FprA receives electrons from NADH oxidase (Far).

aerobic cells (Sies, 1999). Instead, *E. histolytica* uses cysteine as its principal low molecular weight thiol (Ariyanayagam and Fairlamb, 1999; Fahey *et al.*, 1984; Nozaki *et al.*, 1999). As expected, coding sequences for enzymes that use glutathione as a cofactor, such as glutathione-S-transferase, glutathione-dependent peroxidase, glutathione reductase or glutaredoxin, are all absent from the *E. histolytica* genome. In addition, genes encoding catalases and peroxidases are also missing, as previously suggested (Sykes and Band, 1977; Weinbach and Diamond, 1974).

Other genes were identified that code for proteins involved in detoxification of H_2O_2, including one with homology to rubrerythrin. Rubrerythrin is a non-haeme iron protein thought to be able to reduce H_2O_2 as part of an oxidative stress protection system (Weinberg *et al.*, 2004). So far, the nature of its redox partner is unknown in *E. histolytica*, and it remains to be determined whether protection against oxidative stress is indeed its main function. Another group of H_2O_2-detoxifying proteins identified in *E. histolytica* are peroxiredoxins. Peroxiredoxins are known from a wide variety of organisms. They are able to reduce H_2O_2 as well as peroxynitrite with the use of electrons provided by thiols. In addition to involvement in the detoxification of reactive oxygen species, peroxiredoxins seem to play a role in other processes such as signalling and differentiation (Hofmann *et al.*, 2002; Rhee *et al.*, 2005; Wood *et al.*, 2003a,b). All peroxiredoxins contain a conserved cysteine residue that undergoes a cycle of peroxide-dependent oxidation and thiol-dependent reduction during the reaction. The whole protein family can be divided into three classes based on the number and position of active site Cys residues (2-Cys, atypical 2-Cys and 1-Cys peroxiredoxins; Wood *et al.*, 2003a,b). In *E. histolytica* five different genes coding for peroxiredoxins were identified (Prx1–5). They all belong to the 2-Cys peroxiredoxin family. Four of them (Prx1–4) share 98% sequence identity and have an unusual N-terminal Cys-rich repeat (KECCKKECQEKECQEKECCC) of unknown function. In contrast, the fifth peroxiredoxin (Prx5) lacks the cysteine-rich N-terminal extension and shares only 30% identity with Prx1–4. Biochemical studies have shown that *E. histolytica* peroxiredoxins are able to detoxify H_2O_2 and cumene hydroperoxide (Bruchhaus *et al.*, 1997; Poole *et al.*, 1997). Moreover, up-regulation of peroxiredoxin and FeSOD was associated with metronidazole resistance in cultured *E. histolytica* trophozoites (Samarawickrema *et al.*, 1997; Wassmann *et al.*, 1999).

Reactions catalysed by peroxiredoxins are dependent on the presence of physiological thiols like thioredoxin (Rhee *et al.*, 2005; Wood *et al.*, 2003b). Thioredoxins are small proteins involved in thiol-redox processes (Holmgren, 2000). They contain two redox-active site cysteine residues of the motif CXXC (Watson *et al.*, 2004). Five genes coding for classical cytoplasmic thioredoxins were identified in the *E. histolytica* genome (Trx1–5). These thioredoxins have a length of 103–114 amino acids and share 25–47% sequence identity. Trx1–3 have identical active site motifs of the sequence WCGPC, whereas the active sites of Trx4 and Trx5 have the sequences SCPSC and WCKDC, respectively. In addition, another five thioredoxin-related proteins were identified (Trx6–10). All have a signal sequence of 15–19 amino acid residues and the active site motif WCGHC, which is also known from the active site of protein disulphide isomerases. However, in contrast to the latter group of enzymes, the *E. histolytica* thioredoxin-related molecules contain only one rather than two active-site motifs and only two of the proteins have an endoplasmic reticulum (ER) membrane retention signal (Freedman *et al.*, 2002). Thus it remains to be determined whether the thioredoxin-related molecules of *E. histolytica* do constitute protein disulphide isomerases or whether they undertake other functions within the cell.

Thioredoxins are kept in the reduced state by the enzyme thioredoxin reductase, which catalyses the reduction of oxidised thioredoxin by NADPH using FAD and its redox-active disulphide (Nakamura, 2005). Two different genes with homology to thioredoxin reductases have been previously described from *E. histolytica* [thioredoxin reductase (TrxR) and NADPH:flavin oxidoreductase (p34)]. They share about 87% sequence identity and both contain the 2 conserved sequence motifs forming the FAD and NAD(P)H binding domains. p34 was shown to catalyse the NADPH-dependent reduction of oxygen to H_2O_2 as well as of disulphides like DTNB and cystine (Bruchhaus *et al.*, 1998; Lo and Reeves, 1980). Therefore, in addition to disulphide reductase activity the enzyme has H_2O_2-forming NADPH oxidase activity. It was also shown that p34 can transfer reducing equivalents to peroxiredoxin, converting the protein from its non-active, oxidised form back into its active, reduced form (Bruchhaus *et al.*, 1997). However, it is unlikely that peroxiredoxin is directly reduced by p34 *in vivo*. It is more likely that *E. histolytica* contains a classical thioredoxin redox system consisting of thioredoxin reductase, thioredoxin and peroxiredoxin (Poole *et al.*, 1997).

In addition to genes coding for proteins with homology to thioredoxin reductase, four other gene families were identified that encode various flavoproteins. One of these families includes four members that have between 53 and 61% sequence identity to A-type flavoproteins (flavorubredoxin/flavodiiron). A-type flavoproteins belong to a large family of enzymes that are widespread among anaerobic and facultatively anaerobic prokaryotes. In addition to bacteria, homologous genes are also found in the genomes of the pathogenic amitochondriate protistan parasites *T. vaginalis* and *G. intestinalis* (Andersson *et al.*, 2003; Sarti *et al.*, 2004). The A-type flavoproteins are made up of two independent structural modules. The N-terminal region forms a metallo-β-lactamase-like domain, containing a non-haeme di-iron site, whereas the C-terminal region is a flavodoxin-like domain, containing one FMN moiety. These enzymes have significant nitric oxide reductase activity (Gomes *et al.*, 2002; Sarti *et al.*, 2004). For *Escherichia coli* it is known that the nitric oxide reductase (FIRd) receives electrons from a NADH:oxidoreductase (FIRd-red). Consistent with that situation, the *E. histolytica* genome contains a gene encoding an NADH oxidase with 25% sequence identity to several bacterial FIRd-reds.

The three other *E. histolytica* gene families with homology to iron–sulphur flavoproteins (families B–D) are characterised by the presence of a flavodoxin-like domain forming a typical FMN binding site. Family B and family C consist of 3 members each, which share sequence identity of 42 and 46%, respectively. Family D consists of 2 members, which share only 33% sequence identity. At present, the function of the various flavodoxin-like molecules remains to be determined and deserves to be investigated fully, particularly as to whether they do indeed have antioxidant capacity.

4. METABOLISM

Biochemical analysis of *E. histolytica* metabolism has a long history (Reeves, 1984), dating back to shortly after the development of culture media that allowed the generation of substantial numbers of axenic cells. The genome sequence has confirmed most of the predicted metabolic pathways shown biochemically to be present or absent in *E. histolytica* in the past. As with most parasites, secondary loss of biosynthetic pathways is a recurring theme. However, a few surprises have also been uncovered. Every single enzyme involved in metabolism cannot realistically be discussed in this chapter. In this section, only the major energy generating and biosynthetic aspects of metabolism will be covered. Enzyme names, EC numbers and accession numbers are given in the the supplementary table for this section.

4.1. Energy metabolism

4.1.1. Glycolysis

E. histolytica lacks a functional tricarboxylic acid (TCA) cycle and oxidative phosphorylation. It is not able to convert organic substrates such as glucose into H_2O and CO_2, but has to rely on the energy generated by various types of substrate level phosphorylation (Reeves, 1984). Glycolysis is the major pathway of ATP generation, but in addition the genome project has identified a number of genes that could result in more ATP generation through the catabolism of amino acids. These enzymes will be described further below. As *E. histolytica* lacks compartmentalised energy generation, it has been classified as a type I amitochondriate protist (Martin and Müller, 1998) in contrast to the type II amitochondriate protists containing hydrogenosomes such as *T. vaginalis*. Nevertheless, it does contain a mitochondrial remnant, the mitosome (see Section 8).

In *E. histolytica*, glycolysis appears to be localised in the cytosol. This is in contrast to trypanosomes in which a major part is carried out in the glycosomes (Parsons, 2004) and the pathway is regarded as a potential target for chemotherapy (Opperdoes and Michels, 2001). The kinetic properties of recombinant *E. histolytica* glycolysis enzymes have recently been studied by Saavedra *et al.* (2005). Their analysis suggested that fructose-1,6-bisphosphate aldolase, phosphoglycerate mutase, glyceraldehyde-3-phosphate dehydrogenase and pyruvate phosphate dikinase might be regulating the glycolytic flux.

4.1.1.1. Hexokinases Glucose taken up by *E. histolytica* is phosphorylated by two hexokinase (EC 2.7.1.1) isoenzymes (Hxk1 and Hxk2). The two *E. dispar* isoenzymes are shifted towards a slightly more basic pI, which is the basis of the classical biochemical method for distinguishing *E. histolytica* from *E. dispar* by starch gel electrophoresis (Farri *et al.*, 1980). The pI differences among the two *E. histolytica* isoforms (Ortner *et al.*, 1995) and between the two species (Ortner *et al.*, 1997b) are the result of genetic differences that lead to different amino acid sequences and charge differences. Hxk1 phosphorylates glucose and mannose, while Hxk2 phosphorylates mainly glucose and is much less active with mannose as a substrate (Kroschewski *et al.*, 2000).

4.1.1.2. Glucose-6-phosphate isomerase Glucose 6-phosphate is converted to fructose 6-phosphate by glucose-6-phosphate isomerase (EC 5.3.1.9). The genome has two genes for this enzyme, which code for proteins that differ only by a single insertion or deletion of seven amino acid residues. Glucose-6-phosphate isomerase is another of the enzymes for the classical differentiation of *Entamoeba* zymodemes by starch gel electrophoresis (Sargeaunt, 1987).

4.1.1.3. Phosphofructokinases The main phosphofructokinase activity in *E. histolytica* is pyrophosphate (PPi)-dependent (EC 2.7.1.90; Reeves *et al.*, 1976). There is a single gene (Deng *et al.*, 1998) encoding this 60 kDa enzyme. The gene is a candidate for lateral transfer from bacteria (Loftus *et al.*, 2005) (see Section 10). The enzyme is expressed at a 10-fold higher level and displays about 10-fold higher activity than a second phosphofructokinase of 48 kDa (XP_653373) (Chi *et al.*, 2001). The substrate specificity of the smaller enzyme is disputed. Whereas Bruchhaus *et al.* (1996) reported that this minor enzyme also used PPi as phosphate donor, Chi *et al.* (2001) found only an ATP-dependent activity. The 48 and 60 kDa enzymes are highly divergent with $<20\%$ sequence identity. Interestingly, the specificity of the 60 kDa phosphofructokinase can be changed from PPi to ATP by mutation of a single amino acid residue (Chi and Kemp, 2000). The authors concluded that ATP rather than PPi was the primordial high energy compound. In the genome, there are 2 additional genes encoding isoforms of the 48 kDa enzyme, which have not been studied at the protein level.

4.1.1.4. Fructose-1,6-bisphosphate aldolase Fructose 1,6-bisphosphate is cleaved to glyceraldehyde 3-phosphate and dihydroxyacetone 3-phosphate by fructose-1,6-bisphosphate aldolase (EC 4.1.2.13). The enzyme, a class II aldolase (Marsh and Lebherz, 1992), has been cloned (XP_650373) and exhibits strong sequence similarity to eubacterial aldolases (Sanchez *et al.*, 2002). A second gene (XP_655966) encodes a protein differing from the first by a single deletion of 28 amino acids flanked by short divergent stretches. These bacterial-type aldolases are also found in *T. vaginalis*, *G. intestinalis* and other protists (Sanchez *et al.*, 2002). *E. histolytica* has no gene coding for a class I aldolase like those found in animals, which might make aldolase an interesting target for chemotherapy.

*4.1.1.5. **Triose-phosphate isomerase*** Triose-phosphate isomerase (EC 5.3.1.1) converts dihydroxyacetone 3-phosphate into glyceraldehyde 3-phosphate. The gene was previously cloned (Landa *et al.*, 1997), and is highly similar to the annotated gene product. This dimer-forming enzyme represents the first *E. histolytica* protein for which the structure has been solved by X-ray crystallography (Rodriguez-Romero *et al.*, 2002).

*4.1.1.6. **Glyceraldehyde-3-phosphate dehydrogenase*** Glyceraldehyde-3-phosphate dehydrogenase (EC 1.2.1.12) oxidises and phosphorylates glyceraldehyde 3-phosphate to 1,3-bisphosphoglycerate in two coupled reactions using NAD^+ as cofactor (Reeves, 1984). The genome project revealed 5 putative genes, 3 of which encode the identical protein sequence of 36.0 kDa and a predicted pI of 7.04. The fourth gene product, XP_648981, differs from these 3 only by a 13 amino acid deletion, while XP_650370 is a clearly distinct 34.8 kDa isoform with a lower predicted pI of 5.80. Interestingly, the isoforms XP_650356 and XP_650370 of different pI are encoded within the same contig.

*4.1.1.7. **Phosphoglycerate kinase*** Phosphoglycerate kinase has an unusual substrate (Reeves and South, 1974), transferring the high energy phosphate group from 3-phosphoglyceroyl phosphate to GDP leading to the formation of GTP (EC 2.7.2.10). There is one candidate gene encoding a 45 kDa protein.

*4.1.1.8. **Phosphoglycerate mutase*** Phosphoglycerate mutase (Reeves, 1984) isomerises 3-phosphoglycerate to 2-phosphoglycerate (EC 5.4.2.1). Five divergent putative genes for this enzyme are found in the genome. Two gene products of 62 kDa were classified as 2,3-bisphosphoglycerate-independent phosphoglycerate mutases (XP_649031 and XP_654182); they differ only at their C-termini and display significant similarity to bacterial phosphoglycerate mutases. The three other genes are very divergent. XP_651808 was identified as a candidate for lateral gene transfer (LGT) (Loftus *et al.*, 2005) (see Section 10). The remaining two gene products XP_649053 and XP_657284 are related to genes found in both prokaryotes and eukaryotes.

4.1.1.9. Enolase (2-phosphoglycerate dehydratase) Enolase (EC 4.2.1.11) converts 2-phosphoglycerate to phosphoenolpyruvate. The gene has been cloned (Beanan and Bailey, 1995) and the protein characterised (Hidalgo *et al.*, 1997) previously. The 47 kDa gene product is a typical eukaryotic enolase (XP_649161). A carboxy-terminally truncated incomplete ORF is also found.

4.1.1.10. Pyruvate, orthophosphate dikinase and pyruvate kinase In *E. histolytica*, both activities forming ATP and pyruvate from phosphoenolpyruvate have been found. The exergonic pyruvate kinase reaction uses ADP (Saavedra *et al.*, 2004), and the pyruvate, orthophosphate dikinase uses AMP and PPi in a slightly endergonic reaction (Varela-Gomez *et al.*, 2004). The dikinase activity is found in C4 plants where it is involved in phosphoenolpyruvate generation for gluconeogenesis. In *E. histolytica* it was discovered long before the pyruvate kinase (Reeves, 1968).

The cloning of pyruvate, orthophosphate dikinase (EC 2.7.9.1) was reported by two groups. The published sequences (Bruchhaus and Tannich, 1993; Saavedra Lira *et al.*, 1992) are highly similar or identical to XP_657332 and XP_654666. In addition there are two shorter related ORFs.

In the genome three putative pyruvate kinase genes (EC 2.7.1.40) have been identified. The three are identical except for an amino-terminal deletion in XP_648240 and an internal deletion in XP_653635.

4.1.1.11. Pyruvate:ferredoxin oxidoreductase (PFOR) and ferredoxin PFOR (EC 1.2.7.1) is an enzyme of major importance to *E. histolytica*, as the parasite lacks NAD^+-dependent pyruvate dehydrogenase and pyruvate decarboxylase (Reeves, 1984). No evidence for the latter two genes was found in the genome, confirming the biochemical results. PFOR oxidatively decarboxylates pyruvate to acetyl-CoA. The electrons are transferred to ferredoxin which, in its reduced form, can activate and reduce metronidazole, the major anti-amoebic drug (Müller, 1986). The activated form of metronidazole can potentially react with a number of biomolecules and is able to cleave the parasite DNA. In human cells, metronidazole is not activated and is much less toxic. In *T. vaginalis*, down-regulation of PFOR is one mechanism of producing metronidazole resistance (Kulda, 1999); however, PFOR expression appears unaltered in partially resistant *E. histolytica* (Samarawickrema *et al.*, 1997; Wassmann *et al.*, 1999). All eukaryotic PFOR genes, including that of *E. histolytica*, appear to have been acquired during an ancient LGT event from bacteria

(Horner *et al.*, 1999; Rotte *et al.*, 2001). There are two putative PFORs in the *E. histolytica* genome, displaying minor sequence differences.

The genome contains seven ferredoxin genes in total with five quite divergent sequences. All are related to eubacterial and archaeal ferredoxins (Nixon *et al.*, 2002). The gene pairs XP_655183/XP_655182 and XP_654311/XP_652694 are identical. The other three gene products represent more divergent ORFs. The deduced proteins have similar molecular masses, between 6.1 and 8.8 kDa, and different predicted isoelectric points between 4.2 and 8.6 kDa.

4.1.1.12. Acetyl-CoA synthetase (acetate thiokinase) The normal fate of acetyl-CoA in mitochondriate organisms is entry into the tricarboxylic acid cycle. However, this pathway is absent from *E. histolytica*. Instead, the cleavage energy of the thioester bond of acetyl-CoA can be used to generate one ATP molecule. One of the known acetyl-CoA synthetases generates ATP from ADP and Pi (EC 6.2.1.13). Such an enzyme has been characterised by Reeves *et al.* (1977) and cloned (Field *et al.*, 2000), and reported to be a 77 kDa protein. The common acetyl-CoA synthetase activity that produces ATP from AMP and PPi (EC 6.2.1.1) appears to be absent in *E. histolytica*.

4.1.1.13. Aldehyde and alcohol dehydrogenases The *E. histolytica* genome encodes a complex system of alcohol or aldehyde dehydrogenases. In total, there are 25 predicted genes, 3 of which are on the list of LGT candidates.

Alcohol dehydrogenase ADH1 was the first alcohol dehydrogenase to be characterised in *E. histolytica* (Reeves *et al.*, 1971) and is a NADPH-dependent enzyme (EC 1.1.1.2). The gene was previously cloned (Kumar *et al.*, 1992); in the genome 3 genes are almost identical to that sequence, while 1 (XP_652772) has 67% identity.

Fermentation in *E. histolytica* uses the bifunctional NADH-dependent enzyme ADH2, which belongs to the ADHE family and has both alcohol dehydrogenase and aldehyde dehydrogenase activities (Lo and Reeves, 1978). Under anaerobic conditions, reduction of the acetyl-CoA generated by PFOR to ethanol is one way to regenerate the NAD^+ used by glyceraldehyde-3-phosphate dehydrogenase. ADH2 first reduces acetyl-CoA to an enzyme-bound hemiacetal which is then hydrolysed to acetaldehyde (EC 1.2.1.10) and further reduced to ethanol (EC 1.1.1.1). If the enzyme is also able to work in the reverse direction, *E. histolytica* would be able to generate acetyl-CoA and energy from ethanol in the presence of oxygen. This would explain older reports of ethanol stimulated oxygen uptake in *E. histolytica* (Weinbach and Diamond, 1974). The enzyme is closely related to AdhE from *E. coli* and other bacteria (Reid and Fewson, 1994), and there is strong support for its aquisition by LGT (Andersson *et al.*, 2006;

Field *et al.*, 2000; Loftus *et al.*, 2005) (see Section 10). Like its bacterial homologue, ADH2 appears to form helical rods that sediment with membrane fractions (Avila *et al.*, 2002). Two groups have previously cloned ADH2 (Bruchhaus and Tannich, 1994; Yang *et al.*, 1994), and in total the genome contains five full-length ADH2 genes and one that is truncated. All share between 98 and 100% sequence identity.

In total, there are 11 alcohol dehydrogenase ADH3 genes in the genome, 2 of which have been reported previously (Kimura *et al.*, 1996; Rodriguez *et al.*, 1996). The recombinant enzyme characterised by Rodriguez *et al.* (1996) was NADPH-specific, like ADH1. There are five genes similar to these previously reported sequences. The rest of the ADH3 sequences fall into two groups of three similar sequences. All 11 ADH3 sequences are between 44 and 100% identical on the amino acid level. XP_649823 was originally on the list of LGT candidates (Loftus *et al.*, 2005), and a similarity to ADH3 sequences of gram-negative bacteria had been noted before (Nixon *et al.*, 2002). However, a related sequence is now known to exist in *T. vaginalis* also (see Section 10).

The genome encodes three additional distinct alcohol dehydrogenases. XP_656535 is a putative Zn-containing enzyme, and is on the list of LGT candidates. XP_652753 has been annotated as a Fe-containing alcohol dehydrogenase and XP_652262 simply as putative alcohol dehydrogenase.

One NADPH-dependent aldehyde dehydrogenase encoding gene (ALDH1) is present and was reported previously (Zhang *et al.*, 1994).

4.1.2. Energy storage: The glycogen metabolism

E. histolytica uses glycogen as its major energy store. Glycogen is a polymer of α-1,4-linked glucose chains with α-1,6 branch points, which in *E. histolytica* has a compact structure as suggested by branch points every 5–6 glucose residues (Bakker-Grunwald *et al.*, 1995). The cytoplasm of trophozoites contains numerous glycogen granules that were first observed by electron microscopy (Rosenbaum and Wittner, 1970) and later characterised biochemically (Takeuchi *et al.*, 1977). A glycogen phosphorylase activity (EC 2.4.1.1), associated with the glycogen granules, generates glucose 1-phosphate from orthophosphate and the linear portion of various glucopolysaccharides (Werries and Thurn, 1989). The genome contains at least six putative full-length and truncated genes encoding glycogen phosphorylases, two of which were cloned by Wu and Müller (2003). These authors noted a marked sequence divergence in those regions of the enzymes involved in regulation by phosphorylation and concluded that classical regulation by phosphorylation may not occur.

Glycogen phosphorylase degrades the linear chains only down to the α-1,6 branch points. The remaining core molecule is called limit dextrin.

Degradation can proceed further with the help of a debranching enzyme that has been purified (Werries *et al.*, 1990). It exhibits activities of both amylo-1,6-glucosidase (EC 3.2.1.33) and 4-α-glucanotransferase (EC 2.4.1.25). The genome contains two genes putatively encoding a full-length (XP_653608) and a truncated glycogen debranching enzyme. The deduced molecular mass of the large protein is 166 kDa, which corresponds to the biochemical data (Werries *et al.*, 1990).

Glucose 1-phosphate is isomerised to glucose 6-phosphate by phosphoglucomutase (EC 5.4.2.2) before entering the glycolytic pathway. The isoelectric points of the phosphoglucomutases from *E. histolytica* and *E. dispar* differ, and this was exploited for differentiation of the two species by starch gel electrophoresis (Sargeaunt *et al.*, 1978). The migration properties are reproduced by recombinant enzymes and are the result of primary sequence differences (Ortner *et al.*, 1997a). *E. histolytica* has one gene coding for this important enzyme, and in addition there are two distantly related members of the phosphoglucomutase/phosphomannomutase family.

Genes encoding the enzymes involved in glycogen biosynthesis in *E. histolytica* have been identified: a glycogen synthase (EC 2.4.1.11) of 155 kDa and 2 putative branching enzymes (EC 2.4.1.18). The glycogen precursor UDP-glucose is generated from UTP and glucose 1-phosphate by UTP:glucose-1-phosphate uridylyltransferase (EC 2.7.7.9). Two UTP-hexose-1-phosphate uridyltransferases have been characterised biochemically, a larger glucose 1-phosphate-specific enzyme of 45 kDa and a less specific enzyme of 40 kDa reported to use both galactose 1-phosphate and glucose 1-phosphate (Lobelle-Rich and Reeves, 1983). The genome contains one larger ORF encoding a putative UTP:glucose-1-phosphate uridylyltransferase of 54.7 kDa and 2 smaller ones encoding enzymes of 46.3 kDa with high similarity identified as UTP:*N*-acetyl-glucosamine-1-phosphate uridyltransferases. These enzymes are interesting in that they could possibly be involved in the activation of *N*-acetyl-glucosamine 1-phosphate as a precursor of the chitin cyst wall.

4.1.3. Catabolism of sugars other than glucose

4.1.3.1. Activation of fructose and galactose for glycolysis Neither Hxk1 nor Hxk2 can use fructose or galactose as a substrate, but there are 2 genes encoding bacterial-type enzymes that may do so, a 33 kDa fructokinase, which is one of the candidates for LGT to the *E. histolytica* lineage (see Section 10), and a 43 kDa galactokinase. The fructokinase groups with bacterial fructose 6-kinases (EC 2.7.1.4), and the galactokinase groups with galactose 1-kinases (EC 2.7.1.6). This substrate specificity has been noted before (Reeves, 1984). Fructose 6-phosphate enters as an intermediate of the glycolytic pathway (see Section 4.1.1.3). As described earlier (see Section

4.1.2), galactose 1-phosphate can be activated to UDP-galactose (Lobelle-Rich and Reeves, 1983) and then epimerised to UDP-glucose by UDP-glucose 4-epimerase (EC 5.1.3.2) (Reeves, 1984). In the genome, a single candidate 38 kDa ORF for the latter enzyme has been identified. The UDP-bound glucose can then be used either for the synthesis of glycogen or fed into the glycolysis pathway via glucose 1-phosphate and glucose 6-phosphate. This efficient pathway allows *E. histolytica* to grow on galactose instead of glucose (Reeves, 1984).

4.1.3.2. Anomerisation of aldoses The 1-position in the pyranose form of aldoses has a hydroxyl group that can be in either the α- or β-configuration. These forms can be interconverted by means of an aldose 1-epimerase (EC 5.1.3.3), an enzyme that has recently been characterised (Villalobo *et al.*, 2005). There is a single gene encoding this product.

4.1.3.3. Activation of pentoses Two gene candidates encoding pentose-activating enzymes have been identified in the *E. histolytica* genome: a 35 kDa ribokinase (EC 2.7.1.15) and a 56 kDa xylulokinase (EC 2.7.1.17). The latter is another bacterial-type sequence putatively acquired by LGT.

4.1.3.4. Interconversion of hexoses and pentoses The pathway of interconversion between hexoses and pentoses in *E. histolytica* was described many years ago (Reeves, 1984; Susskind *et al.*, 1982). A transketolase (EC 2.2.1.1) converts fructose 6-phosphate and glyceraldehyde 3-phosphate into xylulose 5-phosphate and erythrose 4-phosphate. Erythrose 4-phosphate and dihydroxyacetone phosphate are condensed by the glycolytic enzyme fructose-1,6-bisphosphate aldolase to sedoheptulose 1,7-bisphosphate, an extended substrate specificity of the aldolase. Phosphofructokinase then is able to remove a phosphate group forming diphosphate and sedoheptulose 7-phosphate. This molecule and glyceraldehyde 3-phosphate are then converted by transketolase to the pentoses ribose 5-phosphate and xylulose 5-phosphate. A transaldolase activity is absent (Reeves, 1984) consistent with there being no such gene in the genome. In contrast, 7 gene products were identified as likely transketolases: 3 highly similar proteins of 73 kDa and 4 truncated versions.

4.2. Amino acid catabolism

4.2.1. General features

As discussed earlier, glycolysis under anaerobic conditions can use only part of the energy contained in glucose for ATP generation. *E. histolytica* is capable not only of taking up amino acids (Reeves, 1984), but also using them for the generation of energy, as suggested by Zuo and Coombs (1995). The genome has revealed a number of unusual genes, often with bacterial affinities, coding for enzymes of amino acid catabolism (Anderson and Loftus, 2005).

In many cases, the degradation of amino acids starts with a transamination reaction (EC 2.6.1.-) generating a 2-ketoacid. The *E. histolytica* genome has five ORFs identified as aminotransferases. These ORFs are distinct from each other with the exception of XP_655090 and XP_655099, which differ only by one insertion and are LGT candidates. So far there is no enzymological data on this group of enzymes, so their substrate specificities in *E. histolytica* are unknown.

Both amino acid degradation and glycolysis have 2-ketoacids as intermediates. Pyruvate is one common intermediate, as amino acid degradation can produce either pyruvate or other 2-ketoacids. PFOR (see Section 4.1.1.11) is known to have a relaxed specificity, and in addition to pyruvate it can oxidatively decarboxylate 2-ketobutanoate, oxaloacetate and 2-ketoglutarate (Samarawickrema *et al.*, 1997). The reaction generates CoA-thioesters with the potential of producing one ATP per molecule.

The amino acids asparagine, aspartate, serine, alanine, tryptophan, cysteine, threonine, methionine, glutamine and glutamate can all be transformed into one of these 2-ketoacids in one or very few steps. This underlines the major importance of the PFOR in the energy metabolism of *E. histolytica*. The enzyme is indispensable, and as it always generates reduced ferredoxin it will always activate metronidazole. Consequently, it would be very difficult for *E. histolytica* to become resistant to metronidazole.

4.2.1. Aspartate and asparagine

E. histolytica takes up asparagine and aspartate in the presence or absence of glucose (Zuo and Coombs, 1995). Four putative asparaginases (EC 3.5.1.1) are found in the genome. Three are identical and share only 48% amino acid identity with the fourth (XP_656586). Asparaginase mediates the formation of aspartate from aspargine by releasing ammonia. The predicted sequences appear to possess a signal sequence, as suggested by the TargetP programme (http://www.cbs.dtu.dk/services/TargetP/), which is reminiscent of a periplasmic isotype (EcA, type II) (Swain *et al.*, 1993) that is up-regulated under anaerobic and carbon-restricted conditions (Cedar and Schwartz, 1967).

Aspartate can be converted to fumarate and ammonia by aspartate ammonia-lyase (aspartase, EC 4.3.1.1). Addition of a water molecule by fumarase (EC 4.2.1.2) produces malate. The genome encodes a putative fumarase that is related to bacterial class I fumarases. The aspartase is a member of the bacterial class II fumarase/aspartase protein family (Woods *et al.*, 1988), and also on the list of LGT candidates.

Aspartate is also decomposed into oxaloacetate and ammonia by aspartate aminotransferase, with the concomitant production of glutamate from 2-oxoglutarate. Oxaloacetate is then converted into malate via malate dehydrogenase (EC 1.1.1.37) and, since *E. histolytica* lacks both a functional TCA cycle and a phosphoenolpyruvate carboxykinase, the malate generated can be decarboxylated oxidatively to pyruvate by malic enzyme (EC 1.1.1.39). Both of these enzymes are present in *E. histolytica*. Two very similar genes have been identified as encoding malic enzyme and are LGT candidates.

4.2.2. Serine, threonine

Serine and threonine are also taken up by *E. histolytica* in the presence and absence of glucose (Zuo and Coombs, 1995). Serine can be deaminated by the pyridoxal phosphate-dependent serine dehydratase (L-serine ammonia-lyase, EC 4.3.1.17) to pyruvate and ammonia. The enzyme was characterised by Takeuchi *et al.* (1979) who showed that addition of serine to the culture medium stimulated oxygen consumption. In an analogous reaction, threonine dehydratase (threonine ammonia-lyase, EC 4.3.1.19) breaks down threonine to 2-oxobutanoate. Both ketoacids can then be oxidised by PFOR to acetyl-CoA or propionyl-CoA. Both catabolic reactions can be carried out by the same enzyme, as has been shown in yeast for example (Ramos and Wiame, 1982). In the *E. histolytica* genome annotation, four gene products have been annotated as threonine dehydratases, but none as serine dehydratase. XP_650405 and XP_652480 are identical while XP_655614 and XP_657171 share 95 and 37% identity with the others, respectively. The exact substrate specificities of these four putative serine/threonine dehydratases have not been reported.

Degradation of serine via the non-phosphorylated serine pathway, by the sequential reactions of L-serine: pyruvate aminotransferase (EC 2.6.1.51), D-glycerate dehydrogenase (EC 1.1.1.29) and D-glycerate kinase (EC 2.7.1.31) (Snell, 1986) results in the glycolytic intermediate 3-phosphoglycerate. The genome encodes several putative aminotransferases (see Section 4.2.1), but it is not yet known if serine is among their substrates. An unusual bacterial-type NADPH-dependent D-glycerate dehydrogenase was characterised by Ali *et al.* (2003), and there are two genes encoding D-glycerate dehydrogenases, one of which (XP_648124) is among the weaker LGT candidates (see Section 10). The genome also contains two genes encoding identical glycerate kinases. The enzyme has recently been characterised (V. Ali and T. Nozaki, unpublished data).

4.2.3. Methionine, homocysteine and cysteine

Methionine γ-lyase (EC 4.4.1.11) decomposes methionine to methanethiol (mercaptomethane), ammonia and 2-oxobutanoate. In *E. histolytica*, two methionine γ-lyases, EhMGL1 and EhMGL2, of similar molecular weights have been characterised (Tokoro *et al.*, 2003). These two isoenzymes show marked differences in substrate specificity, isoelectric point, enzymological and biochemical parameters (Tokoro *et al.*, 2003). Both enzymes can also act on other amino acids. In addition to degrading methionine, both EhMGL1 (pI 6.01) and EhMGL2 (pI 6.63) can convert homocysteine to hydrogen sulphide, ammonia and 2-oxobutanoate. EhMGL2 also decomposes cysteine to hydrogen sulphide, ammonia and pyruvate, whereas EhMGL1 is only weakly active against cysteine. Decomposition of homocysteine by methionine γ-lyase is essential since this parasite lacks the other known enzymes capable of destroying this toxic amino acid. In the genome, three ORFs correspond to EhMGL1 and one to EhMGL2. So far, the only eukaryotes known to possess methionine γ-lyases are *E. histolytica* and *T. vaginalis* (Lockwood and Coombs, 1991). As the enzymes are absent from the human host and important for the generation of metabolic energy, they could be targets for chemotherapy (Coombs and Mottram, 2001; Tokoro *et al.*, 2003).

In addition to serving as a source of metabolic energy, another important role of methionine is as a donor of methyl groups via *S*-adenosylmethionine synthetase (synonymous with methionine adenosyltransferase, EC 2.5.1.6). Seven gene candidates were identified, four full-length and three truncated. The *S*-adenosylhomoserine left after the transfer of the activated methyl group can be hydrolysed by *S*-adenosylhomocysteine hydrolase (EC 3.3.1.1), giving adenosine and homocysteine. Two candidate genes with identical sequences and one truncated form are present.

However, *E. histolytica* lacks the remaining enzymes for the reverse transsulphuration pathway (forming cysteine from methionine) (Nozaki *et al.*, 2005), that is cystathionine β-synthase and cystathionine γ-lyase. In addition, *E. histolytica* lacks all enzymes involved in the forward transsulphuration (forming methionine from cysteine) including cobalamin-dependent methionine synthase (EC 2.1.1.13) or cobalamin-independent methionine synthase (EC 2.1.1.14), which suggests that *E. histolytica* is capable of neither converting homocysteine to cystathionine nor recycling homocysteine to methionine.

E. histolytica lacks the methylthioadenosine cycle enzymes except for two, 5′-methylthioadenosine/*S*-adenosyl homocysteine nucleosidase (EC 3.2.2.9) and aspartate aminotransferase (AT, EC 2.6.1.1). The significance of these two enzymes in *E. histolytica* is unknown.

4.2.4. Arginine

In *G. intestinalis* and *T. vaginalis* the arginine deiminase (EC 3.5.3.6) pathway is important for energy generation (Knodler *et al.*, 1994; Linstead and Cranshaw, 1983; Schofield and Edwards, 1994), generating one ATP molecule from the breakdown of arginine to ornithine. In contrast, no arginine deiminase gene or dihydrolase pathway was detected in the *E. histolytica* genome.

In *E. histolytica*, arginine can either be degraded by arginase (EC 3.5.3.1) via ornithine or by arginine decarboxylase (EC 4.1.1.19) via agmatine. The arginine decarboxylase reaction uses up protons and may be involved in the acid resistance needed for the passage of cysts through the human stomach (Anderson and Loftus, 2005). Another function suggested for arginine degradation was that it depletes arginine as a substrate for human macrophages, preventing NO synthesis and amoebicidal activity (Elnekave *et al.*, 2003). Both enzymes could also be important for the generation of the polyamine putrescine (see Section 4.3). The genome contains a single gene encoding a 96 kDa polypeptide annotated as ornithine/arginine/lysine decarboxylase, the substrate specificity of which has not yet been examined on the recombinant protein level. There is a single gene encoding a putative 33 kDa arginase.

4.2.5. Glutamate, glutamine

In aerobic organisms, the 2-oxoglutarate generated from glutamate in a transaminase reaction enters the citric acid cycle for further catabolism. In *E. histolytica*, which also contains transaminases, 2-oxoglutarate can be oxidised by PFOR to give succinyl-CoA from which one molecule of ATP can be generated.

Several other gene products of *E. histolytica* could act on glutamine and glutamate. The genome lacks a glutaminase (EC 3.5.1.2) to carry out the simple hydrolysis of glutamine. Instead there is a putative glucosamine-fructose-6-phosphate aminotransferase (EC 2.6.1.16), which uses the energy in the amide group of glutamine to generate glucosamine 6-phosphate from fructose 6-phosphate. This product may be used for cyst wall biosynthesis.

4.2.6. Tryptophan

Tryptophan can be degraded to indole, pyruvate and ammonia by the PLP-dependent enzyme tryptophanase (EC 4.1.99.1), for which one candidate gene exists. To date, tryptophanase has only been found in bacteria and T. Vaginalis and it is also on the list of LGT candidates.

4.2.7. Alanine: A possible special case

Alanine could potentially be transformed into pyruvate by alanine aminotransferase (synonymous with alanine:pyruvate transaminase, EC 2.6.1.2). However, *E. histolytica* is reported to excrete alanine (Zuo and Coombs, 1995), suggesting that this enzyme is not used under the culture conditions tested. Conceivably, the purpose of the excretion process may be to carry excess nitrogen out of the cell in the absence of a functional urea cycle.

4.2.8. Catabolism of other amino acids

Most of the enzymes for branched-chain amino acid metabolism are missing in *E. histolytica*, but leucine, isoleucine and valine could be transformed into 2-oxoisocaproate, 2-oxo-3-methylvalerate and 2-oxovalerate, respectively, by a putative branched-chain amino acid aminotransferase (EC 2.6.1.42), one of the aminotransferases mentioned earlier (see Section 4.2). This could produce ammonia or transfer the amino group to 2-oxoglutarate to form glutamate. Subsequent oxidative decarboxylation to give the respective CoA-derivatives could be envisaged, but so far no gene candidates for the necessary dehydrogenases have been identified.

One gene encodes a putative histidine ammonia-lyase (EC 4.3.1.3), which is responsible for the decomposition of histidine into urocanate and ammonia. Other than the formation of ammonia, the significance of this enzyme is not clear since the downstream enzymes involved in histidine catabolism from urocanate to glutamate were not found.

Currently, there is little information regarding the fate of the amino acids glycine, proline, phenylalanine, tyrosine and lysine in *E. histolytica*. No genes for the catabolic enzymes necessary were detected except for an LGT candidate bacterial-type 96 kDa broad-specificity ornithine/arginine/lysine decarboxylase that may be acting on lysine.

4.3. Polyamine metabolism

The absence of *S*-adenosyl-L-methionine decarboxylase (EC 4.1.1.50), which converts *S*-adenosyl methionine into decarboxylated *S*-adenosyl methionine, spermidine synthase (EC 2.5.1.16) and spermine synthase (EC 2.5.1.22), suggests a complete lack of polyamine metabolism in this parasite (Anderson and Loftus, 2005). However, as mentioned earlier, *E. histolytica* possesses genes encoding arginase and arginine decarboxylase. Both could be involved in the production of putrescine via agmatine and agmatinase (EC 3.5.3.11) or via ornithine and ornithine decarboxylase (EC 4.1.1.17). The high putrescine concentration in trophozoites demonstrated by NMR spectroscopy (9.5 mM) (Bakker-Grunwald *et al.*, 1995) reinforces the physiological significance of putrescine. However, the fate of putrescine is unknown as neither spermine nor spermidine has been demonstrated in *E. histolytica*.

There is controversy regarding the presence or absence of trypanothione, a spermidine-containing thiol, in *E. histolytica*. Trypanothione is a major thiol in trypanosomes and leishmania (Fairlamb and Cerami, 1992) and contains two molecules of glutathione joined by a spermidine linker. The first reports detected the presence of trypanothione in *E. histolytica* (Ondarza *et al.*, 1997) but were contradicted soon after (Ariyanayagam and Fairlamb, 1999). More recently another study reaffirmed its presence

(Ondarza *et al.*, 2005). However, the gene encoding trypanothione reductase reported from *E. histolytica* strain HK-9 (AF503571) has no homologue in the genome of HM-1:IMSS. Although this matter has not been resolved, there is general agreement that the major thiol in *E. histolytica* is cysteine (Fahey *et al.*, 1984).

The *E. histolytica* genome encodes a 46 kDa ornithine decarboxylase with similarity to both plant and vertebrate enzymes, and there is also the 96 kDa ornithine/arginine/lysine decarboxylase (see Section 4.2.4). Only the former enzyme has been characterised at the biochemical level (Arteaga-Nieto *et al.*, 2002) and has been shown to be insensitive to difluoromethylornithine (DFMO), as is *E. histolytica* (Gillin *et al.*, 1984).

The conversion of arginine into putrescine via agmatine, in a reaction initiated by arginine decarboxylase, is generally present in bacteria and plants. Although arginine decarboxylase is present in *E. histolytica*, agmatinase (EC 3.5.3.11), which further catalyses conversion of agmatine into putrescine and urea, appears absent. However, one gene identified as a 33 kDa arginase also shares 21% sequence identity with human mitochondrial agmatinase and therefore its substrates need to be examined on the biochemical level to see whether the enzyme can act on arginine, agmatine, or both. At present, the role of arginine decarboxylase in *E. histolytica* is not clear, although as mentioned earlier this enzyme may also be involved in acid resistance in *E. histolytica*.

4.4. Biosynthesis of amino acids

4.4.1. Cysteine and serine

One of the areas in which reduction of metabolism is most evident is in amino acid biosynthesis. Biosynthetic pathways for most amino acids other than serine and cysteine (Ali *et al.*, 2003, 2004a; Nozaki *et al.*, 1998a, 1999) have been lost in *E. histolytica*. Similarly, *P. falciparum*, which predominantly acquires amino acids from host haemoglobins, lacks biosynthesis of most amino acids (Gardner *et al.*, 2002). Intracellular concentrations of some amino acids (glutamate, leucine, valine and proline in descending order of abundance) are very high in *E. histolytica* ranging from 6 to 21 mM (Bakker-Grunwald *et al.*, 1995). In particular, the glutamate and proline concentrations are much higher in the cells than in the growth medium (21 and 7.3 mM vs. 5.9 and 1.8 mM, respectively). Glutamate accounts for over one-third of the total amino acid pool (Bakker-Grunwald *et al.*, 1995), and is likely to play a central role in homeostasis not only of amino acids but also of energy metabolism in general. Thus, it is likely that these amino acids are actively taken up by as-yet unidentified amino acid transporters.

Retention of the serine and cysteine biosynthetic pathways when the others have been lost is likely related to the physiological importance of cysteine, which is the major intracellular thiol of this parasite. The cysteine biosynthetic pathway consists of two major steps, catalysed by serine acetyltransferase (EC 2.3.1.30), which produces *O*-acetylserine from serine and acetyl-coenzyme A, and cysteine synthase (EC 2.5.1.47), which subsequently transfers an alanyl moiety from *O*-acetylserine to sulphide to produce cysteine. *E. histolytica* possesses three genes each for cysteine synthase and serine acetyltransferase. Cysteine synthases 1 and 2 were considered to be allelic isotypes (Nozaki *et al.*, 1998b), while cysteine synthase 3 appears to be distinct, with only 83% identity to cysteine synthases 1 and 2. In contrast, all three serine acetyltransferase genes seem to be distinct, showing only 48–73% identity (V. Ali and T. Nozaki., unpublished data). It was previously shown that cysteine synthases 1 and 2 and serine acetyltransferase 1 are unique in that (a) they do not form a heterocomplex, in contrast to other organisms (Bogdanova and Hell, 1997; Droux *et al.*, 1998) and (b) serine acetyltransferase 1 is sensitive to allosteric inhibition by both L-cysteine and L-cystine (Nozaki *et al.*, 1999). Since all variants of these two enzymes lack organelle-targeting sequences, the significance of the multiple isotypes is unknown. It is important to determine subcellular distribution and specific functions of these isotypes to understand the significance of the redundancy. As this pathway is absent in humans, it is a rational target for development of new chemotherapeutic drugs against amoebiasis.

Serine is synthesised *de novo* utilising the glycolytic intermediate 3-phosphoglycerate, in a pathway that includes three sequential reactions catalysed by D-phosphoglycerate dehydrogenase (EC 1.1.1.95), phospho-L-serine aminotransferase (EC 2.6.1.52), and *O*-phospho L-serine phosphatase (EC 3.1.3.3). Although the final enzyme has not yet been enzymologically and functionally analysed, the first two enzymes have been characterised (Ali and Nozaki, 2006; Ali *et al.*, 2004a).

4.4.2. Interconversion of glutamate–glutamine and aspartate–asparagine

The single step interconversions of glutamate and glutamine, catalysed by glutamate synthase (EC 1.4.1.13) and glutamine synthetase (EC 6.3.1.2), and of aspartate and asparagine by asparagine synthase (EC 6.3.5.4) are found in *E. histolytica*. There are two isotypes of glutamine synthetase with 47% amino acid identity and five candidate genes. NADPH-dependent glutamate synthase (EC 1.4.1.13) catalyses the formation of two glutamates from glutamine and 2-oxo-glutarate in bacteria, yeasts and plants, and together with glutamine synthetase is involved in ammonia fixation under ammonia-restricted conditions. NADPH-dependent glutamate synthase is normally composed of two large and two small subunits (Petoukhov *et al.*, 2003). Although three genes encoding the small subunit are present, the large subunit appears to be absent in *E. histolytica*. These putative NADPH-dependent glutamate synthase small subunits share 80% amino acid identity and show 44% amino acid identity to homologues from the Archaea. The similarity to archaeal-type glutamate synthase (Nesbo *et al.*, 2001) suggests that the *E. histolytica* small subunits may function as a glutamate synthase without the large subunit, as shown for gltA from the archaean *Pyrococcus* (Jongsareejit *et al.*, 1997).

The two enzymes that catalyse interconversion between aspartate and asparagine, aspartate ammonia ligase (EC 6.3.1.1) and asparaginase (EC 3.5.1.1; see Section 4.2.1), are present in *E. histolytica*. Two types of aspartate ammonia ligases, AsnA and AsnB, are known from other organisms: the former utilises only ammonia, while the latter uses both ammonia and glutamine as amide donors in a reverse reaction. Mammals possess only AsnA, whereas prokaryotes have both AsnA and AsnB (Boehlein *et al.*, 1996; Nakamura *et al.*, 1981). Interestingly, *E. histolytica* possesses only the AsnB homologue. Thus, the amoebic enzyme is likely involved in the formation of glutamate from glutamine, in addition to asparagine formation from aspartate.

4.4.3. Synthesis of glutamate and aspartate

Glutamate can be formed from 2-oxo-glutarate and ammonia in a reversible reaction catalysed by glutamate dehydrogenase (EC 1.4.1.2), which is present in *E. histolytica*. It is known that this enzyme plays a dominant role in ammonia fixation under ammonia-non-restricted conditions as this reaction consumes no ATP. In addition, glutamate dehydrogenase is also involved in gluconeogenesis from glutamate.

Aspartate ammonia-lyase (synonymous with aspartase, EC 4.3.1.1), which decomposes aspartate into fumarate and ammonia in a reversible reaction, is also present in *E. histolytica* (see Section 4.2.1).

4.5. Lipid metabolism

For *E. histolytica*, the lack of oxidative phosphorylation means that the high energy content of lipids such as fatty acids cannot be exploited. Therefore, lipids such as phospholipids and cholesterol are primarily membrane components in *E. histolytica* (Das *et al.*, 2002; Sawyer *et al.*, 1967). Although these components are mainly acquired from their food or from the human host, *E. histolytica* does have some capability for biosynthesis, as well as extending and remodelling lipids, and for attaching lipids to proteins.

4.5.1. Lipid biosynthetic capabilities

4.5.1.1. Polyisoprene biosynthesis and protein prenylation Cholesterol is an important membrane constituent generated from C_5 isoprene precursors. *E. histolytica* trophozoites in axenic culture need cholesterol in their growth medium (Reeves, 1984), and it is likely that they acquire it from their human host. Reeves (1984) even cites several studies which show that hypercholesteremia in the host increases the damage inflicted by amoebic infection. *E. histolytica* lacks several enzymes for the classical sterol biosynthesis pathway (Schroepfer, 1981). The first stage of sterol biosynthesis is the formation of isopentenyl- or dimethylallyl diphosphate. In the *E. histolytica* genome no candidate genes for the generation of these intermediates were found, neither for the mevalonate pathway nor for the mevalonate-independent methylerythritol 4-phosphate (MEP) pathway that operates in bacteria and plants (Hunter *et al.*, 2003; Rohmer *et al.*, 1993). In a later step towards cholesterol synthesis, two molecules of C_{15} farnesyl diphosphate are dimerised to give C_{30} presqualene diphosphate by squalene synthetase (EC 2.5.1.21). This enzyme activity and those catalysing the subsequent steps also appear to be absent. The genome data thus support the long-standing conclusion that cholesterol biosynthesis is absent from *E. histolytica*.

Unexpectedly, the *E. histolytica* genome appears to encode enzymes involved in the intermediate stages of cholesterol biosynthesis from C_5 isopentenyl diphosphate to C_{15} farnesyl diphosphate. The latter compound, and the larger C_{20} compound geranylgeranyl diphosphate, may serve as precursors for the hydrophobic modification of GTP-binding proteins allowing them to bind to membranes (Grunler *et al.*, 1994). Protein prenylation is a ubiquitous process. It is important in human cell biology, health and disease (McTaggart, 2006), but it is also essential for parasites such that protein farnesylation has been proposed as a potential novel target for anti-parasitic chemotherapy (Maurer-Stroh *et al.*, 2003), including anti-*E. histolytica* chemotherapy (Ghosh *et al.*, 2004).

The first enzyme in this pathway is the isopentenyl-diphosphate δ-isomerase that catalyses the conversion of isopentenyl diphosphate to dimethylallyl diphosphate (EC 5.3.3.2). There is a single gene encoding this enzyme that is of presumed bacterial origin and is on the list of LGT candidates. The two isomeric C_5 isoprenyl diphosphates undergo condensation to C_{10} geranyl diphosphate, catalysed by geranyl-diphosphate synthase (EC 2.5.1.1). Farnesyl-diphosphate synthase (EC 2.5.1.10) then adds another C_5 unit to give C_{15} farnesyl diphosphate. Finally geranylgeranyl-diphosphate synthase (EC 2.5.1.29) adds another C_5 prenyl unit to give C_{20} geranylgeranyl diphosphate. The genome contains five putative prenyl transferase genes, which all have been annotated as geranylgeranyl-diphosphate synthases. Their sequences are highly similar, with the exception that the ORFs are disrupted in two of them (XP_650479 and XP_655958). These prenyl transferases appear to be of bacterial origin as well, and XP_650913 is on the list of LGT candidates. When searching for geranyl-diphosphate synthase or farnesyl-diphosphate synthase in the *E. histolytica* genome, the closest matches are for the same genes, so that the substrate specificity of these enzymes is unclear and needs to be examined biochemically.

The *E. histolytica* genome contains one sequence each for the α and β chains of protein farnesyltransferase (EC 2.5.1.58), which were previously cloned and characterised as recombinant proteins (Kumagai *et al.*, 2004).

In addition to the protein farnesyltransferase, a protein geranylgeranyltransferase I (EC 2.5.1.59) β chain has recently been cloned and expressed together with the protein farnesyltransferase α chain (Makioka *et al.*, 2006). The heterodimeric molecule had protein geranylgeranyltransferase activity of unusually broad substrate specificity. The α and β chains of the protein (Rab-)geranylgeranyltransferase II (EC 2.5.1.60) have also been cloned, as cDNAs (M. Kumagai, A. Makioka, T. Takeuchi and T. Nozaki, unpublished data).

The *E. histolytica* genome encodes candidate enzymes for the modification of prenylated proteins. There are two highly divergent proteins both identified as CAAX prenyl proteases (EC 3.4.24.84). CAAX is the carboxy terminus of the substrate protein in which C is the prenylated cysteine residue, A is an aliphatic amino acid and X is the terminal residue. The proteases cleave after the modified cysteine. After the processing step, a prenylcysteine carboxyl methyltransferase (EC 2.1.1.100) methylates the carboxy-terminal residue; there are two divergent candidate genes for this enzyme.

Taken together, the *E. histolytica* genome contains all the necessary genes to encode the pathway from isopentenyl diphosphate to a processed farnesylated or geranylgeranylated protein. The source of the starting material, isopentenyl diphosphate, remains unknown at this time, but there may be a previously unknown pathway for its synthesis or *E. histolytica* may be able to aquire it from its environment.

4.5.1.2. Fatty acid biosynthesis *E. histolytica* encodes an unusual 138 kDa acetyl-CoA carboxylase with 2 bacterial-type carboxylase domains, an acetyl-CoA carboxylase and a pyruvate carboxylase. Since no biotin carboxylase domain is found in the *E. histolytica* genome, it was proposed that the enzyme removes a carboxyl group from oxaloacetate and transfers it to acetyl-CoA, forming malonyl-CoA and pyruvate (Jordan *et al.*, 2003; Loftus *et al.*, 2005). This fusion protein has not been identified in any organisms other than *Giardia* and *Entamoeba*.

In the classical pathway of fatty acid biosynthesis, starting from acetyl-CoA sequential two-carbon units are added from malonyl-CoA. In each round of extension, the β-keto group is reduced in three steps before a new two-carbon unit is added. The whole pathway is carried out in a large fatty acid synthase complex, where the growing chain is linked to an acyl carrier protein. *E. histolytica* lacks this classical pathway. There are, however, plant homologues of fatty acid chain elongases such as *Arabidopsis thaliana* KCS1 (Todd *et al.*, 1999). There are eight putative fatty acid elongases in the *E. histolytica* genome, and all are very similar to each other. These enzymes could be involved in elongation of fatty acids taken up from the host or food sources, but their function and substrate specificity are unknown at this time.

4.5.2. Phospholipid metabolism

Phospholipids amount to 60–70% of the total lipids in *E. histolytica* (Sawyer *et al.*, 1967). So far little information is available at the biochemical level on how phospholipids are synthesised, acquired or remodelled. The genome project has revealed a number of genes, indicating that the phospholipid metabolism could be more complex than expected.

4.5.2.1. Phospholipid biosynthesis In order to produce phospholipids one has to generate the important intermediate phosphadidate (1,2-diacylglycerol 3-phosphate) by phosphorylation and acylation of glycerol. *E. histolytica* contains one gene for a glycerol kinase (EC 2.7.1.30). The second step would be the transfer of the acyl group to glycerol-3-phosphate by glycerol-3-phosphate *O*-acyltransferase (EC 2.3.1.15), but no candidate gene for this enzyme has been found in the genome. There are, however, two potential 1-acylglycerol-3-phosphate *O*-acyltransferases (EC 2.3.1.51) that could attach the second acyl group. After the attachment of the acyl groups, and in preparation for the attachment of the activated aminoalcohols, the phosphate is removed by phosphadidate phosphatase (EC 3.1.3.4), for which there is one gene, resulting in a diacylglycerol.

The activation of ethanolamine (EC 2.7.1.82) or choline (EC 2.7.1.32) for attachment to the phosphadidate starts with phosphorylation. There are two genes identified as choline/ethanolamine kinases that share 37% amino acid identity. Next, ethanolamine phosphate and choline phosphate are converted into CDP-ethanolamine (EC 2.7.7.14) and CDP-choline (EC 2.7.7.15), respectively. The genome encodes 2 enzymes sharing 57% sequence identity that are identified as ethanolamine-phosphate cytidylyltransferases. The substrate specificity of these enzymes needs to be examined on the biochemical level. Finally, the activated ethanolamine or choline is attached to diacylglycerol by the enzymes ethanolaminephosphotransferase (EC 2.7.8.1) or diacylglycerol choline-phosphotransferase (EC 2.7.8.2) producing phosphatidylethanolamine or phosphatidylcholine, respectively. For these activities a total of eight possible genes are found that share varying degrees of sequence similarity.

In *E. histolytica*, an alternative pathway of phospholipid biosynthesis could involve the biosynthesis of phosphatidylserine. In this pathway, the phosphatidate itself is activated by CTP in a reaction catalysed by phosphatidate cytidylyltransferase (EC 2.7.7.41) resulting in CDP-diacylglycerol. Three genes have been identified. Phosphatidylserine synthase then catalyses the reaction of CDP-diacylglycerol with serine to give phosphatidylserine (EC 2.7.8.8); one gene has been found.

Some organisms can form phosphatidylethanolamine from phosphatidylserine using a decarboxylase, but such an enzyme appears to be absent from the *E. histolytica* genome. There are, however, several candidate methyltransferases of yet unknown substrate specificity, which might be able to generate phosphatidylcholine from phosphatidylethanolamine.

Taken together, large portions of the pathways needed to generate the most important phospholipids can be assembled from genes tentatively identified to date in the *E. histolytica* genome. The first acylation of glycerol 3-phosphate to lysophosphatidate remains an important gap. As *E. histolytica* could potentially aquire all the necessary phospholipids from the host, the functional relevance of the described biosynthetic pathways may not be high.

Finally, two additional interesting enzymes present in *E. histolytica* should be mentioned. The first was previously characterised using cDNA sequences and recombinant proteins as L-*myo*-inositol 1-phosphate synthase (EC 5.5.1.4; Lohia *et al.*, 1999). This enzyme catalyses the complicated isomerisation of glucose 6-phosphate to L-*myo*-inositol 1-phosphate. Inositol is found in phosphatidylinositol (PI) and in GPI-anchors of some membrane proteins, as well as playing a major role in signal transduction via the secondary messenger 1,4,5-inositol trisphosphate. There are three *myo*-inositol 1-phosphate synthase genes, all highly similar to each other and to the previously sequenced cDNA.

The second is phospholipid-cholesterol acyltransferase (EC 2.3.1.43), which transfers an acyl group from phospholipids such as phosphatidylcholine to cholesterol giving a cholesterol ester. The genome contains seven genes for this enzyme. So far nothing is known about the importance of cholesterol esters for *E. histolytica*.

4.5.2.2. Phospholipid degradation Phospholipids are degraded by phospholipases. Whereas phospholipases A1 (EC 3.1.1.32) and A2 (EC 3.1.1.4) cleave the acyl residues in the 1 or 2 position of the glycerol core, phospholipases C (EC 3.1.4.3) and D (EC 3.1.4.4) cleave at the phosphate, phospholipase C on the glycerol side, and phospholipase D on the aminoalcohol side. In *E. histolytica* phospholipase A activity has been implicated in virulence (Ravdin *et al.*, 1985), as it liberates toxic fatty acids and lysophospholipids (Said-Fernandez and Lopez-Revilla, 1988). Phospholipases A have been found in two forms, a membrane-bound Ca-dependent form active at alkaline pH and a soluble Ca-independent form active at acid pH (Long-Krug *et al.*, 1985; Vargas-Villarreal *et al.*, 1998). The genome encodes 11 potential phospholipases A with predicted pI values between 4.8 and 8.8 and various degrees of sequence similarity. In addition, the *E. histolytica* genome encodes three potential phospholipases D.

Finally, there are two highly similar genes for phospholipases C, but these are homologous to PI-specific phospholipases C (EC 3.1.4.11) and most likely do not cleave PI or phosphatidylcholine but GPI-anchors instead. So far there are no studies using individual recombinant phospholipases, and it is not yet known how much these enzymes may contribute to the virulence of *E. histolytica*.

4.6. Coenzyme A biosynthesis and pantothenate metabolism

Analysis of the genome revealed a complete lack of known folate-dependent enzymes and folate transporters, suggesting this cofactor is not utilised by *E. histolytica*. This is at odds with a study on the nutritional requirements of *E. histolytica* in which folate was found to be essential for growth (Diamond and Cunnick, 1991). More experimental research will be needed to resolve this discrepancy. Most organisms require folate as a cofactor for several reactions of amino acid metabolism and for synthesis of thymidylate, a component of DNA. The microsporidian *E. cuniculi*, which possesses the smallest-known eukaryotic genome, still contains a folate transporter and several folate-dependent enzymes (Katinka *et al.*, 2001). In eukaryotes possessing mitochondria or chloroplasts, folate is required for the formylation of methionine on the initiator tRNA used for organelle protein synthesis. Although *E. histolytica* possesses a mitochondrion-derived organelle, the mitosome, there is no organellar genome (Leon-Avila and Tovar, 2004) and so no need for organellar protein synthesis. The most important metabolic consequences of the loss of folate metabolism for *E. histolytica* are therefore the absence of thymidylate synthesis and methionine recycling, although it remains possible that *E. histolytica* possesses folate-independent enzymes carrying out these steps.

Phosphopantothenoyl-cysteine decarboxylase (EC 4.1.1.36) and phosphopantothenoyl-cysteine synthetase (EC 6.3.2.5, synonymous with phosphopantothenate-cysteine ligase) exist as a fusion protein in *E. histolytica*, as in Bacteria and Archaea. The amino- and carboxyl-terminal domains possess decarboxylase and synthetase activity, respectively (Kupke, 2002, 2004; Kupke *et al.*, 2000; Strauss *et al.*, 2001). The role of this enzyme in coenzyme A biosynthesis is not well understood in *E. histolytica* as the other necessary enzymes are absent.

4.7. Nucleic acid metabolism

Like many protistan parasites, *E. histolytica* lacks *de novo* purine synthesis (Reeves, 1984). The genome reveals that nucleic acid metabolism of *E. histolytica* is similar to that of the other lumenal parasites *G. intestinalis* and *T. vaginalis* in lacking pyrimidine synthesis and thymidylate synthase (Aldritt *et al.*, 1985; Wang and Cheng, 1984). In addition, *E. histolytica* appears to lack ribonucleotide reductase, a characteristic shared with *G. intestinalis* (Baum *et al.*, 1989). Ribonucleotide reductase was found, however, in genomic sequences of the species *E. invadens* and *E. moshkovskii*, indicating that the enzyme was lost or replaced relatively recently. Among eukaryotes, the loss of these areas of nucleic acid metabolism is otherwise rare. The enzymes were likely lost during adaptation to living in an organic nutrient rich environment.

4.8. Missing pieces

Several important enzymes and pathways could not be found within the genome and their presumed sequence divergence from known enzymes and pathways labels them as possible drug targets once they are identified. Phosphopyruvate carboxylase, which reversibly converts phosphoenolpyruvate to oxaloacetate, is a central enzyme of carbon metabolism in *E. histolytica* (Reeves, 1970), but could not be identified. Isoprenyl-PP synthesis and aminoethylphosphonate synthesis are also likely to be present, but no candidate genes could be identified.

4.9. Transporters

A total of 174 transporters were identified within the genome, a number intermediate between the 62 transporters of *P. falciparum* and the 286 transporters of *S. cerevisiae* (http://membranetransport.org). *E. histolytica* has a number of ion transporters similar to those of yeast, but fewer identifiable nutrient and organellar transporters. Both *Plasmodium* and *Entamoeba* have reduced metabolisms and take up many complex nutrients. The higher number of transporters in *Entamoeba* suggests that they may be more substrate specific than the *Plasmodium* transporters or that they may have a higher level of redundancy.

Since glucose transport activity has experimentally been characterised in *E. histolytica* and glucose is thought to be the major energy source, it was surprising to find no homologues of known hexose transporters in the genome. Most hexose transporters belong to the sugar porter subfamily of the major facilitator superfamily (TC 2.A.1.1), members of which are found in prokaryotes, animals, fungi, plants and other protists, including *D. discoideum*, but no proteins of this family were found in the

E. histolytica genome. A group of candidate monosaccharide transporters found within the genome are related to the glucose/ribose porter family from prokaryotes (TC 2.A.7.5). These transporters consist of two related domains, and the *Entamoeba* proteins appear to have the N-terminal and C-terminal domains switched relative to the bacterial proteins. Functional characterisation of transporter-encoding genes will be necessary for a more complete picture.

5. THE CYTOSKELETON

The eukaryotic cytoskeleton is composed of three main elements: actin microfilaments, tubulin-based microtubules and intermediate filaments. Despite the fact that *E. histolytica* is very motile and performs phagocytosis very efficiently, its cytoskeletal components are simple. No genes encoding homologues of intermediate filament network proteins, including keratins, desmin and vimentin, have been identified in *E. histolytica*, providing further evidence that these particular cytoskeletal components are rather poorly conserved in evolution. In contrast, microfilament and microtubule components have been readily identified. Additional detail is given in the supplementary tables for this section.

5.1. Actin and microfilaments

Genome information suggests that *E. histolytica* has a greater dependence than other protists on an actin-rich cytoskeletal network. Microfilament proteins are represented by actin and several actin-binding proteins, although there are notable differences with respect to analogous proteins in other eukaryotes. There are eight actin genes in the *E. histolytica* genome, in addition to six others that encode divergent actins. Three divergent actins surprisingly contain an extra N-terminal domain with as yet unknown functional characteristics. Examples of hybrid actins are rather scarce and have been found as ubiquitin fusions (Archibald *et al.*, 2003). The functional significance of these *E. histolytica* hybrid actins is as yet unknown.

Under physiological salt concentrations, monomeric actin assembles into polymers of F-actin, thus building microfilaments. Actin assembles and disassembles in an extremely dynamic and highly controlled process that is dependent on many different actin-binding proteins (Winder and Ayscough, 2005). The *E. histolytica* genome encodes homologues of actin-binding proteins involved in the severing, bundling, cross-linking and capping of filamentous actin. The number and variety of actin-binding proteins support the view that the actin-rich cytoskeleton is very dynamic in *E. histolytica*.

Since the spontaneous polymerisation of actin monomers is inhibited by the action of sequestering proteins such as thymosin $\beta 4$ and profilin, efficient actin polymerisation requires the intervention of an actin polymerisation-promoting factor. The best-described promoting factors are the Arp2/3 complex and the formin protein family.

The Arp2/3 complex is composed of two actin-related proteins (Arp2 and Arp3, which act as a template for new actin filaments) and works in conjunction with five additional subunits: ARPC1 to 5 (Vartiainen and Machesky, 2004). All subunits have clearly been identified in the *E. histolytica* genome, and among these the Arp2 and Arp 3 subunits are the best conserved. The Arp2/3 complex's ability to nucleate new actin filaments is stimulated by its interaction with nucleation promoting factors such as the Wiskott–Aldrich syndrome protein (WASP) or the suppressor of cAMP-receptor (SCAR) factor. Surprisingly, no proteins with homology to WASP/SCAR components were found in the genome, suggesting that actin nucleation depends on the activity of other, as yet unidentified proteins.

In contrast, *E. histolytica* possesses six genes coding for formins, which have emerged as potent regulators of actin dynamics in eukaryotic cells through their ability to increase actin filament assembly (Higgs and Peterson, 2005). Formins control rearrangements of the actin cytoskeleton, especially in the context of cytokinesis and cell polarisation. Members of this family have been found to interact with Rho-GTPases, profilin and other actin-associated proteins. The precise nature of this polymerisation-accelerating activity differs from one formin to another: some nucleate filaments *de novo*, some require profilin for effective nucleation, while yet others seem to use filament severing as their basic mechanism. However, the formin homology 2 domain (FH2, comprising roughly 400 amino acids) is central to formin activity (Otomo *et al.*, 2005; Xu *et al.*, 2004). Actin nucleation by formins is thought to occur by stabilisation of an unfavourable nucleation intermediate, possibly through FH2 domains binding to monomers in the same manner that they bind to barbed ends (an activity influenced by profilin). The formin homologues from *E. histolytica* all contain an FH2 domain, suggesting that they are potential actin nucleation factors.

Once nucleated, actin filaments are able to grow rapidly by addition of monomers at their barbed ends. Filaments are regulated by several mechanisms (Winder and Ayscough, 2005). Filament length is controlled by capping proteins: barbed end cappers (such as capping protein and gelsolin) block addition of new monomers and thus act to decrease the overall length of the filament. In addition, gelsolin severs actin filaments, thereby rapidly increasing actin dynamics. Actin filaments appear to be significantly shorter in *E. histolytica* when compared with those from fibroblasts and stress fibres are not formed in this amoeba. Although

E. histolytica actin has been shown not to bind DNase I (Meza *et al.*, 1983), the inferred amino acid sequence indicates conservation of all the residues likely to participate in this binding event—suggesting that post-translational modifications of actin monomers may prevent DNAse I -actin binding. It remains to be determined whether such modifications of actin participate in the regulation of actin polymerisation. The genome encodes multiple genes associated with filament capping and severing, as well as candidates for proteins that cross-link actin filaments and thus organise them into a supramolecular network. The organisation of actin into networks and higher-order structures is crucial for both cell shape and function. These structures can be responsible for overall cell shape and related processes, such as bundle formation through α-actinin activity, for example. The arrangement of actin filaments into cross-linked arrays is also mediated by proteins with multiple actin-binding domains, which allows a more perpendicular arrangement of actin filaments. Examples of this type of protein are the large, flexible filamin dimer (Vargas *et al.*, 1996) and the spectrin tetramer. Genome analysis has now identified many candidate genes for actin-binding proteins in *E. histolytica*, and additional protein partners of this versatile family responsible for cytoskeleton regulation are likely to emerge from curation of the sequence and cellular studies of cell motility and phagocytosis in this parasite.

5.2. Tubulins and microtubules

E. histolytica has a lower dependence on a tubulin-based cytoskeleton than most other eukaryotic cells. Protein homologues of the basic (α, β and γ) tubulins are present, although other tubulins more characteristic of organisms with basal bodies and flagella (e.g., ε- and δ-tubulins) are absent from *E. histolytica* (Dutcher, 2001). Nine different tubulins (grouped into multigene families) exist in most eukaryotic cells. Microtubules (MTs) composed of α- and β-tubulins are intranuclear in *E. histolytica* (Vayssie *et al.*, 2004), and this raises the question of how such structures are modulated within the nucleus, given that MT dynamics require MT nucleation-based renewal at the minus end and MT capping at the plus end. Proteins involved in MT nucleation act in concert with γ-tubulin (which is also intra-nuclear in *E. histolytica*), and this parasite possesses at least one homologue to the Spc98 factor, a component of the MT-nucleating Tub4p-γ tubulin complex. In contrast, no homologues of EB1, CLIP-170, APC (all involved in MT capping) or centrins (which operate at the MT organising centre) have yet been identified, suggesting that other factors (or mechanical constraints within the MT) may be required in blocking MT growth. *E. histolytica* does encode candidate proteins involved in MT severing or chromosome segregation. All these proteins are good candidates for experimental analysis of the mechanisms of intranuclear

MT localisation and turnover as well as of the trafficking of tubulins between the cytoplasm and nucleus.

There is little information available on the precise organisation of microtubules and F-actin cytoskeleton during *E. histolytica* motility. In many eukaryotic cells, F-actin–microtubule interactions can be observed in lamellipodia at all stages. Interestingly, microtubules preferentially grow along actin bundles in filopodia, suggesting that a physical link between the structures exists (Leung *et al.*, 2002). Multifunctional MT-associated proteins (MAPs, like MAP1B, MAP2 and plakins) are promising candidates for acting as such links, either via dimerisation of MAPs with single microtubule and actin binding sites or via direct bridging of the two cytoskeletons (e.g., via plakins, which contain binding sites for both microtubules and actin within a single molecule). Plakin homologues have not been identified in the *E. histolytica* genome but a MAP is present. Furthermore, proteins with domains that can bind to actin (and potentially to MT) have been described in *E. histolytica*—the actin binding protein ABP-120 gelation factor, for example (Vargas *et al.*, 1996).

5.3. Molecular motors

The distribution of intracellular factors and vesicles is performed using three sets of molecular transporters: myosin along microfilaments and kinesin and dynein along MTs. Although *E. histolytica* is a highly motile cell, stress fibres and cytoplasmic MTs have never been observed. The fluidity of the parasite's cytoplasm may be related to features of its molecular motors some of which are very surprising. The myosin family of actin filament-based molecular motors consists of at least 20 structurally and functionally distinct classes. The human genome contains nearly 40 myosin genes, representing 12 of these classes. Remarkably, *E. histolytica* is the first reported instance of a eukaryote with only one unconventional myosin. This myosin heavy chain (myosin IB) belongs to the type I myosin family, of which 12 are present in the *Dictyostelium* genome (Eichinger *et al.*, 2005).

All members of the myosin family share a common structure composed of three modules: the head, neck and tail domains. The N-terminal region harbours the motor unit, which uses ATP to power movement along the actin filaments. By interacting with specific proteins and 'cargoes' the tail is responsible for the myosin's specific function and location. In particular, the presence of an SH3 domain in the tail region is important for linking these myosin I molecules with the endocytic machinery and the Arp2/3 complex. Protistan class I myosins are able to recruit the Arp2/3 complex towards the CARMIL adapter protein and Acan125. These homologous adapters consist of multiple, leucine-rich repeat sequences and bear two carboxyl-terminal polyproline motifs that are ligands for the myosin I SH3 domains. CARMIL has been shown to bind the Arp2/3 complex via an acidic motif similar to those found in WASP. In view of the fact that *E. histolytica* does not have WASP homologues, the discovery of a CARMIL homologue through proteomic analysis of *E. histolytica* phagosomes (Marion *et al.*, 2005) provides an important clue for understanding actin nucleation in *E. histolytica*. Interestingly, myosin IB in *E. histolytica* plays a structural role in the actin network due to its ability to cross-link filaments (Marion *et al.*, 2004). The cytoskeletal structuring activity of myosin IB regulates the gelation state of cell cytoplasm and the dynamics of cortical F-actin during phagocytosis.

The most-studied myosin has been the conventional or class II myosin. This double-headed molecule is composed of two heavy chains and two pairs of essential and regulatory light chains. The heavy chain tail consists of an α-helical, coiled coil protein able to form a parallel dimer that in turn can self-associate into bipolar, thick filaments. This enables myosin II to operate in huge filament arrays, which drive high speed motility. In addition to myosin IB, *E. histolytica* also has a conventional myosin II heavy chain (very closely related to its homologue in *Dictyostelium*), which has been reported to be involved in crucial phases of parasite motility, surface receptor capping and phagocytosis (Arhets *et al.*, 1998). *E. histolytica*'s sole isoform shapes the actin network and maintains cytoskeletal integrity. Candidate genes for the regulatory and essential light chain activities were also found, and these possess the EF hand domains necessary for Ca^{2+} binding.

Directional transport along the MTs depends on dynein and kinesin, both MT-associated motor proteins that convert the chemical energy from ATP hydrolysis into movement. These motors are unidirectional and move towards either the MT plus- or minus-ends (Mallik and Gross, 2004). Kinesins and dyneins have been implicated in a wide range of functions—principally intracellular organelle transport during interphase and spindle function during mitosis and meiosis. Members of the dynein family are minus-end directed, although this remains to be confirmed for a few uncharacterised, vertebrate, cytoplasmic dynein heavy chains. It has not yet been reliably established that the *E. histolytica* genome contains a dynein heavy chain gene, although a dynein light chain gene is present: improvements in gene assembly should provide us with more information on this high molecular mass protein.

Kinesins are microtubule-dependent molecular motors that play important roles in intracellular transport and cell division. Even though the motor domain is found within the N-terminus in most kinesins (N-type), it is located within the middle or C-terminal domains in some members of the family (M-type and C-type kinesins, respectively) (Asbury, 2005). The position of the motor domain dictates the polarity of the movement of kinesin along the MT: whereas N- and M-type kinesins are plus-end directed, the C-type kinesins are minus-end directed. Humans possess 31 different kinesins and trypanosomes have more than 40. The *E. histolytica* genome sequence predicts only six kinesin-encoding genes (four N-type, two C-type and no M-type homologues have been found). One of the N-kinesins also contains a domain homologous to the HOOK protein required for the correct positioning of microtubular structures within the cell (Walenta *et al.*, 2001). Considering that *E. histolytica* MTs are intranuclear, the study of kinesin function and trafficking should help elucidate what is likely to be a very interesting MT functional mechanism.

6. VESICULAR TRAFFIC

The requirement for nutritional uptake from the extracellular milieu in the host intestine imposes a heavy reliance on endocytic and phagocytic activities in *Entamoeba* (Espinosa-Cantellano and Martínez-Palomo, 2000). Proliferating trophozoites secrete a number of peptides and proteins including cysteine proteases (Que and Reed, 2000) and *amoeba pores* (Leippe, 1999) required for bacterial cell killing and degradation as well as being implicated in virulence (Petri, 2002). During encystation, the cells also secrete substrates used for the formation of the cyst wall (Eichinger, 1997). Electron micrographic studies have revealed a complex membrane organisation. The trophozoites contain numerous vesicles and vacuoles varying in size and shape (Clark *et al.*, 2000; Mazzuco *et al.*, 1997). Intracellular transport of both endocytosed and synthesised molecules between compartments is regulated by the elaborate orchestration of vesicle formation, transport, docking and fusion to the target compartment (Bonifacino and Glick, 2004; Kirchhausen, 2000). More detail is given in the supplementary tables for this section.

6.1. Complexity of vesicle trafficking

Among a number of molecules and structures involved in vesicular trafficking, three types of coated vesicles, named coatomer protein (COP) I, COPII, and clathrin-coated vesicles are the best characterised (Bonifacino and Glick, 2004; Kirchhausen, 2000). COPI vesicles primarily mediate transport from the Golgi to the ER and between the Golgi cisternae, while COPII vesicles are involved in the transport from the ER to the *cis*-Golgi. The clathrin-dependent pathway has a few independent routes: from the plasma membrane to endosomes, from the Golgi to endosomes and from endosomes to the Golgi. It has been well established that certain subfamilies of Ras-like small GTPases, widely conserved among eukaryotes, regulate both the formation of transport vesicles and their docking and fusion to the target organelles. The ARF and secretion-associated

Ras-related protein (Sar) families of GTPases regulate the formation of COPI and COPII vesicles (Memon, 2004), respectively. In contrast, the Rab family of GTPases (Novick and Zerial, 1997) is involved in the targeting and fusion of vesicles to the acceptor organelles together with the tethering machinery SNARE (a soluble *N*-ethylmaleimide-sensitive factor attachment protein receptor) (Chen and Scheller, 2001). Since individual coat proteins, small GTPases, SNAREs and their associated proteins show distinct intracellular distributions in both unicellular and multicellular organisms, they are believed to play a critical role in the determination of membrane trafficking specificity (Chen and Scheller, 2001; Munro, 2004; Novick and Zerial, 1997). It is generally believed that the total number of proteins involved in the membrane traffic reflects the complexity and multiplicity of its organism. The total number of the putative amoebic genes encoding Arf/Sar, Rab, SNARE, and coat proteins together with those from *S. cerevisiae*, *Caenorhabditis elegans*, *Drosophila melanogaster*, *Homo sapiens* and *A. thaliana*, is shown in Table 2.7. *E. histolytica* reveals complexity similar to yeast, fly and worm in the case of Sar/Arf and SNAREs, while the number of genes encoding three coat proteins [COPI, COPII and adapter proteins (APs)] was higher in *E. histolytica* than these organisms and comparable to that in mammals and plants. In contrast, the number of Rab proteins in *E. histolytica* is exceptionally high, exceeding that in mammals and plants.

6.2. Proteins involved in vesicle formation

6.2.1. COPII-coated vesicles and Sar1 GTPase

COPII components were originally discovered in yeast using genetic and biochemical approaches (reviewed in Bonifacino and Glick, 2004). COPII vesicles mediate the transport from the ER to the Golgi and consists of three major cytosolic components and a total of five essential proteins: the Sec23p–Sec24p complex, the Sec13p–Sec31p complex and the small GTPase Sar1p (Barlowe *et al.*, 1994). Sar1p and Sec23p–Sec24p complex are involved in the formation of the membrane–proximal layer of the coat, while Sec13p–Sec31p complex mediates the formation of the second membrane–distal layer (Shaywitz *et al.*, 1997). These proteins are well conserved among various organisms (Table 2.7). *E. histolytica* encodes one each of Sar1, Sec13 and Sec31, 2 of Sec23 and 5 proteins corresponding to Sec24 (Table 2.7). The yeast and human genomes also encode multiple Sec24 isotypes (3 and 4, respectively). Although Sec24 isotypes have been shown to be responsible for the selection of transmembrane cargo proteins in yeast (Peng *et al.*, 2000; Roberg *et al.*, 1999), the significance of the Sec24 redundancy in *E. histolytica* is not clear. Additional regulatory proteins participate in COPII assembly in yeast, including Sec16p, a putative scaffold protein (Espenshade *et al.*, 1995), and Sec12p, a GEF

TABLE 2.7 The number of genes encoding representative proteins involved in vesicular trafficking in *E. histolytica*

Protein	*E. histolytica*	*S. cerevisiae*	*C. elegans*	*D. melanogaster*	*H. sapiens*	*A. thaliana*	References
Sar1	1	1	1	1	2	4	1,2
COPII	9	6	5	4	9	12	1
Arf	10	6	11	11	27	17	3
COPI	11	7	7	7	9	9	1
AP-1		5	7	5	8	9	
AP-2		4	5	5	5	6	
AP-3		4	4	4	7	4	
AP-4		0	0	0	4	4	
AP total	18	13	16	14	24	23	1
Rab	91	11	29	29	60	57	1,4
Qa	8	7	9	7	12	18	
Qb	10(b + c)	6	7	5	9	11	
Qc		8	4	5	8	8	
R	10	5	6	5	9	14	
SNARE total	28	24	23	20	35	54	1,5,6
NSF	1	1	1	2	1	1	1,7
SNAP	1	1	1	3	1	3	
Sec1	5	4	6	5	7	6	8

References: (1): Bock *et al.* (2001); (2): Wennerberg *et al.* (2005); (3): Kahn *et al.* (2006); (4): Pereira-Leal and Seabra (2001); (5): Burri and Lithgow (2004); (6): Uemura *et al.* (2004); (7): Sanderfoot *et al.* (2000); (8): Boehm *et al.* (2001).

for Sar1p (Barlowe and Schekman, 1993). Homologues of Sec12p and Sec16p appear to be absent in *E. histolytica*. The p24 protein is a non-essential component of vesicle formation (Springer *et al.*, 2000), and in yeast it functions as a cargo adaptor through binding to Sec23p (Kaiser, 2000; Schimmoller *et al.*, 1995). *E. histolytica* encodes four p24 proteins, fewer than in yeast and humans, which have eight. GAP Sec23p is also present in *E. histolytica*; this activates the intrinsic GTPase activity of Sar1p after the formation of COPII vesicle and inactivates the function of Sar1p (Yoshihisa *et al.*, 1993), resulting in the uncoating of COPII vesicles.

6.2.2. COPI-coated vesicles and Arf GTPases

COPI-coated vesicles, which mediate transport from the Golgi to the ER and between the Golgi cisternae (Kirchhausen, 2000), consist of seven proteins (α, β, β', γ, δ, ε and ζ-COP) (Hara-Kuge *et al.*, 1994). The number of proteins making up the COPI coat, and thus the complexity of COPI components, varies among organisms (Table 2.7). While human possesses two isotypes of γ-COP and ζ-COP, yeast has a single gene for each. In humans, the two isotypes of γ-COP and ζ-COP form three different COPI complexes ($\gamma1/\zeta1$, $\gamma1/\zeta2$ and $\gamma2/\zeta1$), which have different intracellular distributions (Wegmann *et al.*, 2004). This implies that COPI-coated vesicles are also involved in functions other than Golgi-to-ER transport (Whitney *et al.*, 1995). In *E. histolytica*, the COPI complex appears more heterogeneous: *E. histolytica* encodes two isotypes each of γ-COP, δ-COP and α-COP and three isotypes of β-COP. In contrast, *E. histolytica* lacks ϵ-COP, which is known to stabilise α-COP (Duden *et al.*, 1998). It has been shown in yeast that all genes encoding components of COPI coat except for Sec28p, the yeast ε-COP homologue, are essential for growth (Duden *et al.*, 1998).

Recruitment of COPI to the Golgi membrane requires the association of a GTP-bound GTPase called Arf (Donaldson *et al.*, 1992; Kahn *et al.*, 2006). Arf was initially identified because of its ability to stimulate the ADP-ribosyltransferase activity of cholera toxin A (Kahn and Gilman, 1984). To recruit the COPI coat, Arfs are activated by a Sec7 domain-containing protein, Arf-GEF, which is a target of a fungal metabolite brefeldin A (Helms and Rothman, 1992; Sata *et al.*, 1998). Among Arf family proteins, Arf1 is involved in the formation of COPI-coated vesicles in the retrograde transport from the Golgi to ER, and is also involved in the assembly of clathrin-AP1 (see next section) on the *trans*-Golgi network (TGN) (Stamnes and Rothman, 1993), clathrin-AP3 on endosomes (Ooi *et al.*, 1998) and the recruitment of AP-4 to the TGN (Boehm *et al.*, 2001). The specific roles of Arfs3–5 are less clear, although Arf4 and Arf5 show *in vitro* activities similar to Arf1. Functional cooperativity of Arfs in the vesicular formation has also been demonstrated recently. At least two of four human Arf isotypes (Arf1, Arf3–5) are essential for a retrograde pathway from the Golgi to the ER, in the secretory pathway from the Golgi to the TGN and in the recycling from endosomes to the plasma membrane (Volpicelli-Daley *et al.*, 2005). In contrast to these Arfs, Arf6 regulates the assembly of actin filaments and is involved in endocytosis on the plasma membrane (Radhakrishna and Donaldson, 1997).

GTPases that share significant similarity to Arf but do not either activate cholera toxin A or rescue *S. cerevisiae Arf* mutants are known as Arls (Arf-like GTPases) (Lee *et al.*, 1997a). Arl1 is involved in endosome-to-Golgi trafficking (Lu *et al.*, 2001, 2004). Other Arls (Arls 2–11) and Arf-related proteins (Arp or ArfRP 1–2) have been localised to the cytosol, nucleus, cytoskeleton and mitochondria (Burd *et al.*, 2004; Pasqualato *et al.*, 2002). The number of Arf, Arl and Arf-related proteins varies among organisms (Table 2.7). Among 27 members identified in humans, only about a half dozen Arf/Arl/Arp proteins, including Arf1–6 and Arl1 (Wennerberg *et al.*, 2005), have been shown to function in membrane traffic (Lu *et al.*, 2001). The localisation and function of the remaining Arf/Arl/Arp remains unclear.

E. histolytica encodes 10 Arf/Arl proteins (Table 2.7). Only 2 *E. histolytica* Arfs (A1 and A2) have a high percentage identity to human Arfs 1, 3, 5 and 6 and yeast Arfs 1–3 (57–76% identity), while the remaining 8 Arf/Arl fall into 3 groups (A4–6, B1–3 and C) and are equally divergent from one another and from other organisms. Both the intracellular distributions and the specific steps in vesicular trafficking mediated by these *Entamoeba* Arf/Arl proteins are unknown. It is worth noting that five of these Arfs lack a conserved glycine residue at the second amino acid position of the amino terminus; this glycine is known to be myristylated and essential for membrane association in other organisms (Randazzo *et al.*, 1995). *Eh*ArfA4 also lacks one of the conserved GTP-binding consensus regions (Box2). Similar deletion of GTP-binding domains has also been observed in proteins belonging to the Rab family (see Section 6.3.1).

6.2.3. Clathrin-coated vesicle and its adaptor proteins

Clathrin-coated vesicles and pits, as demonstrated by electron microscopy, are often indicative of clathrin-mediated endocytosis. However, there is no clear ultrastructural evidence for their occurrence in *Entamoeba* (Chavez-Munguia *et al.*, 2000). Interestingly, heavy- but not light-chain clathrin is encoded in the genome. Since a majority of proteins, including adaptor proteins (APs, adaptins), known to be involved in the assembly of clathrin-coated vesicles are encoded in *E. histolytica*, the fundamental mechanisms and components of clathrin-mediated endocytosis are probably present in this organism, but are likely to be divergent from other eukaryotes. AP is a cytosolic heterotetramer that mainly mediates the integration of membrane proteins into clathrin-coated vesicles in the secretory and endocytic pathways (Boehm and Bonifacino, 2001;

Kirchhausen, 2000). AP is composed of two large, one medium and one small subunit (Keen, 1987). Four major types of AP complexes (AP1–4) have been identified (Boehm and Bonifacino, 2001; Nakatsu and Ohno, 2003). AP-2 (consisting of α, $\beta 2$, $\sigma 2$ and $\mu 2$) mediates endocytosis from the plasma membrane (Conner and Schmid, 2003; Motley *et al.*, 2003), while AP-1 (γ, $\beta 1$, $\sigma 1$ and $\mu 1$A) (Meyer *et al.*, 2000), AP-3 (δ, $\beta 3$A $\sigma 3$ and $\mu 3$A) (Le Borgne *et al.*, 2001; Vowels and Payne, 1998) and AP-4 (ϵ, $\beta 4$, $\sigma 4$ and $\mu 4$) (Aguilar *et al.*, 2001) play a role in the Golgi-endosome, endosomal-lysosomal or the Golgi/lysosome sorting pathway, respectively. AP-4, which is present only in mammals and plants (Boehm and Bonifacino, 2001), was also identified in non-clathrin-coated vesicles mediating the transport from TGN to the plasma membrane or endosomes (Hirst *et al.*, 1999). A few isotypes of AP-1 and AP-3, for example, AP-1B (γ, $\beta 1$, $\sigma 1$ and $\mu 1$B) and AP-3B (δ, $\beta 3$B, $\sigma 3$ and $\mu 3$B), showed tissue-specific expression (Faundez *et al.*, 1998; Folsch *et al.*, 1999). *E. histolytica* encodes 10 large subunits (α, β, γ, δ and ϵ), 4 medium subunits (one each of $\mu 1$ and $\mu 2$ and two $\mu 3$) and 4 small subunits ($\epsilon 1-\epsilon 4$). This suggests that *E. histolytica* produces four types of AP complex, as in humans and plants.

6.3. Proteins involved in vesicle fusion

6.3.1. Rab GTPases

The docking and fusion of transport vesicles to a specific target compartment requires the appropriate Rab protein. Specific interaction of a Rab with its effector molecules in conjunction with the interaction between SNAREs plays a central role in vesicle fusion (Zerial and McBride, 2001). In general, the complexity of the Rab gene family correlates with the degree of multicellularity. For example, *S. pombe*, *S. cerevisiae*, *C. elegans*, *D. melanogaster* and *H. sapiens* consist of 1, 1, $\sim 10^3$, 10^9 and 10^{13} cells, and have 7, 11, 29, 29 and 60 Rab genes, respectively (Pereira-Leal and Seabra, 2001). It has also been shown that in multicellular organisms, Rab proteins are expressed in a highly coordinated (i.e., tissue-, organ-, or developmental stage-specific) fashion (Seabra *et al.*, 2002; Zerial and McBride, 2001). *E. histolytica* possesses an extremely high number of Rab genes—91 (Fig. 2.6). Among its 91 Rabs only 22, including *Eh*Rab1, *Eh*Rab2, *Eh*Rab5, *Eh*Rab7, *Eh*Rab8, *Eh*Rab11, *Eh*Rab21, and their isotypes showed >40% identity to Rabs from other organisms. The 69 remaining *E. histolytica* Rab proteins showed only moderate similarity (<40% identity) and represent unique, presumably *Entamoeba*-specific, Rab proteins. Approximately one-third of Rab proteins form 15 subfamilies, including Rab1, Rab2, Rab7, Rab8, Rab11 and RabC-P, each of which contains up to 9 isoforms. Interestingly, ~70% of *E. histolytica* Rab genes contain one or more introns (Saito-Nakano *et al.*, 2005). SNARE genes are also intron-rich whereas the Sar/Arf GTPase and the three coat protein genes have a low

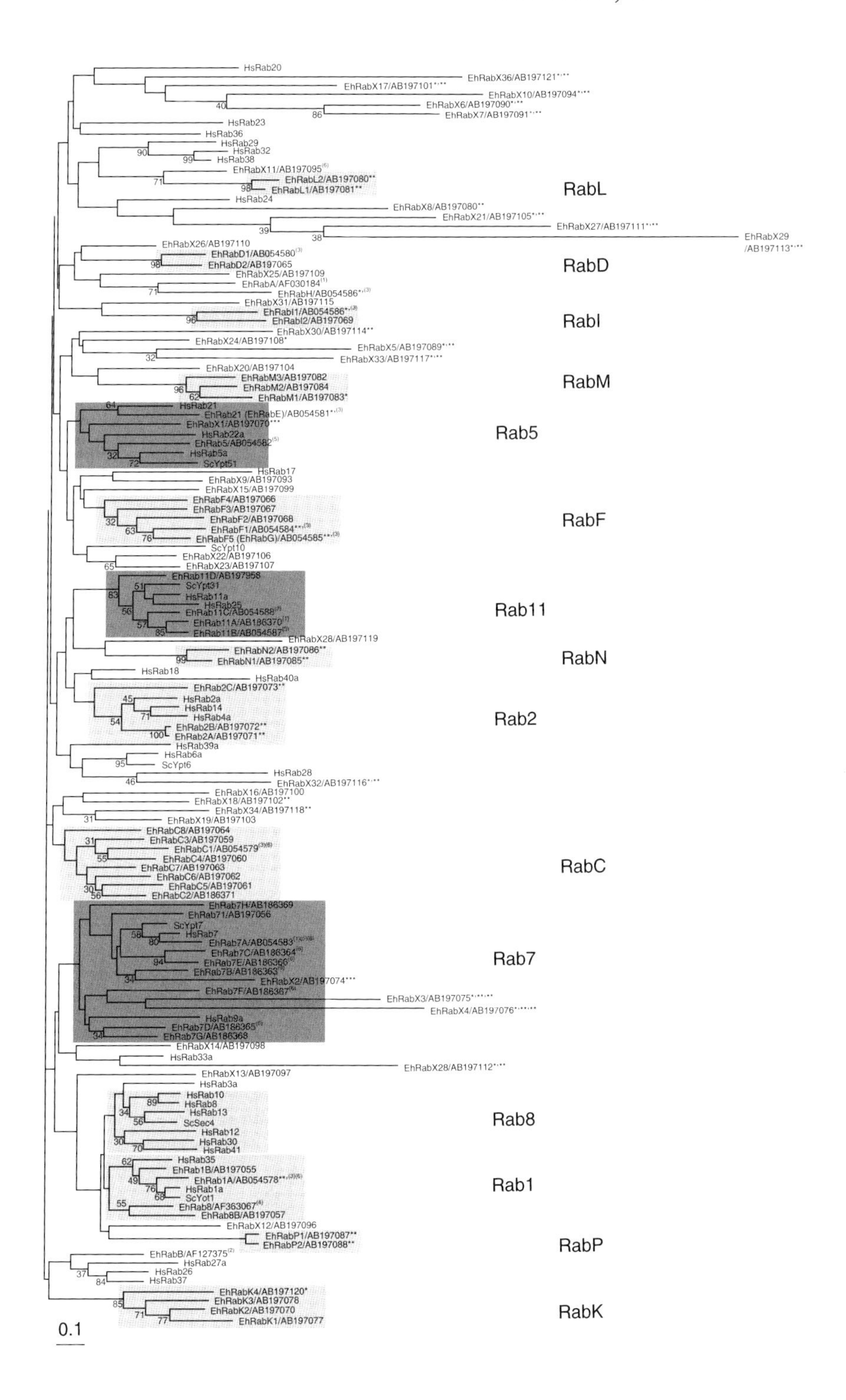
RabL
RabD
RabI
RabM
Rab5
RabF
Rab11
RabN
Rab2
RabC
Rab7
Rab8
Rab1
RabP
RabK
0.1

frequency of introns. The high frequency of introns in the Rab and SNARE gene families may indicate the presence of post-transcriptional regulation of these genes.

Although Rab proteins generally possess a CXC or CC at the carboxyl terminus, 25 *E. histolytica* Rabs have an atypical carboxyl terminus, such as CXXX, XCXX, XXCX, XXXC, or no cysteine at all. The enzyme(s) involved in the lipid modification of these unusual Rab proteins remain poorly understood (see Section 4.5.1.1). It is also worth noting that >20 *E. histolytica* Rabs lack or contain only a degenerate form of the consensus sequence for structural elements such as the GTP-binding regions and the Switch I and II regions, implicated in the binding to GEF, GAP, effectors or guanine nucleotides (Saito-Nakano *et al.*, 2005). These non-conventional *Eh*Rabs are not pseudo-genes since at least some of the genes are known to be expressed as mRNA (Saito-Nakano *et al.*, 2001). It has been shown that neither *Eh*Rab5 nor *Eh*Rab7A rescued the corresponding yeast mutant (Saito-Nakano *et al.*, 2004). Therefore, many, if not all, *E. histolytica* Rabs may have lost functional interchangeability with their homologues in other organisms despite the relatively high percentage of sequence identities. Classification and annotation of the *E. histolytica* Rab proteins have been previously described (Saito-Nakano *et al.*, 2005).

One of the peculiarities of *E. histolytica* Rab proteins was demonstrated by the unprecedented function of *Eh*Rab7A, which plays an important role in the transport of cysteine proteases via interaction with the retromer complex. The *E. histolytica* retromer complex consists of three components, Vps26, Vps29 and Vps35, rather than the 4–5 found in yeast and mammals (Nakada-Tsukui *et al.*, 2005). Homologues of Vps5, Vps17 and sorting nexins are not encoded in the genome. It has been suggested that the *Eh*Rab7A–retromer interaction, mediated by direct binding of *Eh*Rab7A to a unique carboxyl-terminal region of Vps26, regulates intracellular trafficking of cysteine proteases, and possibly other hydrolases as well, by modulating the recycling of a putative cysteine protease receptor

FIGURE 2.6 A phylogenetic tree of Rab proteins from *E. histolytica*, human and yeast. The number on the nodes represents the bootstrap proportions (%) of 1,000 pseudo samples; only bootstrap proportions >30% are shown. *E. histolytica* Rab proteins are indicated in bold. Tentative subfamilies that revealed significant similarity (>40% identity) to their human or yeast counterpart are shaded dark, while *Entamoeba*-specific subfamilies have light shading. The scale bar indicates 0.1 substitutions at each amino acid position. *: *Eh*Rab proteins that lack the conserved effector region, switch regions or GTP-binding boxes. **: *Eh*Rab proteins that possess a non-conventional carboxyl-terminus or lack carboxyl-terminal cysteines. ***: Rab proteins that were not classified as isotypes based on <40% identity to other members of the subfamily. References on tree: (1): Temesvari *et al.* (1999); (2): Rodríguez *et al.* (2000); (3): Saito-Nakano *et al.* (2001); (4): Juarez *et al.* (2001); (5): Saito-Nakano *et al.* (2004); (6): Okada *et al.* (2005).

from lysosomes and phagosomes to the Golgi or post-Golgi compartment (Nakada-Tsukui *et al.*, 2005).

6.3.2. SNARE and their accessory proteins

The final step in membrane trafficking is the fusion of a transport vesicle with its target membrane, which is mediated by the SNARE family of proteins. SNAREs are integral membrane proteins that are present on both donor and acceptor membranes and form a stable complex to tether the two membranes. It is believed that the formation of a SNARE complex pulls the vesicle and target membrane together and provides the energy to drive fusion of the lipid bilayers (Chen and Scheller, 2001; Chen *et al.*, 1999). In a prototypical model, a SNARE complex, which consists of four helices, is formed at each fusion site (Hanson *et al.*, 1997; Poirier *et al.*, 1998). For instance, the fusion of synaptic vesicles with the presynaptic nerve terminus is mediated by the formation of a complex comprising one helix each from syntaxin 1A [Qa-SNARE, also termed target-SNARE (t-SNARE)] and VAMP2 [R-SNARE, vesicular SNARE (v-SNARE)] and two helices from SNAP-25 (Qb- and Qc-SNARE).

The complexity of SNAREs has remained largely unchanged in yeast, fly and worm, but has increased remarkably in mammals and plants (Table 2.7) indicating that although expansion of SNARE repertoires occurs, a set of core SNAREs is sufficient to mediate vesicular fusion of most pathways in multicellular organisms. *E. histolytica* encodes 28 putative SNAREs, 18 Q-SNAREs and 10 R-SNAREs, which is comparable to the complexity in humans and plants. A notable peculiarity of SNAREs in *E. histolytica* is the lack of a group of proteins possessing two helices (Qb and Qc SNAREs) such as SNAP-25. Thus, the prototype model of membrane tethering by a combination of four helices (from Qa, R to Qb/Qc) does not appear to be possible in this organism.

A group of proteins that interact directly with the syntaxin subfamily, including the prototypical member yeast Sec1p and mammalian Munc-18, are essential cytosolic proteins peripherally associated with membranes (Toonen and Verhage, 2003). They are presumed to be chaperones, putting syntaxins into the conformations required for interaction with other SNAREs (Dulubova *et al.*, 1999; Yang *et al.*, 2000). Sec1/Munc-18 proteins are also conserved in *E. histolytica* (there are five *Sec1* genes). Two additional important components involved in the recycling of fusion machinery, *N*-ethylmaleimide sensitive factor (NSF) (Beckers *et al.*, 1989) and soluble NSF attachment protein (SNAP) (Clary *et al.*, 1990; Mayer *et al.*, 1996) are also found in *E. histolytica*.

Other proteins involved in vesicle fusion are the saposin-like proteins mentioned earlier (Section 3.3). The membrane-fusogenic activity of the *E. histolytica* SAPLIPs may play a role in vesicle fusion (Winkelmann *et al.*, 2006), but how they interface with the Rab/SNARE processes remains to be determined.

6.4. Comparisons and implications

While the fundamental machinery of vesicular trafficking is conserved in *E. histolytica*, the high activity of the endocytic and biosynthetic transport pathways in this organism appears to have resulted in the dramatic expansion of the Rab gene repertoire. The diversity and complexity of Rab proteins present in *E. histolytica* likely reflect the vigorous dynamism of membrane transport and the reliance on Rab proteins for the specificity of vesicular trafficking. The high degree of Rab complexity observed in *E. histolytica* (91) has no precedent in other organisms, although the incomplete genome of *T. vaginalis* appears to encode 65 Rabs (Lal *et al.*, 2005) while *Dictyostelium* encodes 50 (Eichinger *et al.*, 2005). Rab proteins have been extensively studied in *T. brucei* and the recent completion of *T. brucei*, *Trypanosoma cruzi* and *Leishmania major* genomes led to identification of all Rab genes in these haemoflagellates (Ackers *et al.*, 2005; Berriman *et al.*, 2005; Quevillon *et al.*, 2003). Among the 16 Rabs present in *T. brucei*, there are only 3 Rab proteins (RabX1-X3) that appear to be unique to kinetoplastids. *T. brucei* encodes 13 Rab proteins homologous to those in humans, suggesting significant conservation of the Rab-dependent core endomembrane systems in kinetoplastids. *P. falciparum* possesses only 11 Rab genes all of which are considered orthologues of yeast and mammalian Rabs, although Rab5a, 5b and 6 revealed unique features (Quevillon *et al.*, 2003). Interestingly, some of these Rabs are expressed in a stage-dependent manner (Quevillon *et al.*, 2003). The comparatively small number of Rabs in these protists reinforces the tremendous diversity and complexity of Rabs seen in *E. histolytica* (Table 2.7).

In marked contrast to the complexity of Rab proteins in *E. histolytica*, the number of SNARE proteins, the other major components of vesicular fusion, is comparable to that in yeast. The apparent disparity in the number of Rab and SNARE proteins suggests one of three possibilities: (1) *Eh*Rab proteins share a single SNARE complex as an interacting partner (Huber *et al.*, 1993; Rowe *et al.*, 2001; Torii *et al.*, 2004), (2) a majority of *Eh*Rabs do not require SNARE proteins for membrane fusion (Demarque *et al.*, 2002), (3) some *Eh*Rabs are primarily involved in cellular functions other than membrane fusion, like Arl GTPases (Burd *et al.*, 2004; Pasqualato *et al.*, 2002). Genome-wide surveys of SNAREs in other protists are not available. The three major types of coatomer protein, which are conserved in *E. histolytica*, are also conserved in kinetoplastids (Berriman *et al.*, 2005). However, in contrast to *E. histolytica*, *T. brucei* does not possess multiple isotypes of COPI and II components except for Sec24, which has two isotypes. *T. cruzi* encodes all four AP complexes while *L. major* and *T. brucei* lack AP-4 or AP-2, respectively, which suggests that the repertoire of AP complexes in kinetoplastids is variable and species-specific. Although low similarity of the *E. histolytica* components to either yeast or mammalian orthologues make unequivocal assignment of *Entamoeba* AP complexes challenging, tentative assignments have been made. It is likely that *E. histolytica* encodes four kinds of AP complex corresponding to APs 1–4.

6.5. Glycosylation and protein folding

6.5.1. Asparagine-linked glycan precursors

Mammals, plants, *Dictyostelium* and most fungi synthesise asparagine-linked glycans (N-glycans) by means of a common 14-sugar precursor dolichol-PP-$Glc_3Man_9GlcNAc_2$ (Figs. 2.7 and 2.8) (Helenius and Aebi, 2004). This lipid-linked precursor is made by at least 14 glycosyltransferases,

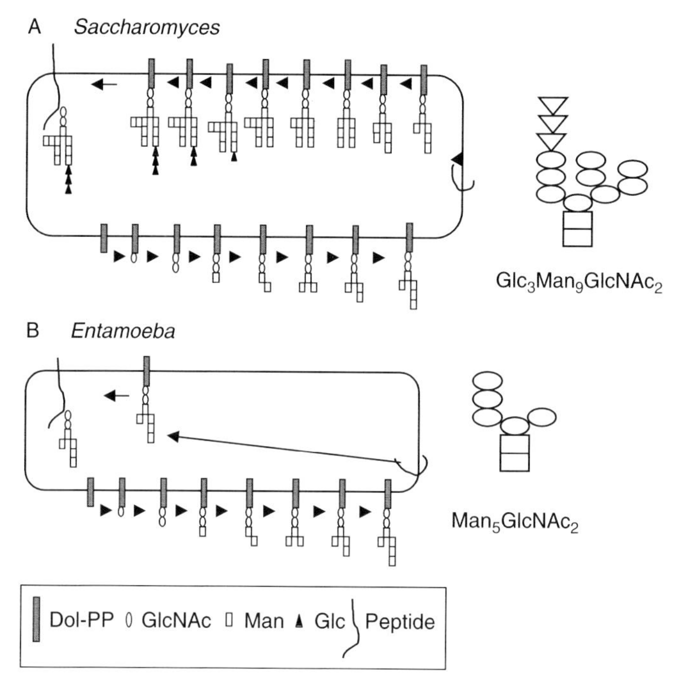

FIGURE 2.7 Synthesis of N-glycan precursors by *S. cerevisiae* (A) and *E. histolytica* (B). The N-glycan precursor of *S. cerevisiae* contains 14 sugars ($Glc_3Man_9GlcNAc_2$), each of which is added by a specific enzyme. The *E. histolytica* N-glycan precursor contains just seven sugars ($Man_5GlcNAc_2$), as the protist is missing enzymes that add mannose and glucose in the lumen of the ER. The figure is modified from Figure 1 of Samuelson *et al.* (2005). Glc = Glucose; GlcNAc = *N*-acetyl glucosamine; Man = Mannose.

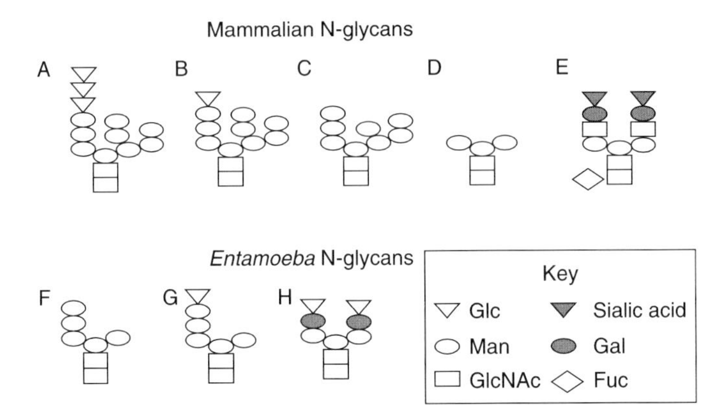

FIGURE 2.8 Selected N-glycans of mammals (A–E) and *Entamoeba* (F–H). Precursors transferred to nascent peptide (A and F). Glycosylated products involved in N-glycan-associated QC of protein folding (B and G). Mannosidase product involved in N-glycan-associated protein degradation (mammals only) (C). Trimmed product that is building block for complex N-glycans (mammals and *Entamoeba*) (D). Complex N-glycans made in the Golgi (E and H). Glc = Glucose; GlcNAc = *N*-acetyl glucosamine; Man = Mannose; Gal = Galactose; Fuc = Fucose.

which are present in the cytosolic aspect or lumen of the ER. The reducing end of the glycan contains two *N*-acetylglucosamines, while nine mannoses are present on three distinct arms. Three glucoses are added to the left arm, which is the same arm that is involved in the quality control (QC) of protein folding (see next section) (Trombetta and Parodi, 2003).

Entamoeba is missing luminal glucosylating and mannosylating enzymes and so makes the truncated, seven-sugar N-glycan precursor dolichol-PP-$Man_5GlcNAc_2$ (Figs. 2.7 and 2.8) (Samuelson *et al.*, 2005). Five mannoses on this N-glycan include the left arm, which is involved in the quality control of protein folding. In contrast, *Entamoeba* is missing the middle and the right arms, which are involved in N-glycan associated QC of protein degradation (see next section). Because *Dictyostelium*, which is phylogenetically related to *Entamoeba*, makes a complete 14-sugar N-glycan precursor, it is likely that *Entamoeba* has lost sets of glycosyltransferases in the ER lumen (Samuelson *et al.*, 2005). Similarly, secondary loss of glycosyltransferases best explains the diversity of N-glycan precursors in fungi, which contain 0–14 sugars, and apicomplexa, which contain 2–10 sugars (Samuelson *et al.*, 2005).

The 14-sugar N-glycan precursor of mammals, plants, *Dictyostelium* and most fungi is transferred to the nascent peptide by an oligosaccharyltransferase (OST), which is composed of a catalytic peptide and six to seven non-catalytic peptides (Kelleher and Gilmore, 2006). In contrast, the *Entamoeba* OST contains a catalytic peptide and just three non-catalytic peptides, while other protists (e.g., *Giardia* and *Trypanosoma*) have an OST

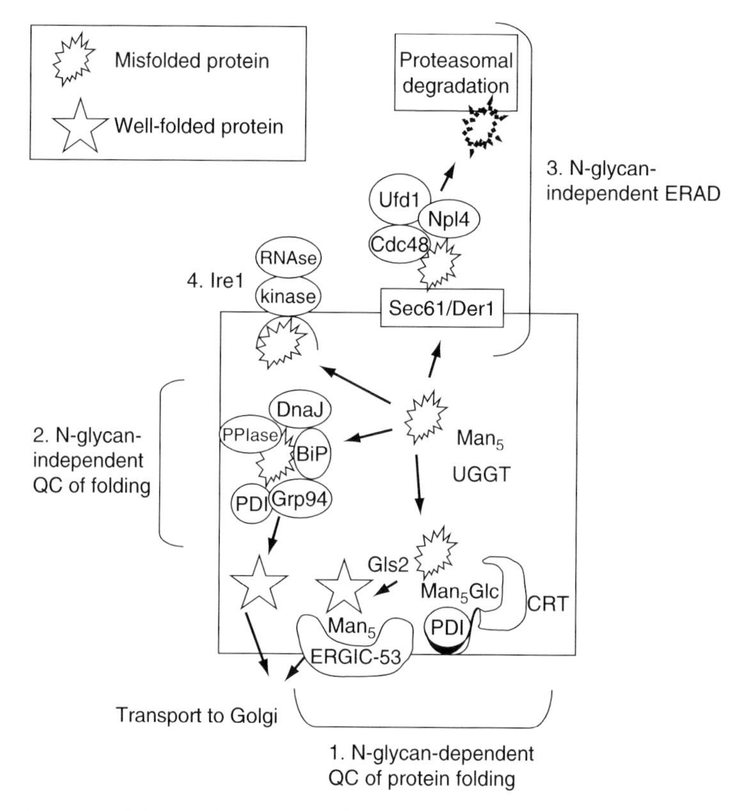

FIGURE 2.9 Model of quality control of protein folding in *Entamoeba*. 1. N-glycan-dependent QC of protein folding. 2. N-glycan-independent QC of protein folding. 3. N-glycan-independent ERAD. 4. Ire1 and unfolded protein response (see text for details).

with a single catalytic peptide. This reduced complexity does not likely affect the site of N-glycan addition to the nascent peptides, which is NxS or NxT (the so-called sequon) (Kornfeld and Kornfeld, 1985).

6.5.2. N-glycans and quality control of protein folding

Protein folding in the lumen of the ER is a complex process that involves N-glycan-dependent and N-glycan-independent QC systems (Helenius and Aebi, 2004; Trombetta and Parodi, 2003). *Entamoeba* has four of five systems present in higher eukaryotes for protein folding (Fig. 2.9).

1. *Entamoeba* has the minimum component parts for N-glycan-dependent QC of protein folding (Helenius and Aebi, 2004; Trombetta and Parodi, 2003). These include a UDP-glucose-dependent glucosyltransferase (UGGT), which adds a single glucose to the

left arm of the N-glycans of misfolded proteins and so forms $GlcMan_5GlcNAc_2$ (Fig. 2.7). The glucosylated N-glycan is then bound and refolded by the lectin calreticulin (CRT), which is a chaperone that works with a protein disulphide isomerase (PDI) to make and break disulphide bonds. A glucosidase (Gls2) removes glucose from the well-folded protein, which is transferred to the Golgi by a mannose-binding lectin (ERGIC-53). The *Entamoeba* system is similar to that of mammals and fungi, which add glucose to the $Man_9GlcNAc_2$ precursor to make $GlcMan_9GlcNAc_2$ (Fig. 2.7). Mammals have a second glucosidase to remove glucose from the $Glc_3Man_9GlcNAc_2$ precursor (Fig. 2.7).

2. *Entamoeba* has N-glycan-independent QC of protein folding within the lumen of the ER, which includes the chaperones Hsp70 and Hsp90 (also known as BiP and Grp94, respectively) (Fig. 2.9) (Helenius and Aebi, 2004; Trombetta and Parodi, 2003). Also involved in this QC system are PDIs; DnaJ proteins that increase the ATPase activity of Hsp70 and Hsp90; and peptidyl-prolyl *cis-trans* isomerases (PPIases). This N-glycan-independent QC system for protein folding is present in all eukaryotes (S. Banerjee, J. Cui, P. W. Robbins, and J. Samuelson, unpublished data).
3. *Entamoeba* and all other eukaryotes have a N-glycan-independent system for ER-associated degradation (ERAD) of misfolded proteins (Fig. 2.9) (Hirsch *et al.*, 2004). This system is composed of proteins (Sec61 and Der1) that dislocate misfolded proteins from the ER lumen to the cytosol. There a complex of proteins (Cdc48, Npl4 and Ufd1) ubiquinate misfolded proteins, which are then degraded in the proteasome. In contrast, *Entamoeba* and the vast majority of eukaryotes are missing an N-glycan-dependent system of ERAD of misfolded proteins (Helenius and Aebi, 2004; Trombetta and Parodi, 2003). In this system, the middle arm of $Man_9GlcNAc_2$ is trimmed to $Man_8GlcNAc_2$, which is recognised by a unique mannose-binding lectin (EDEM) before dislocation into the cytosol for degradation (Fig. 2.9).
4. *Entamoeba* has a transmembrane kinase (Ire1) which recognises misfolded proteins in the lumen of the ER and triggers the unfolded protein response (Fig. 2.9) (Patil and Walter, 2001). The *Entamoeba* unfolded protein response is likely to be different from those of mammals and fungi, because *Entamoeba* is missing an important downstream target, which is a transcription factor called Hac1.

6.5.3. Unique N-glycans

Mammals make complex N-glycans in the Golgi by trimming back the precursor to $Man_3GlcNAc_2$ and then adding *N*-acetyl glucosamine, galactose, sialic acid and fucose (Fig. 2.8) (Hubbard and Ivatt, 1981). In each case, the activated sugars (UDP-GlcNAc, UDP-Gal, CMP-sialic acid and

GDP-fucose) are transferred from the cytosol to the lumen of the Golgi by a specific nucleotide-sugar transporter (NST) (Hirschberg *et al.*, 1998). In turn, each activated sugar is added to the N-glycans by a specific glycosyltransferase.

Entamoeba N-glycans are remarkable for two properties. First, the most abundant N-glycan is unprocessed $Man_5GlcNAc_2$ (Fig. 2.7) (P. E. Magnelli, D. M. Ratner, P. W. Robbins and J. Samuelson, unpublished data). This N-glycan is recognised by the mannose-binding lectin Concanavalin A, which caps glycoproteins on the *Entamoeba* surface (Silva *et al.*, 1975). Unprocessed $Man_5GlcNAc_2$ is also recognised by the anti-retroviral lectin cyanovirin, which binds $Man_9GlcNAc_2$ on the surface of gp120 (Adams *et al.*, 2004). This result suggests the possibility that the anti-retroviral lectin may be active against numerous protists.

Second, complex N-glycans of *Entamoeba*, which are built upon the same $Man_3GlcNAc_2$ core as higher eukaryotes, contain just two additional sugars (galactose and glucose) (Fig. 2.9, D and H) (P. E. Magnelli, D. M. Ratner, P. W. Robbins and J. Samuelson, unpublished data). Galactose is added first to both arms of $Man_3GlcNAc_2$ and then glucose is added to galactose. To make these complex N-glycans, *Entamoeba* has NSTs for glucose (UDP-Glc) and galactose (UDP-Gal) (Bredeston *et al.*, 2005). Glucose is also transferred to N-glycans during the QC of protein folding in the ER, while both galactose and glucose are transferred to proteophosphoglycans (PPGs) (see next section) (Moody-Haupt *et al.*, 2000). Because the complex N-glycans of *Entamoeba* are unique, it is possible that they may be targets of anti-amoebic antibodies.

6.5.4. O-glycans and GPI anchors

The surface of *E. histolytica* trophozoites is rich in glycoconjugates as shown by the ability of many lectins and carbohydrate-specific antibodies to recognise the cell surface (Srivastava *et al.*, 1995; Zhang *et al.*, 2002). Proteophosphoglycans constitute the major glycoconjugate of the *E. histolytica* cell surface. PPG is anchored to the cell surface through a GPI moiety (Bhattacharya *et al.*, 1992). The structure of the PPG GPI has been tentatively determined (Moody-Haupt *et al.*, 2000). In most eukaryotes, PI is glycosidically linked to the reducing end of de-acetylated glucosamine followed by three mannoses that are in turn attached to the ethanolamine that links the protein to the GPI. However, the GPI anchor of *E. histolytica* PPG was found to have a unique backbone that is not observed in other eukaryotes, namely Gal-Man-Man-GlcN-*myo*-inositol. The intermediate and light subunits of the *E. histolytica* Gal/GalNAc lectin, among other cell surface molecules, are anchored to the cell surface through GPI anchors. Though the structure of the GPI anchors is not known, they are thought to be functionally important (Ramakrishnan *et al.*, 2000).

In humans, 23 genes are known to participate in the biosynthesis of GPI anchors. However, only 15 of these were identified in *E. histolytica* (Vats *et al.*, 2005). Interestingly, all the catalytic subunits were identified in *E. histolytica*, the missing genes encoding the accessory subunits suggesting that the biosynthetic pathway may not be significantly different from that in other eukaryotes. The presence of the pathway was also confirmed by detecting the biochemical activities of the first two enzymes—*N*-acetyl glucosamine transferase and deacetylase. In addition, antisense inhibition of the deacetylase blocked GPI anchor biosynthesis and reduced virulence of the parasite (Vats *et al.*, 2005). A novel GIPL (glycosylated inositol phospholipid) was also identified in *E. histolytica* (Vishwakarma *et al.*, 2006). Structural studies indicate that a galactose residue is attached to glucosamine as the terminal sugar instead of mannose. This suggests that *E. histolytica* is capable of synthesising unusual GPI-containing glycoconjugates not observed in other organisms.

In PPG, glycans are attached to a peptide backbone by an O-phosphodiester-linkage (O-P glycans). The *E. histolytica* O-P-glycans have galactose at the reducing end followed by a chain of glucoses. *E. invadens* also has O-P-glycans on its cyst wall proteins but the reducing sugar is a deoxysugar rather than galactose (Van Dellen *et al.*, 2006b). While *Dictyostelium* has also O-P-glycans on glycoproteins in its spore wall, glycoproteins with O-P-glycans are absent from the vast majority of animals and plants (West, 2003).

6.5.5. Significance

The unique glycans of *Entamoeba* lead to three important evolutionary inferences. First, much of the diversity of eukaryotic N-glycans is due to secondary loss of enzymes that make the 14-sugar lipid-linked precursor, which was present in the common ancestor to extant eukaryotes. Despite the truncated N-glycan precursor, *Entamoeba* has conserved the relatively complex N-glycan-dependent QC system for protein folding. Third, the unique N-glycans and O-P-linked glycans are based on a novel set of glycosyltransferases, which are present in *Entamoeba* and remain to be characterised biochemically.

7. PROTEINS INVOLVED IN SIGNALLING

7.1. Phosphatases

The combined actions of protein kinases and phosphatases regulate many cellular activities through reversible phosphorylation of proteins. These activities include such basic functions as growth, motility and metabolism. Although it was once assumed that kinases played the major regulatory role, it is now clear that phosphatases can also be critical participants in some cellular events (Li and Dixon, 2000). There are few publications on the role of phosphatases in *E. histolytica*; however, several investigators have established a role for phosphatases in proliferation and growth. Chaudhuri *et al.* (1999) observed that there was an increase in phosphotyrosine levels in serum starved, growth inhibited, *E. histolytica* cultures. Upon the addition of serum and subsequent growth simulation, an increase in tyrosine phosphatase activity occurred. These investigators also demonstrated that genistein, a tyrosine kinase inhibitor, had no effect on growth, while the addition of sodium orthovanadate, a phosphatase inhibitor, produced a major decrease in cell proliferation. Membrane-bound and secreted acid phosphatase activities have been detected in *E. histolytica* (Aguirre-Garcia *et al.*, 1997; Anaya-Ruiz *et al.*, 1997). The secreted acid phosphatase activity is absent from *E. dispar* (Talamas-Rohana *et al.*, 1999). This secreted acid phosphatase was found to have phosphotyrosine hydrolase activity, and caused cell rounding and detachment of HeLa cells (Anaya-Ruiz *et al.*, 2003), suggesting that phosphatase activity contributes to the virulence of the organism.

There are four families of phosphatases (Stark, 1996). Members of the PPP (protein phosphatase P) family are serine/threonine phosphatases and include PP1, PP2A and PP2B (calcineurin-like) classes. The PPM (protein phosphatase M) family phosphatases also dephosphorylate serine/threonine residues but are unrelated to the PPP family proteins. A third family consists of protein tyrosine phosphatases (PTP) and dual phosphatases. Low molecular weight phosphatases make up the fourth family. In eukaryotic cells, greater than 99% of protein phosphorylation is on serine or threonine residues (Chinkers, 2001). Human cells have about 500 serine/threonine phosphatases and 100 tyrosine phosphatases (Hooft van Huijsduijnen, 1998; Hunter, 1995). *S. cerevisiae* has 31 identified or putative protein phosphatases (Stark, 1996). *E. histolytica* has over 100 putative protein phosphatases. Only a few of these phosphatases have potential transmembrane domains. Some *E. histolytica* phosphatases have varying numbers of LRRs. The LRR domain is thought to be a site for protein–protein interactions (Hsiung *et al.*, 2001; Kobe and Deisenhofer, 1994). LRR domains have been found in a few kinases, but had not been identified in any phosphatases until recently (Gao *et al.*, 2005).

7.1.1. Serine/threonine protein phosphatases

Members of the PPP family of protein phosphatases are closely related metalloenzymes, and complex with regulatory subunits. In contrast, PPM family members are generally monomeric, ranging 42–61 kDa in size. By BLAST analysis, the serine/threonine protein phosphatases of *E. histolytica* are most closely related to PPP phosphatases PP2A, PP2B and PPM phosphatase PP2C.

7.1.1.1. PP2A and PP2B (Calcineurin-like) serine/threonine phosphatases PP2A phosphatases are trimeric enzymes consisting of catalytic, regulatory and variable subunits (Wera and Hemmings, 1995). Calcineurin is a calcium-dependent protein serine/threonine phosphatase (Rusnak and Mertz, 2000). Orthologues of calcineurin are widespread from yeast to mammalian cells. Calcineurin is a heterodimeric complex with catalytic (CaNA) and regulatory (CaNB) subunits. CaNA ranges in size from 58 to 64 kDa. Its conserved domain structure includes a catalytic domain, a CaNB-binding domain, a calmodulin binding domain and an auto-inhibitory (AI) domain. The binding of CaNB and calmodulin activates CaNA. CaNB subunit is 19 kDa, contains 4 EF hand calcium binding motifs, and has similarity to calmodulin. The binding of calmodulin releases the auto-inhibitory domain and results in activation of the phosphatase. Deletion of the AI domain results in a constitutively active protein. Calcineurin is specifically inhibited by cyclosporin A and FK506. Cyclosporin A and FK506 first bind to specific proteins, cyclophilin A and FK506BP, respectively, then bind to CaNA at the CaNB binding site. Cyclophilin A has been identified in *E. histolytica* and treatment with cyclosporin A decreases growth and viability (Carrero *et al.*, 2000, 2004; Ostoa-Saloma *et al.*, 2000).

The *E. histolytica* genome has 51 PP2A and calcineurin-like protein phosphatases. The Pfam motif that classifies proteins as PPP phosphatases is Metallophos (PF00149, calcineurin-like phosphoesterase). This motif is also found in a large number of proteins involved in phosphorylation, including DNA polymerase, exonucleases and other phosphatases. The genome annotation identifies three loci as CaNA orthologues. However, due to the similarity among this family of phosphatases, it is difficult to tell by sequence analyses alone those that are calcium-dependent. Identification of CaNA will have to be confirmed experimentally.

Two of the PPM phosphatases contain a tetratricopeptide repeat (TPR) domain (PF00515). TPR is thought to be involved in protein–protein interactions (Das *et al.*, 1998). Activities that have been ascribed to TPR include regulatory roles, lipid binding and auto-inhibition.

7.1.1.2. PP2C phosphatases PP2C phosphatases are also widespread and are often involved in terminating/attenuating phosphorylation during the cell cycle or in response to environmental stresses such as osmotic and heat shock (Kennelly, 2001). Thirty-five genes were identified as PP2C phosphatases. These proteins can be divided into three broad categories: (1) PP2C domain only small (235–381 amino acids), (2) PP2C domain only large (608–959 amino acids) and (3) PP2C with LRR domains.

7.1.2. Tyrosine phosphatases (PTP)

Tyrosine phosphorylation-dephosphorylation is a key regulatory mechanism for many aspects of cell biology and development (Li and Dixon, 2000). PTPs are a large class of enzymes that have catalytic domains of ~300 amino acids. Forty of these residues are highly conserved (Hooft van Huijsduijnen, 1998). PTPs can be divided into membrane (receptor) and non-membrane (soluble) PTPs (Li and Dixon, 2000). The soluble PTP group includes those that contain conserved SH2, PEST, Ezrin, PDZ or CH2 domains. Two other classes of PTPs are the low molecular weight and dual phosphatases. *S. cerevisiae* lacks classic PTPs but does contain dual phosphatases such as the MAP kinase kinases.

E. histolytica has only four potential PTPs, none of which are receptor PTPs (i.e., PTPs with recognisable transmembrane spanning regions). Two of the PTPs (XM_650778, XM_645883) are 350 and 342 amino acids in length and share 48% identity. Neither of these phosphatases has any other recognisable conserved domain. Non-receptor type 1 PTPs are the closest match to these proteins (Li and Dixon, 2000). Membrane and secreted forms of a PTP that cross-react with anti-human PTP1B have been reported in *E. histolytica* (Aguirre-García *et al.*, 2003; Talamas-Rohana *et al.*, 1999). Both forms have an apparent molecular weight of 55 kDa and disrupt host actin stress fibres. However, since none of the putative PTPs identified by the genome project appear to encode secreted or membrane forms, it is unlikely that these loci represent these previously reported PTP1B cross-reacting proteins.

A third PTP contains a protein tyrosine phosphatase like protein (PTPLA) domain (PF04387). The PTPLA domain is related to the catalytic domains of tyrosine kinases, but it has an arginine for proline substitution at the active site (Uwanogho *et al.*, 1999). It is not yet clear whether this family of proteins actually has phosphatase activity or serves some other regulatory role.

An orthologue of a low molecular weight PTP has also been identified. Low molecular weight protein tyrosine phosphatases have been found in bacteria, yeasts and mammalian cells (Ramponi and Stefani, 1997). They are not similar to other PTPs except in the conserved catalytic domain.

7.1.3. Dual-specificity protein phosphatases

Dual-specificity PTPs (DSP) can hydrolyse both tyrosine and serine/threonine residues, though they hydrolyse phosphorylated tyrosine substrates 40–500-fold faster (Zhang and VanEtten, 1991). In other organisms, DSPs are mostly found in the nucleus and have roles in cell cycle control, nuclear dephosphorylation and inactivation of MAP kinase.

The *E. histolytica* genome has 23 sequences related to DSPs. They fall into three main subclasses: those with the DSP domain only, those with

DSP plus a variable number (one to five) of LRRs and those with the Rhodanese homology domain (RHOD; IPR001763). Rhodanese is a sulphurtransferase involved in cyanide detoxification. Its active site, RHOD, is also found in the catalytic site of the dual-specificity phosphatase CDC25 (Bordo and Bork, 2002).

7.1.4. Leucine rich repeats

LRRs are tandem arrays of 20–29 amino acid, leucine-rich motifs. LRRs have been found in a number of proteins with varied functions including enzyme inhibition, regulation of gene expression, morphology and cytoskeleton formation (Kobe and Deisenhofer, 1994). LRRs are thought to provide versatile sites for protein–protein interactions and have been found linked to a variety of secondary domains. Most LRRs form curved horseshoe-shaped structures with ''a parallel β sheet on the concave side and mostly helical elements on the convex side'' (IPR001611).

The LRR_1 Pfam is the second most abundant Pfam domain found in the *E. histolytica* genome (Table 2.3). The LRR motifs in *E. histolytica* most closely resemble the LRR found in BspA (Section 2.7; Davis *et al.*, 2006). Several *E. histolytica* proteins that contain LRRs are associated with other recognised domains. These include the protein phosphatases PP2C and DSP, as well as protein kinase (PK), F-box (PF00646), gelsolin/villin headpiece (IPR007122), DNA J (IPR001623), Band 41 (B41;IPR000299), WD-40 (IPR001680) and zinc binding (IPR000967) domains. The association of LRRs with phosphatases is unusual. One published example is the phosphatase that dephosphorylates the kinase Akt (Gao *et al.*, 2005). Fungal adenylate cyclases have both LRR and PP2C-like domains, but this is not a widespread feature of adenylate cyclases in other species (Mallet *et al.*, 2000; Yamawaki-Kataoka *et al.*, 1989). The LRR may be a site for interaction with phosphorylated residues in *E. histolytica*. This speculation is supported by the example of the Grr1 protein of yeast, which contains an F-box and an LRR (Hsiung *et al.*, 2001). Grr1 is involved in ubiquitin-dependent proteolysis. The LRR domain of Grr1 binds to phosphorylated targets in the proteasome complex. Another example is the fission yeast phosphatase regulatory subunit, Sds22, which also has LRRs (MacKelvie *et al.*, 1995). The LRR containing phosphatases of *E. histolytica* may represent fusions of regulatory and catalytic subunits.

7.2. Kinases

7.2.1. Cytosolic kinases

Eukaryotic protein kinases are a superfamily of enzymes which are important for signal transduction and cell-cycle regulation. Six families of serine/threonine kinases (STKs), which include AGC, Ste, CK1, CaMK, CMGC and TKL (tyrosine kinase-like), have conserved aspartic acid and lysine amino acids in their active sites and phosphorylate serine or threonine on target proteins (Hanks and Hunter, 1995). Tyrosine kinases (TKs), which lack active site lysine, phosphorylate tyrosine on target proteins. Phosphorylated tyrosine is in turn recognised by Src-homology 2 (SH2) domains that are present on some kinases and other proteins. All seven families of protein kinases are present in metazoa and in *D. discoideum*, while plants lack TK, and *S. cerevisiae* lacks both TK and TKL.

Over 150 predicted *E. histolytica* cytosolic kinases, those that lack signal peptides and *trans*-membrane helices, can be identified, including representatives of each of the 7 groups of kinases (AGC, CAMK, CK1, CMGC, STE, TKL and TK) (Loftus *et al.*, 2005). Two predicted *E. histolytica* TKs, which group with human TKs in phylogenetic trees, contain an AAR peptide in the active site and a Kelch domain at the C-terminus (Gu and Gu, 2003). Four cytosolic protein kinases contain C-terminal SH2 domains, which bind phosphorylated tyrosine residues. Phosphotyrosine has been identified in *E. histolytica* using specific antibodies (Hernandez-Ramirez *et al.*, 2000). The 35 predicted cytosolic *E. histolytica* TKLs include some that contain LRRs and ankyrin repeats at their N-termini. In contrast, the vast majority of *Entamoeba* cytosolic kinases lack accessory domains.

7.2.2. Receptor kinases

Five distinct families of eukaryotic proteins have an N-terminal ectoplasmic domain, a single TMH and a C-terminal cytoplasmic kinase domain (Blume-Jensen and Hunter, 2001). Ire-1 transmembrane kinases, which are present in *S. cerevisiae*, plants and metazoa, detect unfolded proteins in the lumen of the ER and help splice a transcription factor mRNA by means of a unique C-terminal ribonuclease (Patil and Walter, 2001). Receptor tyrosine kinases (RTKs), which include growth hormone and epidermal growth factor (EGF) receptors, are restricted to metazoa and have a diverse set of N-terminal ectoplasmic domains and a conserved C-terminal cytosolic TK (Schlessinger, 2000). Receptor serine/threonine kinases (RSK) of metazoa and receptor-like kinases (RLKs) of plants each contain a C-terminal TKL domain (Massague *et al.*, 2000; McCarty and Chory, 2000; Shiu and Bleecker, 2001). Phylogenetic analyses suggest that plant RLKs, animal RSKs and animal RTKs each form monophyletic groups and that plant RLKs closely resemble cytosolic TKLs of animals called Pelle or IRAK (Shiu and Bleecker, 2001).

E. histolytica contains >80 novel receptor RSKs, each of which has a N-terminal signal sequence, a conserved ectoplasmic domain, a single TMH and a cytosolic kinase domain (Beck *et al.*, 2005). The largest group of *E. histolytica* RSKs has a CXXC-rich ectoplasmic domain with 6–31 internal repeats that each contains 4–6 cysteine residues (Fig. 2.10). Very similar CXXC-rich domains are present in the ectoplasmic domain

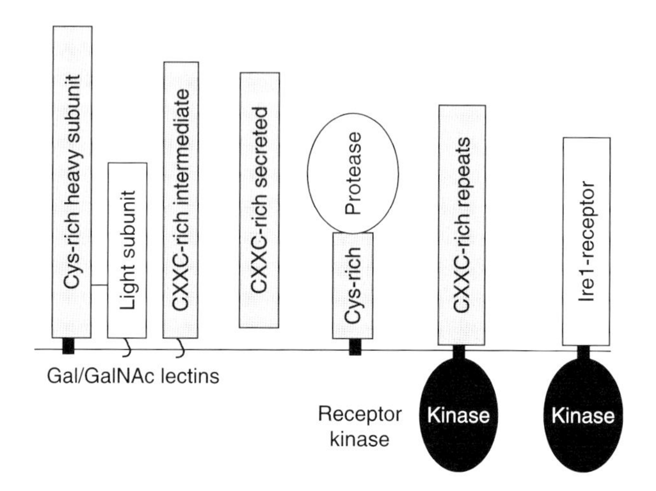

FIGURE 2.10 Structure of cysteine-rich plasma membrane proteins of *E. histolytica*. These proteins include the various subunits of the Gal/GalNAc lectin, a cysteine protease and numerous receptor kinases. Ire1, which is involved in the unfolded protein response, is also a receptor kinase but has no Cys-rich domain.

intermediate subunit of the Gal/GalNAc lectin (see Section 3.1.3). CXXC-rich domains are also present in hypothetical secreted proteins of *E. histolytica*, while cysteine-rich domains are also present in the heavy subunit of the Gal/GalNAc lectin and at the cytosolic aspect of some cysteine proteases (Fig. 2.10).

Ectoplasmic domains of other large families of *Entamoeba* RSKs have one or two 6-Cys domains at the N-terminus and four 6-Cys domains proximal to the plasma membrane. There are no plasma membrane proteins or secreted proteins with similar domains. A minority of RSKs do not contain Cys-rich ectoplasmic domains. Numerous *Entamoeba* RSKs are expressed at the same time, but the specific ligands for the *Entamoeba* RSKs have not been identified (Beck *et al.*, 2005).

As discussed in the section on protein folding (Section 6.5.2), *Entamoeba* has an Ire1 transmembrane kinase, which recognises misfolded proteins in the lumen of the ER and triggers the unfolded protein response (Fig. 2.8).

7.2.3. Significance

While most protists lack TK, TKL, receptor-kinases and Ire1 *E. histolytica* has all four. It is very likely that the *E. histolytica* receptor kinases, which are extensively duplicated, will have important roles in pathogenesis (Beck *et al.*, 2005; Okada *et al.*, 2005). Similarly, trimeric G-proteins and the associated adenyl-cyclases likely have important roles in cyst formation and virulence (Coppi *et al.*, 2002; Frederick and Eichinger, 2004).

7.3. Calcium binding proteins

Ca^{2+}signalling plays a crucial role in the pathogenesis of many protozoan parasites, including *E. histolytica* (Ravdin *et al.*, 1985). Many of the calcium-mediated processes are carried out with the help of calcium binding proteins (CaBPs). CaBPs have been identified and characterised in almost all eukaryotic systems. Some of these, such as calmodulin (CaM) and troponin C, have been studied extensively. A number of CaBPs have also been identified in *E. histolytica*. Among these are two related EF-hand containing proteins, grainin 1 and granin 2, which are likely to be localised in intracellular granules (Nickel *et al.*, 2000). Another protein, URE3-BP, was shown to have a transcription regulatory function (Gilchrist *et al.*, 2001). The CaM-dependent secretion of collagenases from electron dense granules has been demonstrated using *E. histolytica* lysate. However, there is as yet no direct molecular evidence for the presence of CaM in *E. histolytica* (de Muñoz *et al.*, 1991). The CaM-like protein EhCaBP1 has four canonical EF-hand Ca^{2+} binding domains but no functional similarity to CaM (Yadava *et al.*, 1997). Inducible expression of EhCaBP1 antisense RNA demonstrated this protein's role in actin-mediated processes (Sahoo *et al.*, 2004).

Analysis of the whole genome revealed presence of 27 CaBPs with multiple EF-hand calcium binding domains (Bhattacharya *et al.*, 2006). Many of these proteins are architecturally very similar but functionally distinct from CaM. Moreover, functional diversity was also observed among closely related CaBPs such as EhCaBP1 and EhCaBP2 (79% identical at the amino acid level; Chakrabarty *et al.*, 2004). Analysis of partial EST and proteomic databases combined with Northern blots and RT-PCR shows that at least one-third of these genes are expressed in trophozoites, suggesting that many if not all of the 27 are functional genes (Bhattacharya *et al.*, 2006).

What are the roles of these proteins in the context of *E. histolytica* biology? At present the function of only two EhCaBPs are known, EhCaBP1 and URE3-BP. The rest of the proteins are likely to be Ca^{2+} sensors involved in a number of different signal transduction pathways. After binding Ca^{2+} these may undergo conformational changes and the bound form then activates downstream target proteins. It is not clear why *E. histolytica* would need so many Ca^{2+} sensors when many other organisms do not. It is likely that with Ca^{2+} being involved in many functions, some of which are localised in different cellular locations, the various CaBPs may participate in different functions that are spatially and temporally separated.

8. THE MITOSOME

One of the expectations for the *E. histolytica* genome project was that it would identify the function of the mitochondrial remnant known as the mitosome (Tovar *et al.*, 1999) or crypton (Mai *et al.*, 1999). Under the microscope mitosomes are ovoid structures smaller than 0.5 μm in diameter (Leon-Avila and Tovar, 2004). While it is now clear that no mitochondrial genome still persists, from both genome sequencing and cellular localisation data (Leon-Avila and Tovar, 2004), the protein complement of the organelle is still somewhat obscure. The number of identifiable mitosomal proteins remains very small and does not provide great insight into the organelle's function. Genes encoding mitochondrial-type chaperonins (cpn60, hsp10 and mt-hsp70) have been identified and appear to be synthesised with amino-terminal signal sequences. The importation machinery has been shown to be conserved with that in true mitochondria (Mai *et al.*, 1999; Tovar *et al.*, 1999), but none of the proteins involved in mitosomal protein import have been identified with certainty.

Other genes encoding putative mitosomal proteins include pyridine nucleotide transhydrogenase (which moves reducing equivalents between NAD and NADP, and acts as a proton pump (Clark and Roger, 1995); only an incomplete gene is present in the assembly), an ADP/ATP transporter (Chan *et al.*, 2005), a P-glycoprotein-like protein (Pgp6), and a mitochondrial-type thioredoxin, although the latter two are identified based largely on their amino terminal extensions. The only enzymatic pathway that is normally mitochondrial in location is iron–sulphur cluster synthesis. Genes encoding homologues of both IscS/NifS and IscU/NifU proteins are present, but uniquely among eukaryotes the *E. histolytica* homologues are not of mitochondrial origin, having been acquired by distinct LGT from an ε-proteobacterium (Ali *et al.*, 2004b; van der Giezen *et al.*, 2004). The location of these proteins appears to be cytoplasmic as determined by immunofluorescence, using antibodies against both the native proteins as well as detection of epitope-tagged proteins in transformed *E. histolytica* (V. Ali and T. Nozaki, unpublished data). The same pathway has been localised to mitosomes in *Giardia* and is also retained in all other organisms with remnant mitochondria. Given the apparently unique non-compartmentalised nature of iron–sulphur cluster synthesis in *E. histolytica* the location of the proteins needs to be confirmed by immuno-electron-microscopy; such experiments are currently under way (V. Ali and T.Nozaki, unpublished data). The function of the *E. histolytica* mitosome therefore remains an enigma.

9. ENCYSTATION

The infectious stage of *E. histolytica*, and also that most often used for diagnosis, is the quadrinucleate cyst. Because it is not possible to encyst *E. histolytica* in axenic culture, *E. invadens*, which is a reptilian parasite, has been used as a model organism for encystation (Eichinger, 2001; Wang *et al.*, 2003). The *E. invadens* cyst wall is composed of three parts: deacetylated chitin (also known as chitosan), lectins that bind chitin (e.g., Jacob and Jessie) or cyst wall glycoproteins (e.g., plasma membrane Gal/GalNAc lectin), and enzymes that modify chitin or cyst wall proteins (e.g., chitin deacetylase, chitinase and cysteine proteases) (Fig. 2.11).

9.1. Chitin synthases

Chitin fibrils, which are homopolymers of β-1,4-linked *N*-acetyl glucosamine (GlcNAc), are synthesised by chitin synthases. Chitin synthases share common ancestry with cellulose synthases and hyaluronan synthase. They are transmembrane proteins with a catalytic domain in the cytosol (Bulawa, 1993), where UDP-GlcNAc is made into a homopolymer and is threaded through the transmembrane domains into the extracellular space. In *S. cerevisiae*, four accessory peptides, encoded by the Chs4–7 genes, are necessary for the function of its chitin synthases (Trilla *et al.*, 1999). Remarkably, the *E. histolytica* chitin synthase 2 (EhChs2) complements a *S. cerevisiae* chs1/chs3 mutant and the function of EhChs2 is independent of the four accessory peptides (Van Dellen *et al.*, 2006a). This result suggests the possibility that chimaeras of *E. histolytica* and *S. cerevisiae* chitin synthases may be used to map domains in the *S. cerevisiae* chitin synthase that interact with the accessory peptides.

9.2. Chitin deacetylases

Chitin fibrils in the cyst wall are modified by deacetylases and chitinases (see Section 9.3). There are two *E. invadens* chitin deacetylases, which convert chitin into chitosan (Das *et al.*, 2006). Chitosan is a mixture of *N*-acetyl glucosamine and glucosamine and so has a positive charge. It is also present in spore walls of *S. cerevisiae* and in lateral walls of *Mucor* (Kafetzopoulos *et al.*, 1993; Mishra *et al.*, 1997). It is likely that the positive charge of chitosan fibrils contributes to the binding of cyst wall proteins, all of which are acidic (de la Vega *et al.*, 1997; Frisardi *et al.*, 2000; Van Dellen *et al.*, 2002b). Monosaccharide analyses of the *E. invadens* cyst walls following treatment with SDS to remove proteins strongly suggest that chitosan is the only sugar homopolymer present (Das *et al.*, 2006).

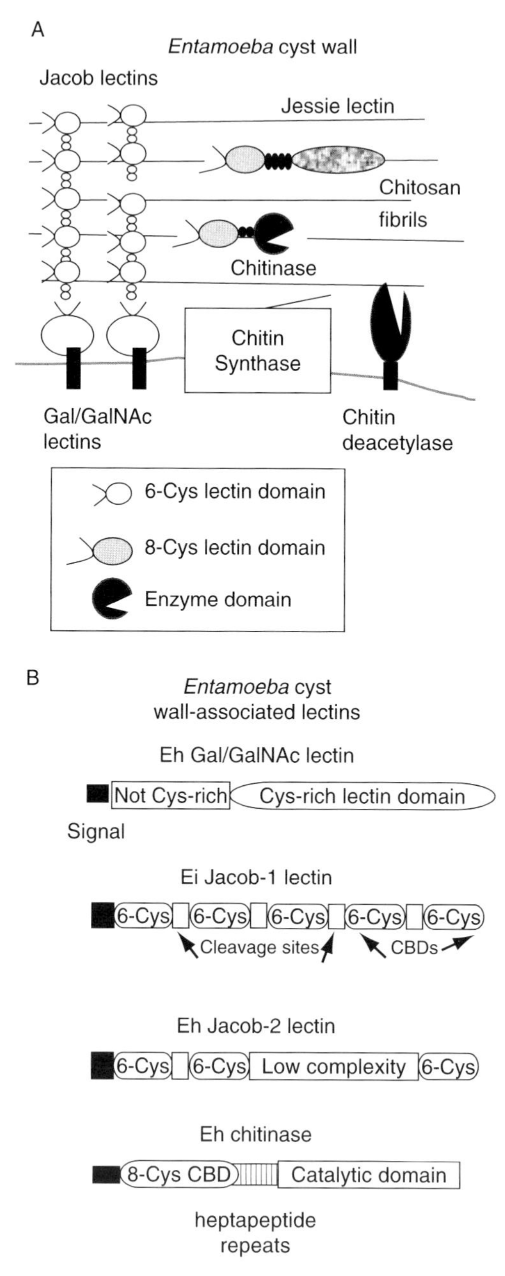

FIGURE 2.11 Model for the *Entamoeba* cyst wall derived primarily from experiments with *E. invadens*. (A) The cyst wall consists of chitosan fibrils, which are made by chitin synthase and chitin deacetylase. Wall proteins include Jacob lectins with tandem arrays of 6-Cys chitin-binding domains (CBDs), as well as chitinase and Jessie lectins that have a single 8-Cys CBD. The Gal/GalNAc lectin in the plasma membrane binds sugars on the Jacob and Jessie lectins. (B) Structures of representative lectins are illustrated in (A).

9.3. Chitinases

Entamoeba species encode numerous chitinases with a conserved type 18 glycohydrolase domain (de la Vega *et al.*, 1997). Recombinant *Entamoeba* chitinases have both endo- and exo-chitinase activities. Two other domains are important in *Entamoeba* chitinases: (1) At the N-terminus is a unique 8-Cys chitin-binding domain (CBD), which is also present as a single domain in *E. histolytica* Jessie lectins (Fig. 2.11) (Van Dellen *et al.*, 2002b). Chitinase and Jessie-3 lectin bind to the *E. invadens* cyst wall by means of this 8-Cys CBD (Van Dellen *et al.*, 2006b). This *E. histolytica* chitinase CBD has the same function as CBDs in chitinases of fungi, nematodes, insects and bacteria, but has no sequence similarity (i.e., it has arisen by convergent evolution) (Shen and Jacobs-Lorena, 1999). (2) Between the CBD and chitinase domains of *Entamoeba* species are low complexity sequences that contain heptapeptide repeats (Ghosh *et al.*, 2000). These polymorphic repeats may be used to distinguish isolates of *E. histolytica* within the same population and may be able to discriminate among isolates from New to Old World (Haghighi *et al.*, 2003). These polymorphic repeats, which are rich in serine and resemble mucin-like domains in other glycoproteins, may also be the sites for addition of O-phosphodiester linked sugars (see Section 6.5.4).

9.4. Jacob lectins

Chitin fibrils in the cyst wall of *E. invadens* are cross-linked by Jacob lectins, which contain three to five unique 6-Cys CBDs (Frisardi *et al.*, 2000). *E. invadens* has at least nine genes encoding Jacob lectins, and the mRNA levels from each gene increase during encystation (Van Dellen *et al.*, 2006b). In addition, at least six Jacob lectin proteins are present in *E. invadens* cyst walls (Van Dellen *et al.*, 2006b). Between the CBDs, Jacob lectins have low complexity sequences that are rich in serine as in the case of chitinase (Van Dellen *et al.*, 2006a). Jacob lectins are post-translationally modified in two ways. First, they are cleaved by cysteine proteases at conserved sites in the serine- and threonine-rich spacers between CBDs. Second, they have O-phosphodiester-linked sugars added to serine and threonine residues. O-phosphodiester-linked glycans are also present in PPGs on the surface of *E. histolytica* trophozoites (Moody-Haupt *et al.*, 2000).

9.5. Gal/GalNAc lectins

The Gal/GalNAc lectins present on the surface of *E. histolytica* trophozoites have been described earlier (see Section 3.1) and in the literature (Mann *et al.*, 1991; Petri *et al.*, 2002). Their possible role in encystation is suggested by two independent experiments. First, the signal for encystation likely depends in part on aggregation of *E. invadens*, which is inhibited by exogenous galactose (Coppi and Eichinger, 1999). Aggregated *E. invadens* secrete catecholamines, which in an autocrine manner stimulate amoebae to encyst (Coppi *et al.*, 2002). Second, in the presence of excess galactose, *E. invadens* forms wall-less cysts that contain four nuclei and makes Jacob lectins and chitinase (Frisardi *et al.*, 2000). Because *E. invadens* trophozoites have a Gal/GalNAc lectin on their surface that is capable of binding sugars on Jacob lectin, and because Jacob lectins have no carboxy-terminal TMH or GPI-anchor, it is likely that the cyst wall is bound to the plasma membrane by the Gal/GalNAc lectin.

9.6. Summary and comparisons

Similar to the cyst wall of *Giardia*, the cyst wall of *E. invadens* is a single homogeneous layer and contains a single homopolymer, chitosan (Fig. 2.11) (Frisardi *et al.*, 2000; Gerwig *et al.*, 2002; Shen and Jacobs-Lorena, 1999). In contrast, *S. cerevisiae* spore walls have multiple layers and contain β-1,3-glucans in addition to chitin, while *Dictyostelium* walls have multiple layers and contain *N*-acetyl galactosamine polymers in addition to cellulose (West, 2003; Yin *et al.*, 2005).

Similar to *Dictyostelium* and in contrast to fungi, the vast majority of *Entamoeba* cyst wall glycoproteins are released by SDS (Frisardi *et al.*, 2000; Van Dellen *et al.*, 2006b; West, 2003; Yin *et al.*, 2005). While some *Dictyostelium* cyst wall proteins have been shown to be cellulose-binding lectins, all of the proteins bound to the cyst wall of *E. invadens* have 6-Cys CBDs (Jacob lectins) or 8-Cys CBDs (Jessie 3 lectin and chitinase) (Frisardi *et al.*, 2000; Van Dellen *et al.*, 2002b). In the same way that *Giardia* cyst wall protein 2 is cleaved by a cysteine protease, Jacob lectins are cleaved by an endogenous cysteine protease at sites between chitin-binding domains (Touz *et al.*, 2002).

Like *Dictyostelium* spore coat proteins and insect peritrophins, cysteine-rich lectin domains of *E. invadens* cyst wall proteins are separated by serine- and threonine-rich domains that are heavily glycosylated (Frisardi *et al.*, 2000; West, 2003; Yin *et al.*, 2005). *S. cerevisiae* cyst wall proteins have also extensive serine- and threonine-rich domains that are heavily glycosylated (Yin *et al.*, 2005). These glycans likely protect proteins in cyst walls or fungal walls from exogenous proteases. While glycoproteins of the *E. invadens* cyst wall and *Dictyostelium* spore coat contain O-phosphodiester-linked glycans, *S. cerevisiae* wall glycoproteins contain O-glycans (Gemmill and Trimble, 1999; West *et al.*, 2005).

Like *S. cerevisiae*, *E. invadens* has enzymes in its wall that modify chitin (Yin *et al.*, 2005). Similar to chitinases of *S. cerevisiae* and bacteria, *E. invadens* chitinase has a CBD in addition to the catalytic domain (Kuranda and Robbins, 1991). It is likely that the CBD is present to localise chitinase to the cyst wall (*E. invadens*) or cell wall (*S. cerevisiae*). Finally, while *E. invadens* uses catecholamines as autocrines for encystation, *Dictyostelium* uses cAMP as an autocrine for sporulation (Coppi *et al.*, 2002; Kriebel and Parent, 2004). An important goal of future research will be to translate what is known about the *E. invadens* cyst wall to that of *E. histolytica*.

10. EVIDENCE OF LATERAL GENE TRANSFER IN THE *E. HISTOLYTICA* GENOME

Lateral (or horizontal) gene transfer (LGT) plays a significant role in prokaryotic genome evolution, contributing up to ~20% of the content of a given genome (Doolittle *et al.*, 2003). LGT has therefore been an important means of acquiring new phenotypes, such as resistance to antibiotics and new physiological and metabolic capabilities, that may permit or facilitate adaptation to new ecological niches (Koonin *et al.*, 2001; Lawrence, 2005a; Ochman *et al.*, 2000). More recently, data from microbial eukaryote genomes suggest that LGT has also played a role in eukaryotic genome evolution, particularly among protists that eat bacteria (Andersson, 2005; Doolittle, 1998; Doolittle *et al.*, 2003; Lawrence, 2005b; Richards *et al.*, 2003). *E. histolytica* lives in the human gut, an environment that is rich in micro-organisms and where LGT is thought to be common between bacteria (Shoemaker *et al.*, 2001). The *E. histolytica* genome thus provides a nice model for investigating prokaryote to eukaryote LGT. In the original genome description (Loftus *et al.*, 2005), 96 putative cases of LGT were identified using phylogenetic analyses of the *E. histolytica* proteome. These have now been reanalysed in the light of more recently published (August, 2005) eukaryotic and prokaryotic genomes. This has allowed evaluation of how previous inferences were influenced by the sparse sampling of eukaryotic and prokaryotic genes and species available at the time of the original analysis. Sparse gene and species sampling is, and is likely to remain, a very serious problem for reconstructing global trees and inferring LGT (Andersson *et al.*, 2001; Richards *et al.*, 2003; Salzberg *et al.*, 2001). Thus, although ecologists differ in their claims for the extent of the unsampled microbial world, they all agree that those species in culture, and the even smaller subset for which genome data exist, represent the smallest tip of a very large iceberg.

10.1. How do the 96 LGT cases stand up?

As before (Loftus *et al.*, 2005), Bayesian and maximum likelihood distance bootstrap phylogenetic analyses were used to identify putative LGT using the following ad hoc conservative criteria: Putative LGT was inferred where either no other eukaryote possessed the gene or where the *E. histolytica* sequence was grouped with bacteria and separated from other eukaryotes by at least two strongly supported nodes (bootstrap support >70%, posterior probabilities >0.95). In cases where tree topologies were more weakly supported but still suggested a possible LGT, bootstrap partition tables were examined for partitions where the *E. histolytica* sequence clustered with another eukaryote. If no such partitions were found that gene was considered to be a putative LGT. Table 2.8 lists the results of the new analyses and also gives BlastP statistics for each sequence.

A total of 41 LGTs remain as strongly supported as before based on the original criteria. For the remaining 55 tree topologies, support for recent LGT into the *Entamoeba* lineage is not as strong as before. For 27 of these 55 trees, 2 strongly supported nodes separating *E. histolytica* from other eukaryotes have been reduced to only 1 well-supported node. However, close scrutiny of the bootstrap partition tables for these trees revealed that, as before, there are no trees in which *E. histolytica* is found together with another eukaryote. Thus, LGT still remains the strongest hypothesis to explain 68 (70%) of the original 96 tree topologies. In a further 14 cases, the position of *E. histolytica* among prokaryotes and eukaryotes was not well supported. The taxonomic sampling of eukaryotes in these trees is very patchy and the trees do not depict consensus eukaryotic relationships. Thus, although the trees do not fulfil the conservative criteria for LGT, they also do not provide strong support for the alternative hypothesis that the *E. histolytica* genes were vertically inherited from a common ancestor shared with all other eukaryotes.

In nine trees *E. histolytica* either clustered with a single newly published eukaryotic sequence, or such a relationship could not be ruled out. In six of these nine trees *E. histolytica* and *T. vaginalis* grouped together, and two trees grouped *E. histolytica* with the diatom *Thalassiosira* (e.g., see Fig. 2.12). Such trees are also not easy to explain within the current consensus for eukaryotic relationships (Baldauf, 2003). Similar topologies have previously been reported for other eukaryotes (Andersson, 2005). The explanations advanced to explain the absence of the gene in other eukaryotes include massive gene loss from multiple eukaryotic lineages, or LGT between the eukaryotic lineages concerned. *Entamoeba* species can ingest both eukaryotes and prokaryotes, and it has been suggested that LGT between eukaryotes, subsequent to one lineage acquiring the gene from a prokaryote, could explain such peculiar tree topologies and sparse distribution (Andersson, 2005). The fact that six of the nine cases recover a relationship between *Entamoeba* and *Trichomonas*, whose relatives often share the same niche, is consistent with this idea. In prokaryotes, recent large-scale analyses support the hypothesis that species from the same

TABLE 2.8 Reassessment of the 96 candidate LGT cases identified in the original genome publication

Acc.[a] RefSeq	Acc.[a]	EhL[b]	Top prokaryotic BLAST hit	PL[c]	% ID[d]	Top eukaryotic BLAST hit	EL[c]	% ID[d]	PE-score[e]	EE-score[f]	P/E ratio[g]
41 LGT cases that remain strongly supported according to our criteria											
EAL43201	XP_648590.1	487	*Treponema denticola*	507	57	*Trichomonas vaginalis*	398	43	1.00E-167	5.00E-88	2.00E-80
EAL43619	XP_649008.1	621	*Vibrio vulnificus*	673	41	*Saccharomyces cerevisiae*	664	40	1.00E-132	1.00E-125	1.00E-07
EAL43678	XP_649067.1	538	*Fusobacterium nucleatum*	562	47	*Trichomonas vaginalis*	477	34	1.00E-135	2.00E-60	5.00E-76
EAL43850	XP_649240.1	880	*Mannheimia succiniciproducens*	898	63	*Mastigamoeba balamuthi*	882	45	0	0	N/A
EAL44182	XP_649570.1	260	*Bacteroides thetaiotaomicron*	273	34	*Yarrowia lipolytica*	298	29	2.00E-35	4.00E-10	5.00E-26
EAL44226	XP_649612.1	262	*Bacteroides thetaiotaomicron*	267	28	*Tetrahymena thermophila*	1476	30	2.00E-25	0.11	1.82E-24
EAL44778	XP_650165.1	188	*Bacteroides thetaiotaomicron*	188	43	*Neurospora crassa*	546	34	8.00E-41	1.8	4.44E-41
EAL45076	XP_650453.1	358	*Bacteroides fragilis*	362	46	*Trichomonas vaginalis*	562	22	1.00E-87	0.24	4.17E-87
EAL45145	XP_650531.1	825	*Staphylococcus aureus*	1036	30	*Trichomonas vaginalis*	2468	20	3.00E-59	0.016	1.88E-57
EAL45220	XP_650606.1	479	*Clostridium tetani*	471	45	*Arabidopsis thaliana*	581	31	1.00E-114	1.00E-54	1.00E-60
EAL44744	XP_650131.1	160	*Bacteroides fragilis*	424	41	*Yarrowia lipolytica*	169	31	3.00E-24	7.00E-11	4.29E-14
EAL46110	XP_651498.1	157	*Bacteroides fragilis*	166	49	*Arabidopsis thaliana*	627	35	5.00E-35	3.2	1.56E-35

(continued)

TABLE 2.8 *(continued)*

Acc.[a]	RefSeq Acc.[a]	EhL[b]	Top prokaryotic BLAST hit	PL[c]	% ID[d]	Top eukaryotic BLAST hit	EL[c]	% ID[d]	PE-score[e]	EE-score[f]	P/E ratio[g]
EAL45378	XP_650765.1	311	*Haloarcula marismortui*	299	43	*Leishmania major*	411	43	3.00E-54	1.00E-32	3.00E-22
EAL45618	XP_651004.1	159	*Bacteroides thetaiotaomicron*	157	46	*Plasmodium vivax*	1275	33	2.00E-28	0.69	2.90E-28
EAL46311	XP_651697.1	248	*Synechococcus elongates*	270	36	*Trichomonas vaginalis*	3075	18	1.00E-30	0.38	2.63E-30
EAL46679	XP_652065.1	218	*Methanosarcina mazei*	230	37	*Candida glabrata*	461	24	8.00E-31	0.079	1.01E-29
EAL46975	XP_652361.1	370	*Bordetella bronchiseptica*	368	46	*Cryptococcus neoformans*	372	40	8.00E-83	3.00E-71	2.67E-12
EAL47525	XP_652912.1	380	*Clostridium perfringens*	296	23	*Plasmodium falciparum*	390	34	2.00E-13	1.3	1.54E-13
EAL47905	XP_653291.1	227	*Clostridium perfringens*	259	33	*Tetrahymena thermophila*	1425	24	4.00E-19	0.32	1.25E-18
EAL48587	XP_653973.1	425	*Desulfovibrio vulgaris*	442	60	*Yarrowia lipolytica*	572	37	1.00E-149	9.00E-57	1.11E-93
EAL48979	XP_654365.1	732	*Thermotoga neapolitana*	740	40	*Cryptococcus neoformans*	735	28	1.00E-135	3.00E-64	3.33E-72
EAL49084	XP_654474.1	350	*Methanococcus jannaschii*	241	29	*Anopheles gambiae*	784	40	1.00E-24	5.00E-06	2.00E-19
EAL49209	XP_654596.1	247	*Bacteroides fragilis*	243	38	*Thalassiosira pseudonana*	269	22	7.00E-43	0.0002	3.50E-39
EAL49277	XP_654665.1	737	*Bacteroides thetaiotaomicron*	781	31	*Cryptococcus neoformans*	935	24	1.00E-111	6.00E-44	1.67E-68
EAL49613	XP_654999.1	168	*Sulfolobus solfataricus*	237	34	*Tetrahymena thermophila*	487	38	1.00E-16	6.00E-06	1.67E-11
EAL49813	XP_655200.1	186	*Escherichia coli*	200	31	*P. brasiliensis*	257	26	2.00E-13	0.47	4.26E-13

EAL49869	XP_655257.1	390	*Campylobacter jejuni*	407	56	*Ashbya gossypii*	490	39	1.00E-124	8.00E-73	1.25E-52
EAL50263	XP_655646.1	390	*Porphyromonas gingivalis*	408	48	*Yarrowia lipolytica*	428	38	1.00E-98	3.00E-60	3.33E-39
EAL50440	XP_655826.1	344	*Bacillus anthracis*	491	54	*Rhizopus oryzae*	510	40	1.00E-101	2.00E-67	5.00E-35
EAL50508	XP_655888.1	348	*Wolinella succinogenes*	340	55	*Mus musculus*	168	40	1.00E-106	2.00E-18	5.00E-89
EAL50603	XP_655988.1	567	*Bacteroides thetaiotaomicron*	622	45	*Trichomonas vaginalis*	632	39	1.00E-141	2.00E-99	5.00E-43
EAL50801	XP_656185.1	499	*Bacteroides thetaiotaomicron*	513	52	*Trichomonas vaginalis*	514	28	1.00E-145	3.00E-40	3.33E-106
EAL50992	XP_656375.1	140	*Archaeoglobus fulgidus*	184	40	*Trichomonas vaginalis*	195	46	1.00E-27	0.018	5.56E-26
EAL50997	XP_656380.1	656	*Bacteroides thetaiotaomicron*	718	53	*Cryptococcus neoformans*	770	32	0	2.00E-69	0.00E + 00
EAL51149	XP_656535.1	343	*Bacteroides fragilis*	359	43	*Pichia ofunaensis*	378	34	8.00E-84	1.00E-53	8.00E-31
EAL51236	XP_656622.1	259	*Symbiobacterium thermophilum*	274	45	*Oryza sativa*	315	21	3.00E-51	0.003	1.00E-48
EAL51348	XP_656749.1	171	*Methanopyrus kandleri*	204	37	*Tetrahymena thermophila*	2872	22	3.00E-21	0.1	3.00E-20
EAL51525	XP_656903.1	316	*Bacteroides thetaiotaomicron*	300	29	*Candida boidinii*	314	32	8.00E-27	0.0007	1.14E-23
EAL51565	XP_656946.1	415	*Clostridium perfringens*	900	43	*Trichomonas vaginalis*	897	40	1.00E-89	5.00E-81	2.00E-09
EAL51925	XP_657304.1	448	*T. tengcongensis*	481	43	*Giardia lamblia*	937	33	3.00E-96	2.00E-60	1.50E-36
EAL52001	XP_657387.1	303	*Oceanobacillus iheyensis*	306	27					2.00E-15	0.00E + 00

(*continued*)

TABLE 2.8 *(continued)*

Acc.[a]	RefSeq Acc.[a]	EhL[b]	Top prokaryotic BLAST hit	PL[c]	% ID[d]	Top eukaryotic BLAST hit	EL[c]	% ID[d]	PE-score[e]	EE-score[f]	P/E ratio[g]
27 LGT cases that are more weakly supported than before according to our criteria											
EAL45152	XP_650539.1	122	*Shewanella oneidensis*	132	34	*Trypanosoma bruzeii*	385	24	5.00E-10	6.6	7.58E-11
EAL43347	XP_648734.1	848	*Burkholderia pseudomallei*	779	38	*Plasmodium falciparum*	2463	32	1.00E-136	4.00E-44	2.50E-93
EAL44257	XP_649643.1	407	*Clostridium acetobutylicum*	406	25	*Homo sapiens*	468	24	6.00E-23	1.00E-14	6.00E-09
EAL45586	XP_650972.1	460	*Clostridium tetani*	476	47	*Xenopus laevis*	513	38	1.00E-116	5.00E-84	2.00E-33
EAL46313	XP_651699.1	118	*Prochlorococcus marinus*	163	42	*Hordeum vulgare*	223	22	2.00E-21	1.4	1.43E-21
EAL46399	XP_651785.1	218	*Clostridium perfringens*	235	65	*Trypanosoma bruzeii*	295	52	3.00E-73	9.00E-54	3.33E-20
EAL46421	XP_651808.1	205	*Clostridium acetobutylicum*	230	40	*Arabidopsis thaliana*	241	33	7.00E-34	6.00E-12	1.17E-22
EAL46701	XP_652087.1	294	*Bacteroides fragilis*	308	45	*Thalassiosira pseudonana*	348	27	4.00E-63	1.00E-14	4.00E-49
EAL46757	XP_652143.1	95	*Lactococcus lactis*	103	31	*Tetrahymena thermophila*	112	32	3.00E-09	1.00E-07	3.00E-02
EAL46858	XP_652245.1	192	*Pseudomonas aeruginosa*	195	41	*Caenorhabditis briggsae*	229	40	6.00E-36	2.00E-17	3.00E-19
EAL47026	XP_652397.1	164	*Bacillus subtilis*	181	30	*Trichomonas vaginalis*	182	26	3.00E-10	2.00E-08	1.50E-02
EAL47464	XP_652839.1	504	*Treponema denticola*	509	39	*Piromyces* sp.	555	27	5.00E-88	2.00E-30	2.50E-58
EAL47648	XP_653034.1	259	*Methanosarcina mazei*	272	36	*Arabidopsis thaliana*	345	25	2.00E-39	3.00E-11	6.67E-29

EAL47787	XP_653173.1	546	*Spirochaeta thermophila*	571	56	*Solanum tuberosum*	552	46	1.00E-175	1.00E-135	1.00E-40
EAL48186	XP_653572.1	232	*Bacillus cereus*	279	34	*Thalassiosira pseudonana*	271	32	2.00E-10	2.00E-08	1.00E-02
EAL49309	XP_654698.1	358	*Methanosarcina mazei*	379	42	*Leishmania major*	373	31	5.00E-77	9.00E-44	5.56E-34
EAL48568	XP_653954.1	113	*Chlamydia pneumoniae*	271	38	*Debaryomyces hansenii*	699	38	5.00E-14	7.00E-16	7.14E + 01
EAL48767	XP_654156.1	165	*Bacteroides fragilis*	177	40	*Trichomonas vaginalis*	189	28	7.00E-28	2.00E-05	3.50E-23
EAL48783	XP_654172.1	217	*Pseudomonas putida*	225	46	*Giardia lamblia*	239	35	2.00E-43	7.00E-24	2.86E-20
EAL49703	XP_655090.1	396	*Clostridium acetobutylicum*	398	34	*Tetrahymena thermophila*	445	29	4.00E-64	3.00E-44	1.33E-20
EAL49996	XP_655383.1	358	*Bacteroides thetaiotaomicron*	368	60	*Brachydanio rerio*	367	43	1.00E-121	5.00E-76	2.00E-46
EAL50325	XP_655711.1	447	*Clostridium tetani*	448	30	*Trichomonas vaginalis*	871	29	4.00E-46	1.00E-37	4.00E-09
EAL50521	XP_655905.1	285	*Streptococcus agalactiae*	323	29	*Leishmania major*	452	24	2.00E-22	3.00E-06	6.67E-17
EAL50620	XP_656005.1	261	*Wolinella succinogenes*	655	27	*Trichomonas vaginalis*	261	28	6.00E-21	1.00E-06	6.00E-15
EAL50838	XP_656225.1	299	*Anabaena* sp.	287	27	*Trichomonas vaginalis*	336	29	4.00E-15	0.0009	4.44E-12
EAL50986	XP_656369.1	219	*Bacteroides thetaiotaomicron*	240	31	*Xenopus laevis*	309	29	2.00E-20	1.00E-12	2.00E-08
EAL52121	XP_657511.1	220	*T. tengcongensis*	222	36	*Caenorhabditis elegans*	255	26	1.00E-30	1.00E-07	1.00E-23

(*continued*)

TABLE 2.8 *(continued)*

Acc.[a]	RefSeq Acc.[a]	EhL[b]	Top prokaryotic BLAST hit	PL[c]	% ID[d]	Top eukaryotic BLAST hit	EL[c]	% ID[d]	PE-score[e]	EE-score[f]	P/E ratio[g]
14 cases where increased sampling has weakened that case for LGT											
EAL42539	XP_647925.1	213	*Bacteroides thetaiotaomicron*	319	47	*Entodinium caudatum*	411	43	3.00E-53	1.00E-32	3.00E-21
EAL42738	XP_648124.1	313	*Campylobacter jejuni*	324	40	*Trichomonas vaginalis*	313	36	1.00E-63	4.00E-42	2.50E-22
EAL44270	XP_649657.1	179	*Methanococcus maripaludis*	193	37	*Anopheles gambiae*	186	21	2.00E-27	2.00E-09	1.00E-18
EAL44593	XP_649979.1	220	*Vibrio vulnificus*	244	24	*Trichomonas vaginalis*	238	21	0.0002	2.6	7.69E-05
EAL45320	XP_650707.1	154	*Geobacillus kaustophilus*	183	53	*Thalassiosira pseudonana*	182	43	8.00E-38	2.00E-32	4.00E-06
EAL45332	XP_650718.1	392	*Methanosarcina acetivorans*	420	48	*Trichomonas vaginalis*	396	47	8.00E-99	2.00E-93	4.00E-06
EAL45528	XP_650913.1	349	*Sulfolobus acidocaldarius*	343	28	*Cyanophora paradoxa*	313	27	1.00E-24	5.00E-17	2.00E-08
EAL45907	XP_651293.1	380	*Streptomyces coelicolor*	603	32	*Dictyostelium discoideum*	457	30	2.00E-39	2.00E-35	1.00E-04
EAL46026	XP_651412.1	176	*Bacteroides fragilis*	184	51	*Tetrahymena thermophila*	323	32	2.00E-44	8.00E-08	2.50E-37
EAL46116	XP_651488.1	662	*Bacillus clausii*	684	48	*Solanum tuberosum*	761	48	0	1.00E-172	0.00E + 00
EAL46656	XP_652044.1	419	*Dictyoglomus thermophilum*	579	30	*S. pombe*	493	41	2.00E-35	2.00E-19	1.00E-16
EAL50605	XP_655990.1	392	*Thermotoga maritima*	417	38	*Cryptococcus neoformans*	445	30	2.00E-69	1.00E-33	2.00E-36
EAL51270	XP_656656.1	251	*Porphyromonas gingivalis*	261	50	*Anopheles gambiae*	272	39	6.00E-53	1.00E-35	6.00E-18

EAL52102	XP_657492.1	345	*Bacteroides thetaiotaomicron*	358	54	*Thalassiosira pseudonana*	354	47	1.00E-105	7.00E-86	1.43E-20
Nine cases where *Entamoeba* is now recovered with a recently sequenced gene from another microbial eukaryote											
EAL44213	XP_649600.1	710	*Bdellovibrio bacteriovorus*	698	37	*Trichomonas vaginalis*	713	35	1.00E-127	1.00E-127	1.00E + 00
EAL44435	XP_649823.1	250	*Bacteroides fragilis*	395	40	*Trichomonas vaginalis*	395	33	1.00E-43	3.00E-35	3.33E-09
EAL44766	XP_650152.1	401	*Porphyromonas gingivalis*	419	36	*Trichomonas vaginalis*	445	32	3.00E-65	1.00E-51	3.00E-14
EAL47785	XP_653171.1	234	*Bacillus anthracis*	242	32	*Trichomonas vaginalis*	256	39	2.00E-30	3.00E-33	6.67E + 02
EAL47859	XP_653246.1	337	*Clostridium acetobutylicum*	322	50	*C. reinhardtii*	352	44	9.00E-74	0	N/A
EAL49158	XP_654544.1	397	*T. tengcongensis*	412	49	*Trichomonas vaginalis*	416	46	1.00E-100	4.00E-99	2.50E-02
EAL49488	XP_654874.1	320	*Geobacter sulfurreducens*	336	34	*Leishmania major*	357	31	1.00E-38	4.00E-30	2.50E-09
EAL49791	XP_655177.1	164	*Oceanobacillus iheyensis*	177	42	*Thalassiosira pseudonana*	96	38	8.00E-30	6.00E-09	1.33E-21
EAL50404	XP_655790.1	718	*T. tengcongensis*	717	37	*Trichomonas vaginalis*	721	34	1.00E-139	1.00E-118	1.00E-21
Five cases where vertical inheritance is now the simplest explanation for the new tree											
EAL44346	XP_649732.1	314	*Oceanobacillus iheyensis*	239	47	*Dictyostelium discoideum*	278	65	1.00E-52	3.00E-95	3.33E + 42

(*continued*)

TABLE 2.8 *(continued)*

Acc.[a] RefSeq	Acc.[a]	EhL[b]	Top prokaryotic BLAST hit	PL[c]	% ID[d]	Top eukaryotic BLAST hit	EL[c]	% ID[d]	PE-score[e]	EE-score[f]	P/E ratio[g]
EAL45466	XP_650849.1	209	*Agrobacterium tumefaciens*	254	31	*Thalassiosira pseudonana*	227	35	3.00E-23	1.00E-27	3.00E + 04
EAL45548	XP_650934.1	259	*Bacillus cereus (strain ZK)*	233	29	*Candida glabrata*	270	30	7.00E-06	5.00E-05	1.40E-01
EAL45595	XP_650981.1	284	*Pyrobaculum aerophilum*	293	27	*Ashbya gossypii*	343	27	1.00E-23	7.00E-16	1.43E-08
EAL50185	XP_655571.1	186	*Aeropyrum pernix*	192	31	*Thalassiosira pseudonana*	149	30	4.00E-13	5.00E-06	8.00E-08

Abbreviated taxon names: *Chlamydomonas reinhardtii: C. reinhardtii; Paracoccidioides brasiliensis: P. brasiliensis; Schizosaccharomyces pombe: S. pombe; Thermoanaerobacter tengcongensis: T. tengcongensis.*

Note: All 96 trees reanalysed here can be downloaded (in pdf format) from the following web site: http://www.ncl.ac.uk/microbial_eukaryotes/

[a] GenBank accession numbers and RefSeq accession numbers, respectively, for the 96 original candidates LGT identified by phylogenetic analysis (Loftus *et al.*, 2005).

[b] EhL, the length of the *E. histolytica* protein.

[c] PL/EL, the protein length of the prokaryotic or eukaryotic top BlastP hit, respectively.

[d] %ID, the percent identity between the *E. histolytica* protein and the top prokaryotic or eukaryotic protein in BlastP alignments (in respective columns).

[e] PE-score, the e-score of the top, prokaryotic hit.

[f] EE-score, the e-score of the top eukaryotic hit.

[g] P/E ratio, the e-score ratio between the top prokaryotic hit and top eukaryotic hit.

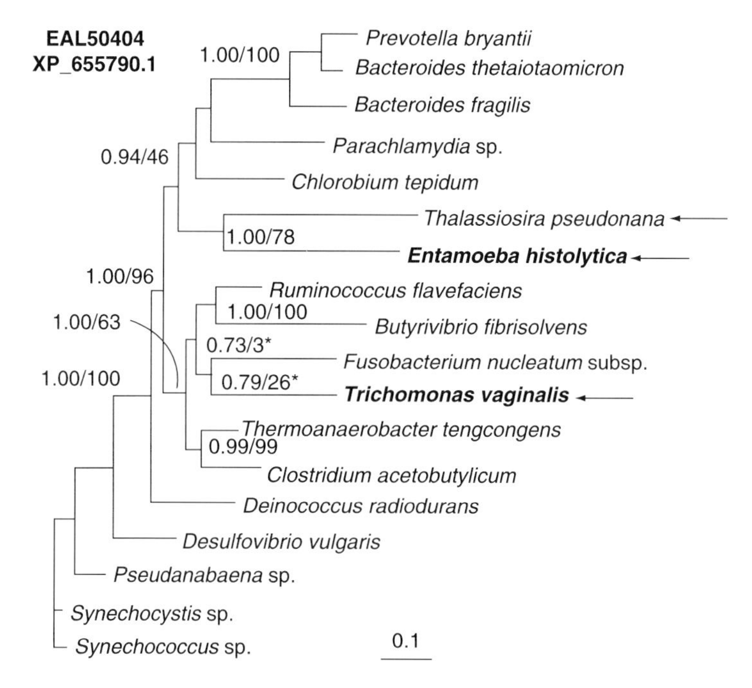

FIGURE 2.12 Phylogenetic relationships of *E. histolytica* glutamine synthase. The gene encoding glutamine synthase (EC 6.3.1.2) is now shared by *E. histolytica* and the diatom *Thalassiosira*. This gene is mainly restricted to prokaryotic genomes (eukaryotes are highlighted by arrows). *T. vaginalis* also contains a homologue but in this case it clusters weakly with *Fusobacterium*. The scale bar represents 10% of inferred sequence divergence. Both the GenBank and RefSeq accession numbers are given for the *E. histolytica* entry. The tree is the consensus Mr Bayes tree with support values corresponding to the posterior probabilities of the Bayesian analysis followed by the bootstrap support value of the equivalent node in maximum likelihood distance analysis. Only a selection of the most relevant support values are shown. A star indicates that the maximum likelihood analysis did not recover the node found in the Bayesian analysis.

environment may share a set of niche-specific genes (Beiko *et al.*, 2005; Mira *et al.*, 2004).

For five trees, the *E. histolytica* gene now appears to be present in eukaryotes from a different taxonomic group and the analysis cannot exclude a common origin for all eukaryotic sequences. Thus, for about 5% of the original 96 cases the simplest explanation is no longer LGT, but vertical inheritance from a common ancestor shared with other eukaryotes.

10.2. Where do the genes come from?

As before, certain prokaryotic groups are favoured as the potential donors of LGT genes in the *E. histolytica* genome (Loftus *et al.*, 2005). In 15 well-resolved trees *E. histolytica* is recovered next to a member of the

Bacteroidetes/Chlorobi group. Bacteroidetes/Chlorobi are abundant members of the intestinal microflora (Shoemaker *et al.*, 2001), providing plenty of opportunities for LGT to occur. Members of the Bacteroidetes/Chlorobi and *Fusobacterium* (one tree) groups are all obligate anaerobes. This bias is consistent with the idea that prokaryotic and eukaryotic cohabitants of the same anaerobic niche are sharing genes (Andersson *et al.*, 2001; Beiko *et al.*, 2005; Lawrence, 2005b). For example, Fig. 2.13 shows an intriguing example where the *T. vaginalis* gene clusters with members of the Bacteroidetes/Chlorobi and *E. histolytica* clusters with *Fusobacterium*.

10.3. What kinds of gene are being transferred?

Most of the 68 laterally transferred genes that can be assigned to a functional category encode enzymes involved in metabolism (Fig. 2.14). This is consistent with the complexity hypothesis, which posits that LGT of genes involved in processing a single substrate are more likely to be transferred than those genes encoding proteins that interact with many other cellular components, such as ribosomal proteins for example (Jain *et al.*, 1999). Mapping the LGT enzymes on the *E. histolytica* metabolic pathway (Loftus *et al.*, 2005) indicates that LGT has affected some important pathways, including iron–sulphur cluster biosynthesis, amino acid metabolism and nucleotide metabolism. Since only 8 of the 68 LGT have obvious homologues in the human genome, the proteins are potentially specific to the parasite and may thus be worth exploring as potential drug targets. The rest of the LGT cases involve hypothetical or unclassified proteins.

11. MICROARRAY ANALYSIS

Microarray-based analyses can be utilised in conjunction with genome sequencing to assign functional roles to annotated genes and to clarify genomic architecture. A number of groups have utilised DNA microarrays in *E. histolytica* (made from random genomic DNA fragments or long or short oligonucleotides based on annotated genes) to successfully study transcriptional differences between virulent and avirulent *E. histolytica* as well as transcriptional responses to heat shock, collagen and calcium exposure, tissue invasion and cyst development (Debnath *et al.*, 2004; Gilchrist *et al.*, 2006; MacFarlane and Singh, 2006; Weber *et al.*, 2006; Ehrenkaufer *et al.*, 2007). Additionally, using a genomic DNA microarray, comparative genomic hybridisations (CGH) between strains and species of *Entamoeba* have been performed (Shah *et al.*, 2005).

Some interesting aspects of amoebic biology have been uncovered using DNA microarray-based expression profiling. To investigate the hypothesis that virulence determinants will be more highly expressed in

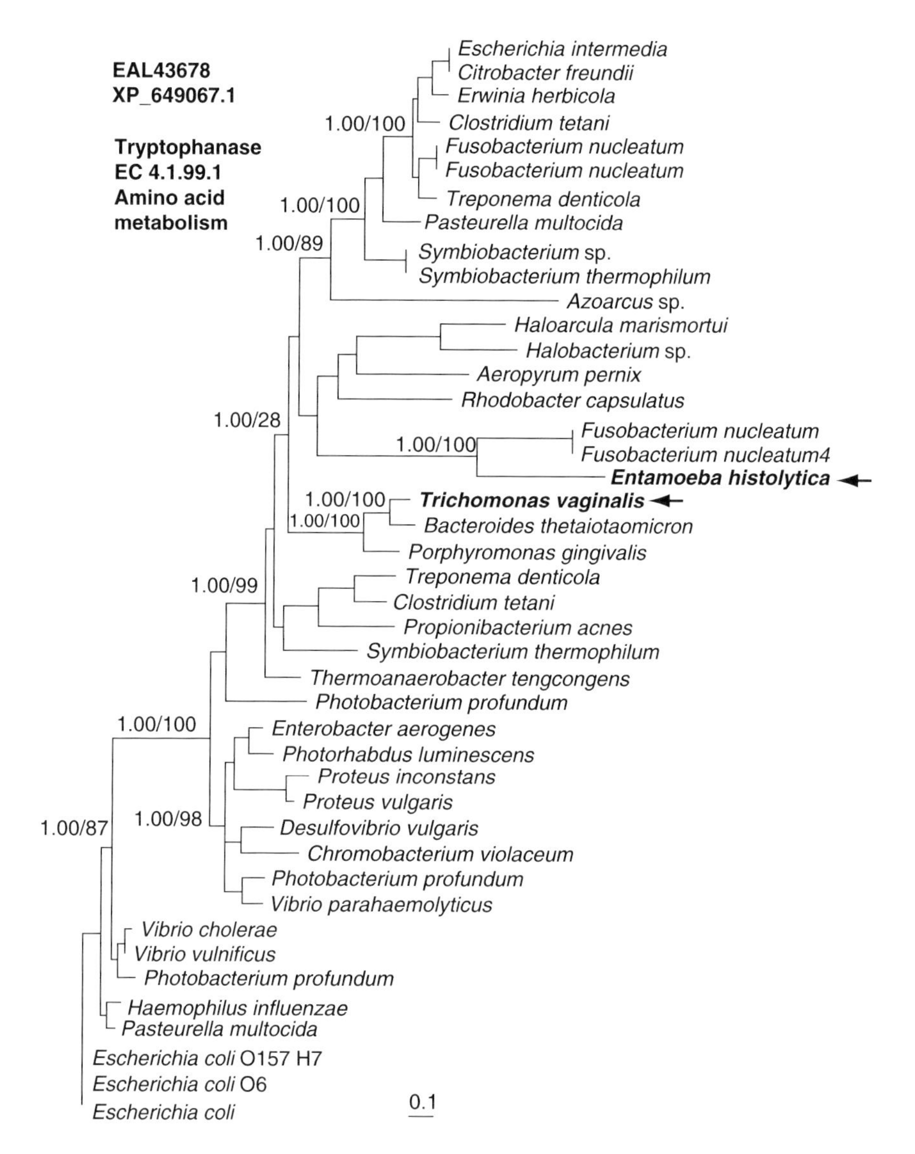

FIGURE 2.13 Phylogenetic relationships of *E. histolytica* tryptophanase. This tree suggests that the *E. histolytica* gene encoding a tryptophanase was acquired by LGT from a relative of the anaerobic bacterium *Fusobacterium*. In contrast, the *T. vaginalis* gene appears to have a separate origin with an LGT from a relative of the anaerobic *Bacteroides* group. The scale bar represents 10% of inferred sequence divergence. Both the GenBank and RefSeq accession numbers are given for the *E. histolytica* entry. The EC number is also shown. Analysis details as for figure 2.12.

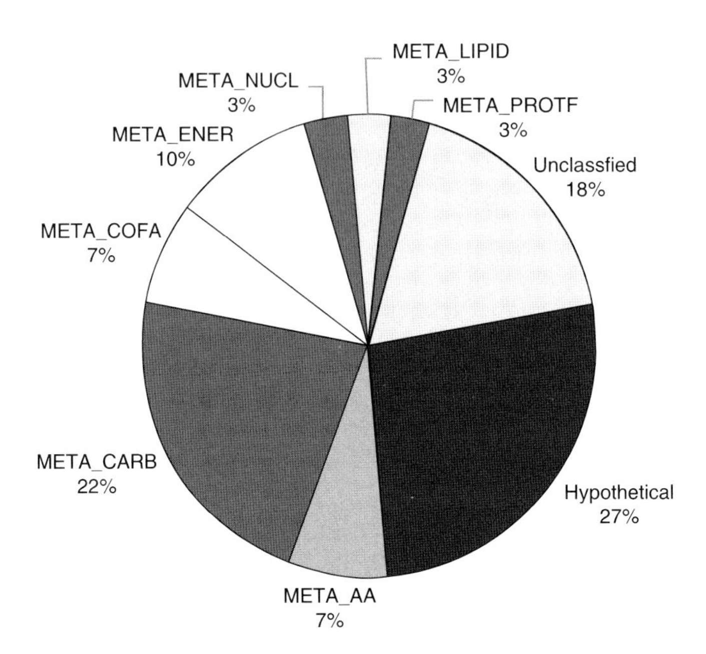

FIGURE 2.14 Pie chart of functional categories for the 68 strongest LGT cases. The cases are those discussed in the text and listed in Table 2.8. Most entries encode metabolic enzymes (KEGG annotation).

virulent strains, the transcriptomes of virulent and avirulent *Entamoeba* species and strains have been studied. It has been confirmed that a number of known virulence determinants have decreased expression in avirulent *Entamoeba* (Davis *et al.*, 2007; MacFarlane and Singh, 2006). A genomic DNA microarray composed of 2,110 genes identified 29 genes with decreased expression in both an attenuated *E. histolytica* strain (Rahman) and the avirulent *E. dispar* (strain SAW760) (MacFarlane and Singh, 2006), while an oligonucleotide microarray composed of 6,242 genes identified 152 genes with a higher level of expression in the virulent *E. histolytica* HM-1:IMSS than in the attenuated Rahman strain (Davis *et al.*, 2007). A majority of these genes are annotated as hypothetical and whether these genes encode novel virulence factors will require genetic analysis of their functions. A peroxiredoxin gene identified as having decreased expression in *E. histolytica* Rahman has been shown to be a virulence factor (Davis *et al.*, 2006), indicating that these comparisons between virulent and avirulent strains are likely to be a fruitful avenue of investigation.

In other microarray-based studies, the large family of transmembrane receptor kinases identified in *E. histolytica* has been found to be differentially expressed under *in vitro* trophozoite culture conditions (Beck *et al.*, 2005). One can easily envision that these kinases may have roles in signalling, allowing the parasite to adapt to its ever-changing environmental milieu. A substantial transcriptional response to heat shock has been demonstrated (Weber *et al.*, 2006), and interestingly lectin gene family members were identified as being differentially regulated under heat shock conditions.

The most comprehensive microarray data to date used a whole genome short oligonucleotide microarray (based on the Affymetrix platform) to profile the transcriptional changes that occur as the parasite colonises and invades the host colon (Gilchrist *et al.*, 2006). Using a mouse model of colitis, in which the microscopic features replicate human disease and substantial pathology can be seen, the transcriptional response of parasites was assayed soon after colonisation (one day after injection into the caecum) and in a long-term (29 days) disease state. Overall, 326 genes were modulated at day 1 after infection, 109 at 29 days after infection, and 88 at both time points. A number of the well-characterised 'virulence determinants' in *E. histolytica* were highly expressed under all conditions tested and not transcriptionally modulated, although some members of the cysteine protease gene family were highly regulated during tissue invasion. A summary of the genes and gene families that have been identified as being transcriptionally active under the conditions mentioned above is listed in Table 2.9.

The life cycle of *E. histolytica* involves transition between the trophozoite stage, responsible for colonisation as well as invasive disease, and the cyst, responsible for infection transmission. Despite its central role, little is known about cyst development in *E. histolytica*, largely due to our inability to generate *E. histolytica* cysts in axenic culture. Using a whole genome microarray and xenic cultures of recently isolated *E. histolytica* strains that contained spontaneously produced cysts, a cyst transcriptome was developed that identified 1,439 developmentally regulated genes (672 cyst-specific and 767 trophozoite-specific genes; Ehrenkaufer *et al.*, 2007). This first large-scale insight into encystation indicates that ~15% of *E. histolytica* genes are transcriptionally controlled in this developmental pathway. Among the genes identified were a number of stage-specific cysteine proteases, transmembrane kinases, transcriptional regulators and other potential initiators of the developmental cascade. Future characterisation of these genes and pathways will provide important insights into developmental processes in this parasite.

The above microarray studies used expression data to identify interesting genes and pathways potentially involved in amoebic pathogenesis or development. In another application of microarrays, CGH identified a number of interesting genomic characteristics of *Entamoeba* (Shah *et al.*, 2005). The *E. histolytica* genome project revealed that a large number of genes are multi-copy or members of highly similar gene families. Due to the repetitive nature of the genome there has been difficulty with genome assembly and thus the large number of gene duplications could have

TABLE 2.9 Examples of microarray-detected transcriptional changes in some gene families and the conditions tested

Gene family	Total number of genes family	Number of genes transcriptionally regulated under condition tested	
		Heat shock[a] (1,131 genes on array)	Host colonisation and invasion[b] (9,435 genes on array)
Cysteine proteases	29[c]	2 up-regulated (CPs 6, 4); 7 down-regulated (CPs 1, 2, 3, 8, 13, 17),	21 genes on array; 4 up-regulated (CPs 1, 9, 4, 6); 1 down-regulated (CP8)
Lectin (heavy, light, and intermediate subunits)	12	1 up-regulated (Hgl-2); 5 down-regulated (Lgl-1 and 3, Igl-1 and -2, Hgl-3)	No change in heavy or intermediate subunits; Light subunit lgl-2 and lgl-3 down-regulated)
Amoeba pore	3	1 down-regulated (*amoeba pore* C)	No substantial changes
Transmembrane receptor kinases	>80	NA	6 up-regulated (TMKs 69, 53, 95, 105, 63, 56); 2 down-regulated (TMKs 03 and 17)
AIG-1 (similar to plant antibacterial proteins)	15	NA	5 up-regulated at day 1; 6 down-regulated at day 29 (all non-overlapping)

[a] Adapted from Weber *et al.* (2006).
[b] Adapted from Gilchrist *et al.* (2006).
[c] Number of cysteine protease gene families in genome annotation at time studies were performed.

represented an assembly artefact. The data from CGH confirmed the high copy number of a significant portion (~14%) of the genome and validated the genome assembly. Additionally, genome-wide genetic diversity was demonstrated among strains of *E. histolytica* (Shah *et al.*, 2005), including the observation that the attenuated *E. histolytica* strain Rahman had a unique genetic pattern suggesting the possibility that a genomic signature may correlate with invasive potential. Since genome sequencing for different *E. histolytica* strains, including clinical isolates, is unlikely the promise of CGH to study genetic diversity and identify genotype-phenotype associations is substantial.

E. dispar, the closely related but avirulent species, had been identified early on as having some genetic divergence from the virulent *E. histolytica*. CGH analysis of *E. histolytica* and *E. dispar* revealed a significant amount of difference between the two species. Whether the genetic drift in these genes is responsible for the non-invasive phenotype of *E. dispar* is not known, but the work has highlighted a number of genes for further functional analyses.

Taken together the DNA microarray analyses of *Entamoeba* have been useful to begin to dissect the genome of this parasite and provide functional context to the genes identified in the genome sequencing effort. Future directions will include analysis of the parasite transcriptome in invasive hepatic disease as well as further characterisation of the developmental conversion to the cyst form. Those data may be useful in the development of novel diagnostic and therapeutic options. Additionally, genetic approaches can now be applied to definitively assign a role for these genes in amoebic biology and pathogenesis.

12. FUTURE PROSPECTS FOR THE *E. HISTOLYTICA* GENOME

Although the genome of *E. histolytica* is not yet complete, it has already revealed much about the biology of the parasite. There appear to be forces acting to compact the genome, leading to a reduction in the coding region and intron length of genes, and resulting in the loss of numerous metabolic pathways. However, there are also opposing evolutionary forces as many gene families have expanded. This applies particularly to genes involved in signalling and trafficking that allow the parasite to sense and respond to its environment, a necessary adaptation for a predatory protist. Unfortunately, it is difficult at present to understand the genome structure on a macro scale due to the fragmented nature of the current assembly. In other parasites, genome structure has been vital to unravelling important biological processes, such as antigenic variation in *T. brucei* and identification of rifin genes in *P. falciparum*. Until the *E. histolytica* genome is complete we will not know what else remains to be uncovered. Efforts are already under way to complete the genome by first generating a HAPPY map (Dear and Cook, 1993). Over 2000 markers are being designed at ~25-kb intervals across all contigs. Using PCR, co-segregation analysis allows the identification of contigs that are physically linked in the genome. This will allow the ordering and orientation of the contigs and will facilitate gap closure. Shotgun genome sequencing projects of *E. invadens* and *E. dispar* are under way (Loftus and Hall, 2005). At present the *E. invadens* genome appears to assemble with fewer problems than were encountered with that of *E. histolytica*. It is anticipated that an essentially complete *E. invadens* genome sequence will be obtained, enabling extensive comparative analyses to be made, and facilitating the study of pathogenicity, host interaction and the evolutionary forces acting on the genome.

ACKNOWLEDGEMENTS

This chapter is dedicated to Louis S. Diamond without whose pioneering studies the sequencing of the genome would have been impossible. The genome sequencing was supported by grants from the Wellcome Trust (064057) and the National Institute of Allergy and Infectious Disease (5R01AI046516–03). UCMA is supported by a Wellcome Trust project grant (075796) awarded to TME and RPH. Analyses by MH and MD were supported by grant P15960 from the Austrian Science Fund, Vienna, Austria. Analyses by CM and AL were supported by grant 1RO3 TW007314–01 from the Fogarty International Center of the National Institutes of Health, USA. Analyses by NG were supported by an INCO-DEV grant in the fifth framework programme of the European Union. Analyses by IB were supported by the Deutsche Forschungsgemeinschaft (DFG), BR 1744/7–1. N.H. thank Lynn Schriml and Regina Cer for preparing the web pages with the tables and M.L. thank Julia Winkelmann for preparing Table 2.6.

REFERENCES

Ackers, J. P., Dhir, V., and Field, M. C. (2005). A bioinformatic analysis of the RAB genes of *Trypanosoma brucei*. *Mol. Biochem. Parasitol.* **141,** 89–97.

Adams, E. W., Ratner, D. M., Bokesch, H. R., McMahon, J. B., O'Keefe, B. R., and Seeberger, P. H. (2004). Oligosaccharide and glycoprotein microarrays as tools in HIV glycobiology; glycan-dependent gp120/protein interactions. *Chem. Biol.* **11,** 875–881.

Adams, M. D., Celniker, S. E., Holt, R. A., Evans, C. A., Gocayne, J. D., Amanatides, P. G., Scherer, S. E., Li, P. W., Hoskins, R. A., Galle, R. F., George, R. A., Lewis, S. E., *et al.* (2000). The genome sequence of *Drosophila melanogaster*. *Science* **287,** 2185–2195.

Aguilar, R. C., Boehm, M., Gorshkova, I., Crouch, R. J., Tomita, K., Saito, T., Ohno, H., and Bonifacino, J. S. (2001). Signal-binding specificity of the mu4 subunit of the adaptor protein complex AP-4. *J. Biol. Chem.* **276,** 13145–13152.

Aguirre-Garcia, M. M., Rosales-Encina, J. L., and Talamas-Rohana, P. (1997). Secreted *Entamoeba histolytica* acid phosphatase (SAP). *Arch. Med. Res.* **28**(Spec No), 184–185.

Aguirre-García, M. M., Anaya-Ruiz, M., and Talamás-Rohana, P. (2003). Membrane-bound acid phosphatase (MAP) from *Entamoeba histolytica* has phosphotyrosine phosphatase activity and disrupts the actin cytoskeleton of host cells. *Parasitology* **126,** 195–202.

Aldritt, S. M., Tien, P., and Wang, C. C. (1985). Pyrimidine salvage in *Giardia lamblia*. *J. Exp. Med.* **161,** 437–445.

Ali, I. K., Zaki, M., and Clark, C. G. (2005). Use of PCR amplification of tRNA gene-linked short tandem repeats for genotyping *Entamoeba histolytica. J. Clin. Microbiol.* **43,** 5842–5847.

Ali, V., Shigeta, Y., and Nozaki, T. (2003). Molecular and structural characterization of NADPH-dependent D-glycerate dehydrogenase from the enteric parasitic protist *Entamoeba histolytica. Biochem. J.* **375,** 729–736.

Ali, V., Hashimoto, T., Shigeta, Y., and Nozaki, T. (2004a). Molecular and biochemical characterization of D-phosphoglycerate dehydrogenase from *Entamoeba histolytica.* A unique enteric protozoan parasite that possesses both phosphorylated and nonphosphorylated serine metabolic pathways. *Eur. J. Biochem.* **271,** 2670–2681.

Ali, V., Shigeta, Y., Tokumoto, U., Takahashi, Y., and Nozaki, T. (2004b). An intestinal parasitic protist, *Entamoeba histolytica,* possesses a non-redundant nitrogen fixation-like system for iron-sulfur cluster assembly under anaerobic conditions. *J. Biol. Chem.* **279,** 2670–2681.

Ali, V., and Nozaki, T. (2006). Biochemical and functional characterization of phosphoserine aminotransferase from *Entamoeba histolytica,* which possesses both phosphorylated and non-phosphorylated serine metabolic pathways. *Mol. Biochem. Parasitol.* **145,** 71–83.

Anaya-Ruiz, M., Rosales-Encina, J. L., and Talamas-Rohana, P. (1997). Membrane acid phosphatase (MAP) from *Entamoeba histolytica. Arch. Med. Res.* **28**(Spec No), 182–183.

Anaya-Ruiz, M., Perez-Santos, J. L., and Talamas-Rohana, P. (2003). An ecto-protein tyrosine phosphatase of *Entamoeba histolytica* induces cellular detachment by disruption of actin filaments in HeLa cells. *Int. J. Parasitol.* **33,** 663–670.

Anderson, D. H., Sawaya, M. R., Cascio, D., Ernst, W., Modlin, R., Krensky, A., and Eisenberg, D. (2003). Granulysin crystal structure and a structure-derived lytic mechanism. *J. Mol. Biol.* **325,** 355–365.

Anderson, I. J., and Loftus, B. J. (2005). *Entamoeba histolytica*: Observations on metabolism based on the genome sequence. *Exp. Parasitol.* **110,** 173–177.

Andersson, J. O., Doolittle, W. F., and Nesbo, C. L. (2001). Genomics. Are there bugs in our genome? *Science* **292,** 1848–1850.

Andersson, J. O., Sjogren, A. M., Davis, L. A., Embley, T. M., and Roger, A. J. (2003). Phylogenetic analyses of diplomonad genes reveal frequent lateral gene transfers affecting eukaryotes. *Curr. Biol.* **13,** 94–104.

Andersson, J. O. (2005). Lateral gene transfer in eukaryotes. *Cell. Mol. Life Sci.* **62,** 1182–1197.

Andersson, J. O., Hirt, R. P., Foster, P. G., and Roger, A. J. (2006). Evolution of four gene families with patchy phylogenetic distributions: Influx of genes into protist genomes. *BMC Evol. Biol.* **6,** 27.

Andrä, J., and Leippe, M. (1994). Pore-forming peptide of *Entamoeba histolytica.* Significance of positively charged amino acid residues for its mode of action. *FEBS Lett.* **354,** 97–102.

Andrä, J., Herbst, R., and Leippe, M. (2003). *Amoeba pores,* archaic effector peptides of protozoan origin, are discharged into phagosomes and kill bacteria by permeabilizing their membranes. *Dev. Comp. Immunol.* **27,** 291–304.

Ankri, S., Stolarsky, T., Bracha, R., Padilla-Vaca, F., and Mirelman, D. (1999). Antisense inhibition of expression of cysteine proteinases affects *Entamoeba histolytica*-induced formation of liver abscess in hamsters. *Infect. Immun.* **67,** 421–422.

Archibald, J. M., Teh, E. M., and Keeling, P. J. (2003). Novel ubiquitin fusion proteins: Ribosomal protein P1 and actin. *J. Mol. Biol.* **328,** 771–778.

Arhets, P., Olivo, J. C., Gounon, P., Sansonetti, P., and Guillen, N. (1998). Virulence and functions of myosin II are inhibited by overexpression of light meromyosin in *Entamoeba histolytica. Mol. Biol. Cell.* **6,** 1537–1547.

Ariyanayagam, M. R., and Fairlamb, A. H. (1999). *Entamoeba histolytica* lacks trypanothione metabolism. *Mol. Biochem. Parasitol.* **103,** 61–69.

Arteaga-Nieto, P., López-Romero, E., Teran-Figueroa, Y., Cano-Canchola, C., Luna Arias, J. P., Flores-Carreón, A., and Calvo-Méndez, C. (2002). *Entamoeba histolytica*: Purification and characterization of ornithine decarboxylase. *Exp. Parasitol.* **101,** 215–222.

Asbury, C. L. (2005). Kinesin: World's tiniest biped. *Curr. Opin. Cell Biol.* **17,** 89–97.

Avila, E. E., Martinez-Alcaraz, E. R., Barbosa-Sabanero, G., Rivera-Baron, E. I., Arias-Negrete, S., and Zazueta-Sandova, L. R. (2002). Subcellular localization of the NAD^+-dependent alcohol dehydrogenase in *Entamoeba histolytica* trophozoites. *J. Parasitol.* **88,** 217–222.

Bailey, G. B., Gilmour, J. R., and McCoomer, N. E. (1990). Roles of target cell membrane carbohydrate and lipid in *Entamoeba histolytica* interaction with mammalian cells. *Infect. Immun.* **58,** 2389–2391.

Bakker-Grunwald, T., Martin, J. B., and Klein, G. (1995). Characterization of glycogen and amino acid pool of *Entamoeba histolytica* by C^{13}-NMR spectroscopy. *J. Eukaryot. Microbiol.* **42,** 346–349.

Bakre, A. A., Rawal, K., Ramaswamy, R., Bhattacharya, A., and Bhattacharya, S. (2005). The LINEs and SINEs of *Entamoeba histolytica*: Comparative analysis and genomic distribution. *Exp. Parasitol.* **110,** 207–213.

Baldauf, S. L. (2003). The deep roots of eukaryotes. *Science* **300,** 1703–1706.

Band, R. N., and Cirrito, H. (1979). Growth response of axenic *Entamoeba histolytica* to hydrogen, carbon dioxide, and oxygen. *J. Protozool.* **26,** 282–286.

Barlowe, C., and Schekman, R. (1993). SEC12 encodes a guanine-nucleotide-exchange factor essential for transport vesicle budding from the ER. *Nature* **365,** 347–349.

Barlowe, C., Orci, L., Yeung, T., Hosobuchi, M., Hamamoto, S., Salama, N., Rexach, M. F., Ravazzola, M., Amherdt, M., and Schekman, R. (1994). COPII: A membrane coat formed by Sec proteins that drive vesicle budding from the endoplasmic reticulum. *Cell* **77,** 895–907.

Barrett, A. J. (1998). Cysteine peptidase. *In* "Handbook of Proteolytic Enzymes" (A. J. Barrett, N. D. Rawlings, and J. F. Woessner, eds.), pp. 543–798. Academic Press, San Diego, CA.

Baum, K. F., Berens, R. L., Marr, J. J., Harrington, J. A., and Spector, T. (1989). Purine deoxynucleoside salvage in *Giardia lamblia*. *J. Biol. Chem.* **264,** 21087–21090.

Beanan, M. J., and Bailey, G. B. (1995). The primary structure of an *Entamoeba histolytica* enolase. *Mol. Biochem. Parasitol.* **69,** 119–121.

Beck, D. L., Boettner, D. R., Dragulev, B., Ready, K., Nozaki, T., and Petri, W. A. J. (2005). Identification and gene expression analysis of a large family of transmembrane kinases related to the Gal/GalNAc lectin in *Entamoeba histolytica*. *Eukaryot. Cell* **4,** 722–732.

Beckers, C. J., Block, M. R., Glick, B. S., Rothman, J. E., and Balch, W. E. (1989). Vesicular transport between the endoplasmic reticulum and the Golgi stack requires the NEM-sensitive fusion protein. *Nature* **339,** 397–398.

Beiko, R. G., Harlow, T. J., and Ragan, M. A. (2005). Highways of gene sharing in prokaryotes. *Proc. Natl. Acad. Sci. USA* **102,** 14332–14337.

Bell, S. P., and Dutta, A. (2002). DNA replication in eukaryotic cells. *Annu. Rev. Biochem.* **71,** 333–374.

Berninghausen, O., and Leippe, M. (1997). Calcium-independent cytolysis of target cells induced by *Entamoeba histolytica*. *Arch. Med. Res.* **28**(Spec No), 158–160.

Berriman, M., Ghedin, E., Hertz-Fowler, C., Blandin, G., Renauld, H., Bartholomeu, D. C., Lennard, N. J., Caler, E., Hamlin, N. E., Haas, B., Bohme, U., Hannick, L., *et al.* (2005). The genome of the African trypanosome *Trypanosoma brucei*. *Science* **309,** 416–422.

Bhattacharya, A., Prasad, R., and Sacks, D. L. (1992). Identification and partial characterization of a lipophosphoglycan from a pathogenic strain of *Entamoeba histolytica*. *Mol. Biochem. Parasitol.* **56,** 161–168.

Bhattacharya, A., Padhan, N., Jain, R., and Bhattacharya, S. (2006). Calcium-binding proteins of *Entamoeba histolytica*. *Arch. Med. Res.* **37,** 221–225.

Bhattacharya, S., Som, I., and Bhattacharya, A. (1998). The ribosomal DNA plasmids of *Entamoeba*. *Parasitol. Today* **14,** 181–185.

Blume-Jensen, P., and Hunter, T. (2001). Oncogenic kinase signalling. *Nature* **411,** 355–365.

Bock, J. B., Matern, H. T., Peden, A. A., and Scheller, R. H. (2001). A genomic perspective on membrane compartment organization. *Nature* **409,** 839–841.

Boehlein, S. K., Schuster, S. M., and Richards, N. G. (1996). Glutamic acid gamma-monohydroxamate and hydroxylamine are alternate substrates for *Escherichia coli* asparagine synthetase B. *Biochemistry* **35,** 3031–3037.

Boehm, M., Aguilar, R. C., and Bonifacino, J. S. (2001). Functional and physical interactions of the adaptor protein complex AP-4 with ADP-ribosylation factors (ARFs). *EMBO J.* **20,** 6265–6276.

Boehm, M., and Bonifacino, J. S. (2001). Adaptins: The final recount. *Mol. Biol. Cell* **12,** 2907–2920.

Bogdanova, N., and Hell, R. (1997). Cysteine synthesis in plants: Protein–protein interactions of serine acetyltransferase from *Arabidopsis thaliana*. *Plant J.* **11,** 251–262.

Bonifacino, J. S., and Glick, B. S. (2004). The mechanisms of vesicle budding and fusion. *Cell* **116,** 153–166.

Bordo, D., and Bork, P. (2002). The rhodanese/Cdc25 phosphatase superfamily. Sequence-structure-function relations. *EMBO Rep.* **3,** 741–746.

Bracha, R., Nuchamowitz, Y., Leippe, M., and Mirelman, D. (1999). Antisense inhibition of *amoeba pore* expression in *Entamoeba histolytica* causes a decrease in amoebic virulence. *Mol. Microbiol.* **34,** 463–472.

Bracha, R., Nuchamowitz, Y., and Mirelman, D. (2003). Transcriptional silencing of an *amoeba pore* gene in *Entamoeba histolytica*: Molecular analysis and effect on pathogenicity. *Eukaryot. Cell* **2,** 295–305.

Braga, L. L., Ninomiya, H., McCoy, J. J., Eacker, S., Wiedmer, T., Pham, C., Wood, S., Sims, P. J., and Petri, W. A. (1992). Inhibition of the complement membrane attack complex by the galactose-specific adhesin of *Entamoeba histolytica*. *J. Clin. Invest.* **90,** 1131–1137.

Bredeston, L. M., Caffaro, C. E., Samuelson, J., and Hirschberg, C. B. (2005). Golgi and endoplasmic reticulum functions take place in different subcellular compartments of *Entamoeba histolytica*. *J. Biol. Chem.* **280,** 32168–32176.

Bruchhaus, I., Leippe, M., Lioutas, C., and Tannich, E. (1993). Unusual gene organization in the protozoan parasite *Entamoeba histolytica*. *DNA Cell Biol.* **12,** 925–933.

Bruchhaus, I., and Tannich, E. (1993). Primary structure of the pyruvate phosphate dikinase in *Entamoeba histolytica*. *Mol. Biochem. Parasitol.* **62,** 153–156.

Bruchhaus, I., and Tannich, E. (1994). Purification and molecular characterization of the NAD^+-dependent acetaldehyde/alcohol dehydrogenase from *Entamoeba histolytica*. *Biochem. J.* **303,** 743–748.

Bruchhaus, I., Jacobs, T., Denart, M., and Tannich, E. (1996). Pyrophosphate-dependent phosphofructokinase of *Entamoeba histolytica:* Molecular cloning, recombinant expression and inhibition by pyrophosphate analogues. *Biochem. J.* **316,** 57–63.

Bruchhaus, I., Richter, S., and Tannich, E. (1997). Removal of hydrogen peroxide by the 29 kDa protein of *Entamoeba histolytica*. *Biochem. J.* **326,** 785–789.

Bruchhaus, I., Richter, S., and Tannich, E. (1998). Recombinant expression and biochemical characterization of an NADPH:flavin oxidoreductase from *Entamoeba histolytica*. *Biochem. J.* **330,** 1217–1221.

Bruchhaus, I., Loftus, B. J., Hall, N., and Tannich, E. (2003). The intestinal protozoan parasite *Entamoeba histolytica* contains 20 cysteine protease genes, of which only a small subset is expressed during in vitro cultivation. *Eukaryot. Cell* **2,** 501–509.

Bruhn, H., Riekens, B., Berninghausen, O., and Leippe, M. (2003). *Amoeba pores* and NK-lysin, members of a class of structurally distinct antimicrobial and cytolytic peptides from protozoa and mammals: A comparative functional analysis. *Biochem. J.* **375,** 737–744.

Bulawa, C. E. (1993). Genetics and molecular biology of chitin synthesis in fungi. *Annu. Rev. Microbiol.* **47,** 505–534.

Burch, D. J., Li, E., Reed, S., Jackson, T. F. H. G., and Stanley, S. L., Jr. (1991). Isolation of a strain-specific *Entamoeba histolytica* cDNA clone. *J. Clin. Microbiol.* **29,** 696–701.

Burchard, G. D., and Bilke, R. (1992). Adherence of pathogenic and non-pathogenic *Entamoeba histolytica* strains to neutrophils. *Parasitol. Res.* **78,** 146–153.

Burchard, G. D., Moslein, C., and Brattig, N. W. (1992a). Adherence between *Entamoeba histolytica* trophozoites and undifferentiated or DMSO-induced HL-60 cells. *Parasitol. Res.* **78,** 336–340.

Burchard, G. D., Prange, G., and Mirelman, D. (1992b). Interaction of various *Entamoeba histolytica* strains with human intestinal cell lines. *Arch. Med. Res.* **23,** 193–195.

Burd, C. G., Strochlic, T. I., and Gangi Setty, S. R. (2004). Arf-like GTPases: Not so Arf-like after all. *Trends Cell Biol.* **14,** 687–694.

Burri, L., and Lithgow, T. (2004). A complete set of SNAREs in yeast. *Traffic* **5,** 45–52.

Capaldi, S. A., and Berger, J. M. (2004). Biochemical characterization of Cdc6/Orc1 binding to the replication origin of the euryarchaeon Methanothermobacter thermoautotrophicus. *Nucleic Acids Res.* **32,** 4821–4832.

Carrero, J. C., Petrossian, P., Olivos, A., Sanchez-Zerpa, M., Ostoa-Soloma, P., and Laclette, J. P. (2000). Effect of cyclosporine A on *Entamoeba histolytica. Arch. Med. Res.* **31,** S8–S9.

Carrero, J. C., Lugo, H., Perez, D. G., Ortiz-Martinez, C., and Laclette, J. P. (2004). Cyclosporin A inhibits calcineurin (phosphatase 2B) and P-glycoprotein activity in *Entamoeba histolytica. Int. J. Parasitol.* **34,** 1091–1097.

Carrozza, M. J., Utley, R. T., Workman, J. L., and Cote, J. (2003). The diverse functions of histone acetyltransferase complexes. *Trends Genet.* **19,** 321–329.

Cawley, S. E., Wirth, A. I., and Speed, T. P. (2001). Phat: A gene finding program for *Plasmodium falciparum. Mol. Biochem. Parasitol.* **118,** 167–174.

Cedar, H., and Schwartz, J. H. (1967). Localization of the two-L-asparaginases in anaerobically grown *Escherichia coli. J. Biol. Chem.* **242,** 3753–3755.

Chakrabarty, P., Sethi, D. K., Padhan, N., Kaur, K. J., Salunke, D. M., Bhattacharya, S., and Bhattacharya, A. (2004). Identification and characterization of EhCaBP2. A second member of the calcium-binding protein family of the protozoan parasite *Entamoeba histolytica. J. Biol. Chem.* **279,** 12898–12908.

Chan, K. W., Slotboom, D. J., Cox, S., Embley, T. M., Fabre, O., van der Giezen, M., Harding, M., Horner, D. S., Kunji, E. R., Leon-Avila, G., and Tovar, J. (2005). A novel ADP/ATP transporter in the mitosome of the microaerophilic human parasite *Entamoeba histolytica. Curr. Biol.* **15,** 737–742.

Chaudhuri, S., Choudhury, N., and Raha, S. (1999). Growth stimulation by serum in *Entamoeba histolytica* is associated with protein tyrosine dephosphorylation. *FEMS Microbiol. Lett.* **178,** 241–249.

Chavez-Munguia, B., Espinosa-Cantellano, M., Castanon, G., and Martinez-Palomo, A. (2000). Ultrastructural evidence of smooth endoplasmic reticulum and golgi-like elements in *Entamoeba histolytica* and *Entamoeba dispar. Arch. Med. Res.* **31,** S165–S167.

Chen, Y. A., Scales, S. J., Patel, S. M., Doung, Y. C., and Scheller, R. H. (1999). SNARE complex formation is triggered by Ca^{2+} and drives membrane fusion. *Cell* **97,** 165–174.

Chen, Y. A., and Scheller, R. H. (2001). SNARE-mediated membrane fusion. *Nat. Rev. Mol. Cell Biol.* **2,** 98–106.

Cheng, X. J., Tsukamoto, H., Kaneda, Y., and Tachibana, H. (1998). Identification of the 150-kDa surface antigen of *Entamoeba histolytica* as a galactose- and *N*-acetyl-D-galactosamine-inhibitable lectin. *Parasitol. Res.* **84,** 632–639.

Cheng, X. J., Hughes, M. A., Huston, C. D., Loftus, B., Gilchrist, C. A., Lockhart, L. A., Ghosh, S., Miller-Sims, V., Mann, B. J., Petri, W. A., Jr., and Tachibana, H. (2001). Intermediate subunit of the Gal/GalNAc lectin of *Entamoeba histolytica* is a member of a gene family containing multiple CXXC sequence motifs. *Infect. Immun.* **69,** 5892–5898.

Cheng, X. J., and Tachibana, H. (2001). Protection of hamsters from amebic liver abscess formation by immunization with the 150- and 170-kDa surface antigens of *Entamoeba histolytica*. *Parasitol. Res.* **87,** 126–130.

Chi, A., and Kemp, R. G. (2000). The primordial high energy compound: ATP or inorganic pyrophosphate? *J. Biol. Chem.* **275,** 35677–35679.

Chi, A. S., Deng, Z., Albach, R. A., and Kemp, R. G. (2001). The two phosphofructokinase gene products of *Entamoeba histolytica*. *J. Biol. Chem.* **276,** 19974–19981.

Chinkers, M. (2001). Protein phosphatase 5 in signal transduction. *Trends Endocrinol. Metab.* **12,** 28–32.

Clark, C. G., and Roger, A. J. (1995). Direct evidence for secondary loss of mitochondria in *Entamoeba histolytica*. *Proc. Natl. Acad. Sci. USA* **92,** 6518–6521.

Clark, C. G., Espinosa Cantellano, M., and Bhattacharya, A. (2000). *Entamoeba histolytica*: An overview of the biology of the organism. *In* "Amebiasis" (J. I. Ravdin, ed.), pp. 1–45. Imperial College Press, London.

Clark, C. G., Ali, I. K., Zaki, M., Loftus, B. J., and Hall, N. (2006a). Unique organisation of tRNA genes in *Entamoeba histolytica*. *Mol. Biochem. Parasitol.* **146,** 24–29.

Clark, C. G., Kaffashian, F., Tawari, B., Windsor, J. J., Twigg-Flesner, A., Davies-Morel, M. C. G., Blessmann, J., Ebert, F., Peschel, B., Le Van, A., Jackson, C. J., Macfarlane, L., *et al.* (2006b). New insights into the phylogeny of *Entamoeba* species provided by analysis of four new small-subunit rRNA genes. *Int. J. Syst. Evol. Microbiol.* **56,** 2235–2239.

Clary, D. O., Griff, I. C., and Rothman, J. E. (1990). SNAPs, a family of NSF attachment proteins involved in intracellular membrane fusion in animals and yeast. *Cell* **61,** 709–721.

Conner, S. D., and Schmid, S. L. (2003). Differential requirements for AP-2 in clathrin-mediated endocytosis. *J. Cell Biol.* **162,** 773–779.

Coombs, G. H., and Mottram, J. C. (2001). Trifluoromethionine, a prodrug designed against methionine gamma-lyase-containing pathogens, has efficacy *in vitro* and *in vivo* against *Trichomonas vaginalis*. *Antimicrob. Agents Chemother.* **45,** 1743–1745.

Coppi, A., and Eichinger, D. (1999). Regulation of *Entamoeba invadens* encystation and gene expression with galactose and *N*-acetylglucosamine. *Mol. Biochem. Parasitol.* **102,** 67–77.

Coppi, A., Merali, S., and Eichinger, D. (2002). The enteric parasite *Entamoeba* uses an autocrine catecholamine system during differentiation into the infectious cyst stage. *J. Biol. Chem.* **277,** 8083–8090.

Cruz-Reyes, J., ur-Rehman, T., Spice, W. M., and Ackers, J. P. (1995). A novel transcribed repeat element from *Entamoeba histolytica*. *Gene* **166,** 183–184.

Cruz-Reyes, J. A., and Ackers, J. P. (1992). A DNA probe specific to pathogenic *Entamoeba histolytica*. *Arch. Med. Res.* **23,** 271–275.

Das, A. K., Cohen, P. W., and Barford, D. (1998). The structure of the tetratricopeptide repeats of protein phosphatase 5: Implications for TPR-mediated protein–protein interactions. *EMBO J.* **17,** 1192–1199.

Das, S., and Lohia, A. (2002). Delinking of S phase and cytokinesis in the protozoan parasite *Entamoeba histolytica*. *Cell. Microbiol.* **4,** 55–60.

Das, S., Stevens, T., Castillo, C., Villasenor, A., Arredondo, H., and Reddy, K. (2002). Lipid metabolism in mucous-dwelling amitochondriate protozoa. *Int. J. Parasitol.* **32,** 655–675.

Das, S., Van Dellen, K., Bulik, D., Magnelli, P., Cui, J., Head, J., Robbins, P. W., and Samuelson, J. (2006). The cyst wall of *Entamoeba invadens* contains chitosan (deacetylated chitin). *Mol. Biochem. Parasitol.* **148,** 86–92.

Davis, J. A., Takagi, Y., Kornberg, R. D., and Asturias, F. A. (2002). Structure of the yeast RNA polymerase II holoenzyme: Mediator conformation and polymerase interaction. *Mol. Cell* **10,** 409–415.

Davis, P. H., Zhang, Z., Chen, M., Zhang, X., Chakraborty, S., and Stanley, S. L., Jr. (2006). Identification of a family of BspA like surface proteins of *Entamoeba histolytica* with novel leucine rich repeats. *Mol. Biochem. Parasitol.* **145,** 111–116.

Davis, P. H., Schulze, J., and Stanley, S. L., Jr. (2007). Transcriptomic comparison of two *Entamoeba histolytica* strains with defined virulence phenotypes identifies new virulence factor candidates and key differences in the expression patterns of cysteine proteases, lectin light chains, and calmodulin. *Mol. Biochem. Parasitol.* **151,** 118–128.

de Muñoz, M. L., Moreno, M. A., Pérez-Garcia, J. N., Tovar, G. R., and Hernandez, V. I. (1991). Possible role of calmodulin in the secretion of *Entamoeba histolytica* electron-dense granules containing collagenase. *Mol. Microbiol.* **5,** 1707–1714.

de la Vega, H., Specht, C. A., Semino, C. E., Robbins, P. W., Eichinger, D., Caplivski, D., Ghosh, S., and Samuelson, J. (1997). Cloning and expression of chitinases of *Entamoebae*. *Mol. Biochem. Parasitol.* **85,** 139–147.

Dear, P. H., and Cook, P. R. (1993). Happy mapping: Linkage mapping using a physical analogue of meiosis. *Nucl. Acids Res.* **21,** 13–20.

Debnath, A., Das, P., Sajid, M., and McKerrow, J. H. (2004). Identification of genomic responses to collagen binding by trophozoites of *Entamoeba histolytica*. *J. Infect. Dis.* **190,** 448–457.

Demarque, M., Represa, A., Becq, H., Khalilov, I., Ben-Ari, Y., and Aniksztejn, L. (2002). Paracrine intercellular communication by a Ca^{2+}—and SNARE-independent release of GABA and glutamate prior to synapse formation. *Neuron* **36,** 1051–1061.

Deng, Z., Huang, M., Singh, K., Albach, R. A., Latshaw, S. P., Chang, K. P., and Kemp, R. G. (1998). Cloning and expression of the gene for the active PPi-dependent phosphofructokinase of *Entamoeba histolytica*. *Biochem. J.* **329,** 659–664.

Dhar, S. K., Choudhury, N. R., Bhattacharya, A., and Bhattacharya, S. (1995). A multitude of circular DNAs exist in the nucleus of *Entamoeba histolytica*. *Mol. Biochem. Parasitol.* **70,** 203–206.

Diamond, L. S., and Cunnick, C. C. (1991). A serum-free, partly defined medium, PDM-805, for the axenic cultivation of *Entamoeba histolytica* Schaudinn, 1903 and other *Entamoeba*. *J. Protozool.* **38,** 211–216.

Dodson, J. M., Clark, C. G., Lockhart, L. A., Leo, B. M., Schroeder, J. W., and Mann, B. J. (1997). Comparison of adherence, cytotoxicity, and Gal/GalNAc lectin gene structure in *Entamoeba histolytica* and *Entamoeba dispar*. *Parasitol. Int.* **46,** 225–235.

Donaldson, J. G., Cassel, D., Kahn, R. A., and Klausner, R. D. (1992). ADP-ribosylation factor, a small GTP-binding protein, is required for binding of the coatomer protein beta-COP to Golgi membranes. *Proc. Natl. Acad. Sci. USA* **89,** 6408–6412.

Doolittle, W. F. (1998). You are what you eat: A gene transfer ratchet could account for bacterial genes in eukaryotic nuclear genomes. *Trends Genet.* **14,** 307–311.

Doolittle, W. F., Boucher, Y., Nesbo, C. L., Douady, C. J., Andersson, J. O., and Roger, A. J. (2003). How big is the iceberg of which organellar genes in nuclear genomes are but the tip? *Phil. Trans. R. Soc. Lond. Ser. B, Biol. Sci.* **358,** 39–57; discussion 57–38.

Dresios, J., Panopoulos, P., Suzuki, K., and Synetos, D. (2003). A dispensable yeast ribosomal protein optimizes peptidyltransferase activity and affects translocation. *J. Biol. Chem.* **278,** 3314–3322.

Droux, M., Ruffet, M. L., Douce, R., and Job, D. (1998). Interactions between serine acetyltransferase and *O*-acetylserine (thiol) lyase in higher plants: Structural and kinetic properties of the free and bound enzymes. *Eur. J. Biochem.* **255,** 235–245.

Duden, R., Kajikawa, L., Wuestehube, L., and Schekman, R. (1998). epsilon-COP is a structural component of coatomer that functions to stabilize alpha-COP. *EMBO J.* **17,** 985–995.

Dulubova, I., Sugita, S., Hill, S., Hosaka, M., Fernandez, I., Sudhof, T. C., and Rizo, J. (1999). A conformational switch in syntaxin during exocytosis: Role of munc18. *EMBO J.* **18,** 4372–4382.

Dhrenkaufer, G. M., Haque, R., Hackney, J. A., Eichinger, D. J., and Singh, U. (2007). Identification of developmentally regulated genes in *Entamoeba histolytica*: Insights into mechanisms of stage conversion in a protozoan parasite. *Cell. Microbiol.* **9,** 1426–1444.

Dutcher, S. K. (2001). The tubulin fraternity: Alpha to eta. *Curr. Opin. Cell Biol.* **13,** 49–54.

Ehrenkaufer, G. M., Haque, R., Hackney, J. A., Eichinger, D. J., and Singh, U. (2007). Identification of developmentally regulated genes in *Entamoeba histolytica:* Insights into mechanisms of stage conversion in a protozoan parasite. *Cell. Microbiol.* **9,** 1426–1444.

Eichinger, D. (1997). Encystation of *Entamoeba* parasites. *Bioessays* **19,** 633–639.

Eichinger, D. (2001). Encystation in parasitic protozoa. *Curr. Opin. Microbiol.* **4,** 421–426.

Eichinger, L., Pachebat, J. A., Glockner, G., Rajandream, M. A., Sucgang, R., Berriman, M., Song, J., Olsen, R., Szafranski, K., Xu, Q., Tunggal, B., Kummerfeld, S., *et al.* (2005). The genome of the social amoeba *Dictyostelium discoideum. Nature* **435,** 43–57.

Elnekave, K., Siman-Tov, R., and Ankri, S. (2003). Consumption of L-arginine mediated by *Entamoeba histolytica* L-arginase (EhArg) inhibits amoebicidal activity and nitric oxide production by activated macrophages. *Parasite Immunol.* **25,** 597–608.

Espenshade, P., Gimeno, R. E., Holzmacher, E., Teung, P., and Kaiser, C. A. (1995). Yeast SEC16 gene encodes a multidomain vesicle coat protein that interacts with Sec23p. *J. Cell Biol.* **131,** 311–324.

Espinosa-Cantellano, M., and Martínez-Palomo, A. (2000). Pathogenesis of intestinal amebiasis: From molecules to disease. *Clin. Microbiol. Rev.* **13,** 318–331.

Fahey, R. C., Newton, G. L., Arrick, B., Overdank-Bogart, T., and Aley, S. B. (1984). *Entamoeba histolytica*: A eukaryote without glutathione metabolism. *Science* **224,** 70–72.

Fairlamb, A. H., and Cerami, A. (1992). Metabolism and functions of trypanothione in the Kinetoplastida. *Annu. Rev. Microbiol.* **46,** 695–729.

Farri, T. A., Sargeaunt, P. G., Warhurst, D. C., Williams, J. E., and Bhojnani, R. (1980). Electrophoretic studies of the hexokinase of *Entamoeba histolytica* groups I to IV. *Trans. R. Soc. Trop. Med. Hyg.* **74,** 672–673.

Faundez, V., Horng, J. T., and Kelly, R. B. (1998). A function for the AP3 coat complex in synaptic vesicle formation from endosomes. *Cell* **93,** 423–432.

Field, J., Rosenthal, B., and Samuelson, J. (2000). Early lateral transfer of genes encoding malic enzyme, acetyl-CoA synthetase and alcohol dehydrogenases from anaerobic prokaryotes to *Entamoeba histolytica. Mol. Microbiol.* **38,** 446–455.

Folsch, H., Ohno, H., Bonifacino, J. S., and Mellman, I. (1999). A novel clathrin adaptor complex mediates basolateral targeting in polarized epithelial cells. *Cell* **99,** 189–198.

Frederick, J., and Eichinger, D. (2004). *Entamoeba invadens* contains the components of a classical adrenergic signaling system. *Mol. Biochem. Parasitol.* **137,** 339–343.

Freedman, R. B., Klappa, P., and Ruddock, L. W. (2002). Protein disulfide isomerases exploit synergy between catalytic and specific binding domains. *EMBO Rep.* **3,** 136–140.

Fridovich, I. (1995). Superoxide radical and superoxide dismutases. *Annu. Rev. Biochem.* **64,** 97–112.

Frisardi, M., Ghosh, S. K., Field, J., Van Dellen, K., Rogers, R., Robbins, P., and Samuelson, J. (2000). The most abundant glycoprotein of amebic cyst walls (Jacob) is a lectin with five Cys-rich, chitin-binding domains. *Infect. Immun.* **68,** 4217–4224.

Gadasi, H., and Kessler, E. (1983). Correlation of virulence and collagenolytic activity in *Entamoeba histolytica. Infect. Immun.* **39,** 528–531.

Gangopadhyay, S. S., Ray, S. S., Kennady, K., Pande, G., and Lohia, A. (1997). Heterogeneity of DNA content and expression of cell cycle genes in axenically growing *Entamoeba histolytica* HM1:IMSS clone A. *Mol. Biochem. Parasitol.* **90,** 9–20.

Gao, T., Furnari, F., and Newton, A. C. (2005). PHLPP: A phosphatase that directly dephosphorylates Akt, promotes apoptosis, and suppresses tumor growth. *Mol. Cell* **18,** 13–24.

Garcia-Rivera, G., Rodriguez, M. A., Ocadiz, R., Martinez, L. P. M. C., Arroyo, R., Gonzalez-Robles, A., and Orozco, E. (1999). *Entamoeba histolytica*: A novel cysteine protease and an adhesin form the 112 kDa surface protein. *Mol. Microbiol.* **33,** 556–568.

Gardner, M. J., Hall, N., Fung, E., White, O., Berriman, M., Hyman, R. W., Carlton, J. M., Pain, A., Nelson, K. E., Bowman, S., Paulsen, I. T., James, K., *et al.* (2002). Genome sequence of the human malaria parasite *Plasmodium falciparum. Nature* **419,** 498–511.

Gelderman, A. H., Bartgis, I. L., Keister, D. B., and Diamond, L. S. (1971a). A comparison of genome sizes and thermal denaturation-derived base composition of DNAs from several members of *Entamoeba* (*histolytica* group). *J. Parasitol.* **57,** 912–916.

Gelderman, A. H., Keister, D. B., Bartgis, I. L., and Diamond, L. S. (1971b). Characterization of the deoxyribonucleic acid of representative strains of *Entamoeba histolytica*, *E. histoytica*-like amebae, and *E. moshkovskii*. *J. Parasitol.* **57,** 906–911.

Gemmill, T. R., and Trimble, R. B. (1999). Overview of N- and O-linked oligosaccharide structures found in various yeast species. *Biochim. Biophys. Acta* **1426,** 227–237.

Gerwig, G. J., van Kuik, J. A., Leeflang, B. R., Kamerling, J. P., Vliegenthart, J. F., Karr, C. D., and Jarroll, E. L. (2002). The *Giardia intestinalis* filamentous cyst wall contains a novel beta (1–3)-*N*-acetyl-D-galactosamine polymer: A structural and conformational study. *Glycobiology* **12,** 499–505.

Ghosh, S., Frisardi, M., Ramirez-Avila, L., Descoteaux, S., Sturm-Ramirez, K., Newton-Sanchez, O. A., Santos-Preciado, J. I., Ganguly, C., Lohia, A., Reed, S., and Samuelson, J. (2000). Molecular epidemiology of *Entamoeba* spp.: Evidence of a bottleneck (Demographic sweep) and transcontinental spread of diploid parasites. *J. Clin. Microbiol.* **38,** 3815–3821.

Ghosh, S., Chan, J. M., Lea, C. R., Meints, G. A., Lewis, J. C., Tovian, Z. S., Flessner, R. M., Loftus, T. C., Bruchhaus, I., Kendrick, H., Croft, S. L., Kemp, R. G., *et al.* (2004). Effects of bisphosphonates on the growth of *Entamoeba histolytica* and *Plasmodium* species *in vitro* and *in vivo*. *J. Med. Chem.* **47,** 175–187.

Gilchrist, C. A., Holm, C. F., Hughes, M. A., Schaenman, J. M., Mann, B. J., and Petri, W. A., Jr. (2001). Identification and characterization of an *Entamoeba histolytica* upstream regulatory element 3 sequence-specific DNA-binding protein containing EF-hand motifs. *J. Biol. Chem.* **276,** 11838–11843.

Gilchrist, C. A., Houpt, E., Trapaidze, N., Fei, Z., Crasta, O., Asgharpour, A., Evans, C., Martino-Catt, S., Baba, D. J., Stroup, S., Hamano, S., Ehrenkaufer, G., *et al.* (2006). Impact of intestinal colonization and invasion on the *Entamoeba histolytica* transcriptome. *Mol. Biochem. Parasitol.* **147,** 163–176.

Gillin, F. D., Reiner, D. S., and McCann, P. P. (1984). Inhibition of growth of *Giardia lamblia* by difluoromethylornithine, a specific inhibitor of polyamine biosynthesis. *J. Protozool.* **31,** 161–163.

Goffeau, A., Barrell, B. G., Bussey, H., Davis, R. W., Dujon, B., Feldmann, H., Galibert, F., Hoheisel, J. D., Jacq, C., Johnston, M., Louis, E. J., Mewes, H. W., *et al.* (1996). Life with 6000 genes. *Science* **274,** 546–567.

Gomes, C. M., Giuffre, A., Forte, E., Vicente, J. B., Saraiva, L. M., Brunori, M., and Teixeira, M. (2002). A novel type of nitric-oxide reductase. *Escherichia coli* flavorubredoxin. *J. Biol. Chem.* **277,** 25273–25276.

Gregan, J., Lindner, K., Brimage, L., Franklin, R., Namdar, M., Hart, E. A., Aves, S. J., and Kearsey, S. E. (2003). Fission yeast Cdc23/Mcm10 functions after pre-replicative complex formation to promote Cdc45 chromatin binding. *Mol. Biol. Cell* **14,** 3876–3887.

Grunler, J., Ericsson, J., and Dallner, G. (1994). Branch-point reactions in the biosynthesis of cholesterol, dolichol, ubiquinone and prenylated proteins. *Biochim. Biophys. Acta* **1212,** 259–277.

Gu, J., and Gu, X. (2003). Natural history and functional divergence of protein tyrosine kinases. *Gene* **317,** 49–57.

Guerrant, R. L., Brush, J., Ravdin, J. I., Sullivan, J. A., and Mandell, G. L. (1981). Interaction between *Entamoeba histolytica* and human polymorphonuclear neutrophils. *J. Infect. Dis.* **143,** 83–93.

Gutsmann, T., Riekens, B., Bruhn, H., Wiese, A., Seydel, U., and Leippe, M. (2003). Interaction of *amoeba pores* and NK-lysin with symmetric phospholipid and asymmetric lipopolysaccharide/phospholipid bilayers. *Biochemistry* **42,** 9804–9812.

Haghighi, A., Kobayashi, S., Takeuchi, T., Thammapalerd, N., and Nozaki, T. (2003). Geographic diversity among genotypes of *Entamoeba histolytica* field isolates. *J. Clin. Microbiol.* **41,** 3748–3756.

Hanks, S. K., and Hunter, T. (1995). Protein kinases 6. The eukaryotic protein kinase superfamily: Kinase (catalytic) domain structure and classification. *FASEB J.* **9,** 576–596.

Hanson, P. I., Roth, R., Morisaki, H., Jahn, R., and Heuser, J. E. (1997). Structure and conformational changes in NSF and its membrane receptor complexes visualized by quick-freeze/deep-etch electron microscopy. *Cell* **90,** 523–535.

Hara-Kuge, S., Kuge, O., Orci, L., Amherdt, M., Ravazzola, M., Wieland, F. T., and Rothman, J. E. (1994). En bloc incorporation of coatomer subunits during the assembly of COP-coated vesicles. *J. Cell Biol.* **124,** 883–892.

Hartwell, L. H., and Weinert, T. A. (1989). Checkpoints: Controls that ensure the order of cell cycle events. *Science* **246,** 629–634.

Hausmann, S., Ho, C. K., Schwer, B., and Shuman, S. (2001). An essential function of *Saccharomyces cerevisiae* RNA triphosphatase Cet1 is to stabilize RNA guanylyltransferase Ceg1 against thermal inactivation. *J. Biol. Chem.* **276,** 36116–36124.

Hecht, O., Van Nuland, N. A., Schleinkofer, K., Dingley, A. J., Bruhn, H., Leippe, M., and Grotzinger, J. (2004). Solution structure of the pore-forming protein of *Entamoeba histolytica*. *J. Biol. Chem.* **279,** 17834–17841.

Helenius, A., and Aebi, M. (2004). Roles of N-linked glycans in the endoplasmic reticulum. *Annu. Rev. Biochem.* **73,** 1019–1049.

Helms, J. B., and Rothman, J. E. (1992). Inhibition by brefeldin A of a Golgi membrane enzyme that catalyses exchange of guanine nucleotide bound to ARF. *Nature* **360,** 352–354.

Hemming, S. A., and Edwards, A. M. (2000). Yeast RNA polymerase II subunit RPB9. Mapping of domains required for transcription elongation. *J. Biol. Chem.* **275,** 2288–2294.

Hernandez-Ramirez, V. I., Anaya-Ruiz, M., Rios, A., and Talamas-Rohana, P. (2000). *Entamoeba histolytica*: Tyrosine kinase activity induced by fibronectin through the beta1-integrin-like molecule. *Exp. Parasitol.* **95,** 85–95.

Hernandez, R., Luna-Arias, J. P., and Orozco, E. (1997). Comparison of the *Entamoeba histolytica* TATA-binding protein (TBP) structure with other TBP. *Arch. Med. Res.* **28** (Spec No), 43–45.

Hidalgo, M. E., Sanchez, R., Perez, D. G., Rodriguez, M. A., Garcia, J., and Orozco, E. (1997). Molecular characterization of the *Entamoeba histolytica* enolase gene and modelling of the predicted protein. *FEMS Microbiol. Lett.* **148,** 123–129.

Higgs, H. N., and Peterson, K. J. (2005). Phylogenetic analysis of the formin homology 2 domain. *Mol. Biol. Cell* **16,** 1–13.

Hirsch, C., Jarosch, E., Sommer, T., and Wolf, D. H. (2004). Endoplasmic reticulum-associated protein degradation: One model fits all? *Biochim. Biophys. Acta* **1695,** 215–223.

Hirschberg, C. B., Robbins, P. W., and Abeijon, C. (1998). Transporters of nucleotide sugars, ATP, and nucleotide sulfate in the endoplasmic reticulum and Golgi apparatus. *Annu. Rev. Biochem.* **67,** 49–69.

Hirst, J., Bright, N. A., Rous, B., and Robinson, M. S. (1999). Characterization of a fourth adaptor-related protein complex. *Mol. Biol. Cell* **10,** 2787–2802.

Hirt, R. P., Harriman, N., Kajava, A. V., and Embley, T. M. (2002). A novel potential surface protein in *Trichomonas vaginalis* contains a leucine-rich repeat shared by micro-organisms from all three domains of life. *Mol. Biochem. Parasitol.* **125,** 195–199.

Hofmann, B., Hecht, H. J., and Flohe, L. (2002). Peroxiredoxins. *Biol. Chem.* **383,** 347–364.

Holmgren, A. (2000). Antioxidant function of thioredoxin and glutaredoxin systems. *Antioxid. Redox. Signal.* **2,** 811–820.

Hooft van Huijsduijnen, R. (1998). Protein tyrosine phosphatases: Counting the trees in the forest. *Gene* **225,** 1–8.

Horner, D. S., Hirt, R. P., and Embley, T. M. (1999). A single eubacterial origin of eukaryotic pyruvate: Ferredoxin oxidoreductase genes: Implications for the evolution of anaerobic eukaryotes. *Mol. Biol. Evol.* **16,** 1280–1291.

Hsiung, Y. G., Chang, H. C., Pellequer, J. L., La Valle, R., Lanker, S., and Wittenberg, C. (2001). F-box protein Grr1 interacts with phosphorylated targets via the cationic surface of its leucine-rich repeat. *Mol. Cell. Biol.* **21,** 2506–2520.

Hubbard, S. C., and Ivatt, R. J. (1981). Synthesis and processing of asparagine-linked oligosaccharides. *Annu. Rev. Biochem.* **50,** 555–583.

Huber, L. A., Pimplikar, S., Parton, R. G., Virta, H., Zerial, M., and Simons, K. (1993). Rab8, a small GTPase involved in vesicular traffic between the TGN and the basolateral plasma membrane. *J. Cell Biol.* **123,** 35–45.

Hunter, T. (1995). Protein kinases and phosphatases: The yin and yang of protein phosphorylation and signaling. *Cell* **80,** 225–236.

Hunter, W. N., Bond, C. S., Gabrielsen, M., and Kemp, L. E. (2003). Structure and reactivity in the non-mevalonate pathway of isoprenoid biosynthesis. *Biochem. Soc. Trans.* **31,** 537–542.

Ito, M. (2005). Conservation and diversification of three-repeat Myb transcription factors in plants. *J. Plant Res.* **118,** 61–69.

Iwashita, J., Sato, Y., Kobayashi, S., Takeuchi, T., and Abe, T. (2005). Isolation and functional analysis of a chk2 homologue from *Entamoeba histolytica. Parasitol. Int.* **54,** 21–27.

Jacobs, T., Bruchhaus, I., Dandekar, T., Tannich, E., and Leippe, M. (1998). Isolation and molecular characterization of a surface-bound proteinase of *Entamoeba histolytica. Mol. Microbiol.* **27,** 269–276.

Jain, R., Rivera, M. C., and Lake, J. A. (1999). Horizontal gene transfer among genomes: The complexity hypothesis. *Proc. Natl. Acad. Sci. USA* **96,** 3801–3806.

Jeffrey, P. D., Russo, A. A., Polyak, K., Gibbs, E., Hurwitz, J., Massague, J., and Pavletich, N. P. (1995). Mechanism of CDK activation revealed by the structure of a cyclinA-CDK2 complex. *Nature* **376,** 313–320.

Jongsareejit, B., Rahman, R. N., Fujiwara, S., and Imanaka, T. (1997). Gene cloning, sequencing and enzymatic properties of glutamate synthase from the hyperthermophilic archaeon Pyrococcus sp. KOD1. *Mol. Gen. Genet.* **254,** 635–642.

Jordan, I. K., Henze, K., Fedorova, N. D., Koonin, E. V., and Galperin, M. Y. (2003). Phylogenomic analysis of the *Giardia intestinalis* transcarboxylase reveals multiple instances of domain fusion and fission in the evolution of biotin-dependent enzymes. *J. Mol. Microbiol. Biotechnol.* **5,** 172–189.

Juarez, P., Sanchez-Lopez, R., Stock, R. P., Olvera, A., Ramos, M. A., and Alagon, A. (2001). Characterization of the Ehrab8 gene, a marker of the late stages of the secretory pathway of *Entamoeba histolytica. Mol. Biochem. Parasitol.* **116,** 223–228.

Jurica, M. S., and Moore, M. J. (2003). Pre-mRNA splicing: Awash in a sea of proteins. *Mol. Cell* **12,** 5–14.

Kafetzopoulos, D., Thireos, G., Vournakis, J. N., and Bouriotis, V. (1993). The primary structure of a fungal chitin deacetylase reveals the function for two bacterial gene products. *Proc. Natl. Acad. Sci. USA* **90,** 8005–8008.

Kahn, R. A., and Gilman, A. G. (1984). Purification of a protein cofactor required for ADP-ribosylation of the stimulatory regulatory component of adenylate cyclase by cholera toxin. *J. Biol. Chem.* **259,** 6228–6234.

Kahn, R. A., Cherfils, J., Elias, M., Lovering, R. C., Munro, S., and Schurmann, A. (2006). Nomenclature for the human Arf family of GTP-binding proteins: ARF, ARL, and SAR proteins. *J. Cell Biol.* **172,** 645–650.

Kaiser, C. (2000). Thinking about p24 proteins and how transport vesicles select their cargo. *Proc. Natl. Acad. Sci. USA* **97,** 3783–3785.

Kang, J. S., Kim, S. H., Hwang, M. S., Han, S. J., Lee, Y. C., and Kim, Y. J. (2001). The structural and functional organization of the yeast mediator complex. *J. Biol. Chem.* **276,** 42003–42010.

Katinka, M. D., Duprat, S., Cornillot, E., Metenier, G., Thomarat, F., Prensier, G., Barbe, V., Peyretaillade, E., Brottier, P., Wincker, P., Delbac, F., El Alaoui, H., *et al.* (2001). Genome sequence and gene compaction of the eukaryote parasite *Encephalitozoon cuniculi. Nature* **414,** 450–453.

Keen, J. H. (1987). Clathrin assembly proteins: Affinity purification and a model for coat assembly. *J. Cell Biol.* **105,** 1989–1998.

Keene, W. E., Hidalgo, M. E., Orozco, E., and McKerrow, J. H. (1990). *Entamoeba histolytica*: Correlation of the cytopathic effect of virulent trophozoites with secretion of a cysteine proteinase. *Exp. Parasitol.* **71,** 199–206.

Kelleher, D. J., and Gilmore, R. (2006). An evolving view of the eukaryotic oligosaccharyltransferase. *Glycobiology* **16,** 47R–62R.

Kelman, L. M., and Kelman, Z. (2004). Multiple origins of replication in archaea. *Trends Microbiol.* **12,** 399–401.

Kennelly, P. J. (2001). Protein phosphatases: A phylogenetic perspective. *Chem. Rev.* **101,** 2291–2312.

Kimura, A., Hara, Y., Kimoto, T., Okuno, Y., Minekawa, Y., and Nakabayashi, T. (1996). Cloning and expression of a putative alcohol dehydrogenase gene of *Entamoeba histolytica* and its application to immunological examination. *Clin. Diagn. Lab. Immunol.* **3,** 270–274.

Kirchhausen, T. (2000). Three ways to make a vesicle. *Nat. Rev. Mol. Cell Biol.* **1,** 187–198.

Kirisako, T., Ichimura, Y., Okada, H., Kabeya, Y., Mizushima, N., Yoshimori, T., Ohsumi, M., Takao, T., Noda, T., and Ohsumi, Y. (2000). The reversible modification regulates the membrane-binding state of Apg8/Aut7 essential for autophagy and the cytoplasm to vacuole targeting pathway. *J. Cell Biol.* **151,** 263–276.

Knodler, L. A., Edwards, M. R., and Schofield, P. J. (1994). The intracellular amino acid pools of *Giardia intestinalis, Trichomonas vaginalis,* and *Crithidia luciliae. Exp. Parasitol.* **79,** 117–125.

Kobe, B., and Deisenhofer, J. (1994). The leucine-rich repeat: A versatile binding motif. *Trends Biochem. Sci.* **19,** 415–421.

Koonin, E. V., Makarova, K. S., and Aravind, L. (2001). Horizontal gene transfer in prokaryotes: Quantification and classification. *Annu. Rev. Microbiol.* **55,** 709–742.

Kornberg, R. D. (2001). The eukaryotic gene transcription machinery. *Biol. Chem.* **382,** 1103–1107.

Kornfeld, R., and Kornfeld, S. (1985). Assembly of asparagine-linked oligosaccharides. *Annu. Rev. Biochem.* **54,** 631–664.

Kriebel, P. W., and Parent, C. A. (2004). Adenylyl cyclase expression and regulation during the differentiation of *Dictyostelium discoideum. IUBMB Life* **56,** 541–546.

Kroschewski, H., Ortner, S., Steipe, B., Scheiner, O., Wiedermann, G., and Duchêne, M. (2000). Differences in substrate specificity and kinetic properties of the recombinant hexokinases HXK1 and HXK2 from *Entamoeba histolytica. Mol. Biochem. Parasitol.* **105,** 71–80.

Kulda, J. (1999). Trichomonads, hydrogenosomes and drug resistance. *Int. J. Parasitol.* **29,** 199–212.

Kumagai, M., Makioka, A., Takeuchi, T., and Nozaki, T. (2004). Molecular cloning and characterization of a protein farnesyltransferase from the enteric protozoan parasite *Entamoeba histolytica. J. Biol. Chem.* **279,** 2316–2323.

Kumar, A., Shen, P. S., Descoteaux, S., Pohl, J., Bailey, G., and Samuelson, J. (1992). Cloning and expression of an $NADP^{+}$-dependent alcohol dehydrogenase gene of *Entamoeba histolytica. Proc. Natl. Acad. Sci. USA* **89,** 10188–10192.

Kupke, T., Uebele, M., Schmid, D., Jung, G., Blaesse, M., and Steinbacher, S. (2000). Molecular characterization of lantibiotic-synthesizing enzyme EpiD reveals a function for bacterial Dfp proteins in coenzyme A biosynthesis. *J. Biol. Chem.* **275,** 31838–31846.

Kupke, T. (2002). Molecular characterization of the 4′-phosphopantothenoylcysteine synthetase domain of bacterial dfp flavoproteins. *J. Biol. Chem.* **277,** 36137–36145.

Kupke, T. (2004). Active-site residues and amino acid specificity of the bacterial 4′-phosphopantothenoylcysteine synthetase CoaB. *Eur. J. Biochem.* **271,** 163–172.
Kuranda, M. J., and Robbins, P. W. (1991). Chitinase is required for cell separation during growth of *Saccharomyces cerevisiae*. *J. Biol. Chem.* **266,** 19758–19767.
Lal, K., Field, M. C., Carlton, J. M., Warwicker, J., and Hirt, R. P. (2005). Identification of a very large Rab GTPase family in the parasitic protozoan *Trichomonas vaginalis*. *Mol. Biochem. Parasitol.* **143,** 226–235.
Landa, A., Rojo-Dominguez, A., Jimenez, L., and Fernandez-Velasco, D. A. (1997). Sequencing, expression and properties of triosephosphate isomerase from *Entamoeba histolytica*. *Eur. J. Biochem.* **247,** 348–355.
Lander, E. S., Linton, L. M., Birren, B., Nusbaum, C., Zody, M. C., Baldwin, J., Devon, K., Dewar, K., Doyle, M., FitzHugh, W., Funke, R., Gage, D., *et al.* (2001). Initial sequencing and analysis of the human genome. *Nature* **409,** 860–921.
Lawrence, J. G. (2005a). Horizontal and vertical gene transfer: The life history of pathogens. *Contrib. Microbiol.* **12,** 255–271.
Lawrence, J. G. (2005b). Common themes in the genome strategies of pathogens. *Curr. Opin. Genet. Dev.* **15,** 584–588.
Le Borgne, R., Planque, N., Martin, P., Dewitte, F., Saule, S., and Hoflack, B. (2001). The AP-3-dependent targeting of the melanosomal glycoprotein QNR-71 requires a di-leucine-based sorting signal. *J. Cell Sci.* **114,** 2831–2841.
Lee, F. J., Huang, C. F., Yu, W. L., Buu, L. M., Lin, C. Y., Huang, M. C., Moss, J., and Vaughan, M. (1997a). Characterization of an ADP-ribosylation factor-like 1 protein in *Saccharomyces cerevisiae*. *J. Biol. Chem.* **272,** 30998–31005.
Lee, J. H., Kim, J. M., Kim, M. S., Lee, Y. T., Marshak, D. R., and Bae, Y. S. (1997b). The highly basic ribosomal protein L41 interacts with the beta subunit of protein kinase CKII and stimulates phosphorylation of DNA topoisomerase IIalpha by CKII. *Biochem. Biophys. Res. Commun.* **238,** 462–467.
Leippe, M., Ebel, S., Schoenberger, O. L., Horstmann, R. D., and Müller-Eberhard, H. J. (1991). Pore-forming peptide of pathogenic *Entamoeba histolytica*. *Proc. Natl. Acad. Sci. USA* **88,** 7659–7663.
Leippe, M., Tannich, E., Nickel, R., van der Goot, G., Pattus, F., Horstmann, R. D., and Müller-Eberhard, H. J. (1992). Primary and secondary structure of the pore-forming peptide of pathogenic *Entamoeba histolytica*. *EMBO J.* **11,** 3501–3506.
Leippe, M., Andrä, J., and Müller-Eberhard, H. J. (1994a). Cytolytic and antibacterial activity of synthetic peptides derived from *amoeba pore*, the pore-forming peptide of *Entamoeba histolytica*. *Proc. Natl. Acad. Sci. USA* **91,** 2602–2606.
Leippe, M., Andrä, J., Nickel, R., Tannich, E., and Müller-Eberhard, H. J. (1994b). *Amoeba pores*, a family of membranolytic peptides from cytoplasmic granules of *Entamoeba histolytica*: Isolation, primary structure, and pore formation in bacterial cytoplasmic membranes. *Mol. Microbiol.* **14,** 895–904.
Leippe, M., Sievertsen, H. J., Tannich, E., and Horstmann, R. D. (1995). Spontaneous release of cysteine proteinases but not of pore-forming peptides by viable *Entamoeba histolytica*. *Parasitology* **111,** 569–574.
Leippe, M. (1997). *Amoeba pores*. *Parasitol. Today* **13,** 178–183.
Leippe, M. (1999). Antimicrobial and cytolytic polypeptides of amoeboid protozoa: Effector molecules of primitive phagocytes. *Dev. Comp. Immunol.* **23,** 267–279.
Leippe, M., Bruhn, H., Hecht, O., and Grotzinger, J. (2005). Ancient weapons: The three-dimensional structure of *amoeba pore* A. *Trends Parasitol.* **21,** 5–7.
Leon-Avila, G., and Tovar, J. (2004). Mitosomes of *Entamoeba histolytica* are abundant mitochondrion-related remnant organelles that lack a detectable organellar genome. *Microbiology* **150,** 1245–1250.
Leung, C. L., Green, K. J., and Liem, R. K. (2002). Plakins: A family of versatile cytolinker proteins. *Trends Cell Biol.* **12,** 37–45.

Li, E., Yang, W. G., Zhang, T., and Stanley, S. L., Jr. (1995). Interaction of laminin with *Entamoeba histolytica* cysteine proteinases and its effect on amebic pathogenesis. *Infect. Immun.* **63,** 4150–4153.

Li, L., and Dixon, J. E. (2000). Form, function, and regulation of protein tyrosine phosphatases and their involvement in human diseases. *Semin. Immunol.* **12,** 75–84.

Li, Y., Chen, Z. Y., Wang, W., Baker, C. C., and Krug, R. M. (2001). The 3′-end-processing factor CPSF is required for the splicing of single-intron pre-mRNAs *in vivo*. *RNA* **7,** 920–931.

Liepinsh, E., Andersson, M., Ruysschaert, J. M., and Otting, G. (1997). Saposin fold revealed by the NMR structure of NK-lysin. *Nat. Struct. Biol.* **4,** 793–795.

Linstead, D., and Cranshaw, M. A. (1983). The pathway of arginine catabolism in the parasitic flagellate *Trichomonas vaginalis*. *Mol. Biochem. Parasitol.* **8,** 241–252.

Lioutas, C., Schmetz, C., and Tannich, E. (1995). Identification of various circular DNA molecules in *Entamoeba histolytica*. *Exp. Parasitol.* **80,** 349–352.

Lioutas, C., and Tannich, E. (1995). Transcription of protein-coding genes in *Entamoeba histolytica* is insensitive to high concentrations of α-amanitin. *Mol. Biochem. Parasitol.* **73,** 259–261.

Lo, H., and Reeves, R. E. (1980). Purification and properties of NADPH:flavin oxidoreductase from *Entamoeba histolytica*. *Mol. Biochem. Parasitol.* **2,** 23–30.

Lo, H. S., and Reeves, R. E. (1978). Pyruvate-to-ethanol pathway in *Entamoeba histolytica*. *Biochem. J.* **171,** 225–230.

Lobelle-Rich, P. A., and Reeves, R. E. (1983). Separation and characterization of two UTP-utilizing hexose phosphate uridylyltransferases from *Entamoeba histolytica*. *Mol. Biochem. Parasitol.* **7,** 173–182.

Lockwood, B. C., and Coombs, G. H. (1991). Purification and characterization of methionine gamma-lyase from *Trichomonas vaginalis*. *Biochem. J.* **279,** 675–682.

Loftus, B., Anderson, I., Davies, R., Alsmark, U. C., Samuelson, J., Amedeo, P., Roncaglia, P., Berriman, M., Hirt, R. P., Mann, B. J., Nozaki, T., Suh, B., *et al.* (2005). The genome of the protist parasite *Entamoeba histolytica*. *Nature* **433,** 865–868.

Loftus, B. J., and Hall, N. (2005). *Entamoeba*: Still more to be learned from the genome. *Trends Parasitol.* **21,** 453.

Lohia, A., and Samuelson, J. (1993). Cloning of the *Eh cdc2* gene from *Entamoeba histolytica* encoding a protein kinase p34^{cdc2} homologue. *Gene* **127,** 203–207.

Lohia, A., Hait, N. C., and Lahiri Majumder, A. (1999). L-*myo*-inositol 1-phosphate synthase from *Entamoeba histolytica*. *Mol. Biochem. Parasitol.* **98,** 67–79.

Long-Krug, S. A., Fischer, K. J., Hysmith, R. M., and Ravdin, J. I. (1985). Phospholipase A enzymes of *Entamoeba histolytica*: Description and subcellular localization. *J. Infect. Dis.* **152,** 536–541.

Lu, L., Horstmann, H., Ng, C., and Hong, W. (2001). Regulation of Golgi structure and function by ARF-like protein 1 (Arl1). *J. Cell Sci.* **114,** 4543–4555.

Lu, L., Tai, G., and Hong, W. (2004). Autoantigen Golgin-97, an effector of Arl1 GTPase, participates in traffic from the endosome to the trans-golgi network. *Mol. Biol. Cell* **15,** 4426–4443.

Luaces, A. L., and Barrett, A. J. (1988). Affinity purification and biochemical characterization of histolysin, the major cysteine proteinase of *Entamoeba histolytica*. *Biochem. J.* **250,** 903–909.

Luaces, A. L., Pico, T., and Barrett, A. J. (1992). The ENZYMEBA test: Detection of intestinal *Entamoeba histolytica* infection by immuno-enzymatic detection of histolysain. *Parasitology* **105,** 203–205.

Lushbaugh, W. B., Hofbauer, A. F., and Pittman, F. E. (1984). Proteinase activities of *Entamoeba histolytica* cytotoxin. *Gastroenterology* **87,** 17–27.

Lushbaugh, W. B., Hofbauer, A. F., and Pittman, F. E. (1985). *Entamoeba histolytica*: Purification of cathepsin B. *Exp. Parasitol.* **59,** 328–336.

MacFarlane, R. C., and Singh, U. (2006). Identification of differentially expressed genes in virulent and nonvirulent *Entamoeba* species: Potential implications for amebic pathogenesis. *Infect. Immun.* **74,** 340–351.
Machida, Y. J., Hamlin, J. L., and Dutta, A. (2005). Right place, right time, and only once: Replication initiation in metazoans. *Cell* **123,** 13–24.
MacKelvie, S. H., Andrews, P. D., and Stark, M. J. (1995). The *Saccharomyces cerevisiae* gene SDS22 encodes a potential regulator of the mitotic function of yeast type 1 protein phosphatase. *Mol. Cell. Biol.* **15,** 3777–3785.
Mai, Z., Ghosh, S., Frisardi, M., Rosenthal, B., Rogers, R., and Samuelson, J. (1999). Hsp60 is targeted to a cryptic mitochondrion-derived organelle (''Crypton'') in the microaerophilic protozoan parasite *Entamoeba histolytica. Mol. Cell. Biol.* **19,** 2198–2205.
Maiorano, D., Lutzmann, M., and Mechali, M. (2006). MCM proteins and DNA replication. *Curr. Opin. Cell Biol.* **18,** 130–136.
Majoros, W. H., Pertea, M., and Salzberg, S. L. (2004). TigrScan and GlimmerHMM: Two open source ab initio eukaryotic gene-finders. *Bioinformatics* **20,** 2878–2879.
Makioka, A., Kumagai, M., Takeuchi, T., and Nozaki, T. (2006). Characterization of protein geranylgeranyltransferase I from the enteric protist *Entamoeba histolytica. Mol. Biochem. Parasitol.* **145,** 216–225.
Mallet, L., Renault, G., and Jacquet, M. (2000). Functional cloning of the adenylate cyclase gene of *Candida albicans* in *Saccharomyces cerevisiae* within a genomic fragment containing five other genes, including homologues of CHS6 and SAP185. *Yeast* **16,** 959–966.
Mallik, R., and Gross, S. P. (2004). Molecular motors: Strategies to get along. *Curr. Biol.* **14,** R971–R982.
Mandal, P. K., Bagchi, A., Bhattacharya, A., and Bhattacharya, S. (2004). An *Entamoeba histolytica* LINE/SINE pair inserts at common target sites cleaved by the restriction enzyme-like LINE-encoded endonuclease. *Eukaryot. Cell* **3,** 170–179.
Mann, B. J., Torian, B. E., Vedvick, T. S., and Petri, W. A., Jr. (1991). Sequence of a cysteine-rich galactose-specific lectin of *Entamoeba histolytica. Proc. Natl. Acad. Sci. USA* **88,** 3248–3252.
Marino, G., Uria, J. A., Puente, X. S., Quesada, V., Bordallo, J., and Lopez-Otin, C. (2003). Human autophagins, a family of cysteine proteinases potentially implicated in cell degradation by autophagy. *J. Biol. Chem.* **278,** 3671–3678.
Marion, S., Wilhelm, C., Voigt, H., Bacri, J. C., and Guillen, N. (2004). Overexpression of myosin IB in living *Entamoeba histolytica* enhances cytoplasm viscosity and reduces phagocytosis. *J. Cell Sci.* **117,** 3271–3279.
Marion, S., Laurent, C., and Guillen, N. (2005). Signalization and cytoskeleton activity through myosin IB during the early steps of phagocytosis in *Entamoeba histolytica*: A proteomic approach. *Cell. Microbiol.* **7,** 1504–1518.
Marsh, J. J., and Lebherz, H. G. (1992). Fructose-bisphosphate aldolases: An evolutionary history. *Trends Biochem. Sci.* **17,** 110–113.
Martin, W., and Müller, M. (1998). The hydrogen hypothesis for the first eukaryote. *Nature* **392,** 37–41.
Masai, H., and Arai, K. (2002). Cdc7 kinase complex: A key regulator in the initiation of DNA replication. *J. Cell. Physiol.* **190,** 287–296.
Massague, J., Blain, S. W., and Lo, R. S. (2000). TGFbeta signaling in growth control, cancer, and heritable disorders. *Cell* **103,** 295–309.
Maurer-Stroh, S., Washietl, S., and Eisenhaber, F. (2003). Protein prenyltransferases: Anchor size, pseudogenes and parasites. *Biol. Chem.* **384,** 977–989.
Mayer, A., Wickner, W., and Haas, A. (1996). Sec18p (NSF)-driven release of Sec17p (alpha-SNAP) can precede docking and fusion of yeast vacuoles. *Cell* **85,** 83–94.
Mazzuco, A., Benchimol, M., and De Souza, W. (1997). Endoplasmic reticulum and Golgi-like elements in *Entamoeba*. *Micron* **28,** 241–247.

McCarty, D. R., and Chory, J. (2000). Conservation and innovation in plant signaling pathways. *Cell* **103,** 201–209.
McCoy, J. J., Mann, B. J., Vedvick, T. S., Pak, Y., Heimark, D. B., and Petri, W. A., Jr. (1993). Structural analysis of the light subunit of the *Entamoeba histolytica* galactose-specific adherence lectin. *J. Biol. Chem.* **268,** 24223–24231.
McLaughlin, J., and Faubert, G. (1977). Partial purification and some properties of a neutral sulfhydryl and an acid proteinase from *Entamoeba histolytica*. *Can. J. Microbiol.* **23,** 420–425.
McTaggart, S. J. (2006). Isoprenylated proteins. *Cell. Mol. Life Sci.* **63,** 255–267.
Mehlotra, R. K. (1996). Antioxidant defense mechanisms in parasitic protozoa. *Crit. Rev. Microbiol.* **22,** 295–314.
Memon, A. R. (2004). The role of ADP-ribosylation factor and SAR1 in vesicular trafficking in plants. *Biochim. Biophys. Acta* **1664,** 9–30.
Merchant, A. M., Kawasaki, Y., Chen, Y., Lei, M., and Tye, B. K. (1997). A lesion in the DNA replication initiation factor Mcm10 induces pausing of elongation forks through chromosomal replication origins in *Saccharomyces cerevisiae*. *Mol. Cell. Biol.* **17,** 3261–3271.
Meyer, C., Zizioli, D., Lausmann, S., Eskelinen, E. L., Hamann, J., Saftig, P., von Figura, K., and Schu, P. (2000). mu1A-adaptin-deficient mice: Lethality, loss of AP-1 binding and rerouting of mannose 6-phosphate receptors. *EMBO J.* **19,** 2193–2203.
Meza, I., Sabanero, M., Cázares, F., and Bryan, J. (1983). Isolation and characterization of actin from *Entamoeba histolyica*. *J. Biol. Chem.* **258,** 3936–3941.
Mira, A., Pushker, R., Legault, B. A., Moreira, D., and Rodriguez-Valera, F. (2004). Evolutionary relationships of *Fusobacterium nucleatum* based on phylogenetic analysis and comparative genomics. *BMC Evol. Biol.* **4,** 50.
Mishra, C., Semino, C. E., McCreath, K. J., de la Vega, H., Jones, B. J., Specht, C. A., and Robbins, P. W. (1997). Cloning and expression of two chitin deacetylase genes of *Saccharomyces cerevisiae*. *Yeast* **13,** 327–336.
Moody-Haupt, S., Patterson, J. H., Mirelman, D., and McConville, M. J. (2000). The major surface antigens of *Entamoeba histolytica* trophozoites are GPI-anchored proteophosphoglycans. *J. Mol. Biol.* **297,** 409–420.
Morgan, D. O. (1995). Principles of CDK regulation. *Nature* **374,** 131–134.
Morgan, D. O. (1996). The dynamics of cyclin dependent kinase structure. *Curr. Opin. Cell Biol.* **8,** 767–772.
Motley, A., Bright, N. A., Seaman, M. N., and Robinson, M. S. (2003). Clathrin-mediated endocytosis in AP-2-depleted cells. *J. Cell Biol.* **162,** 909–918.
Müller, M. (1986). Reductive activation of nitroimidazoles in anaerobic microorganisms. *Biochem. Pharmacol.* **35,** 37–41.
Mullikin, J. C., and Ning, Z. (2003). The phusion assembler. *Genome Res.* **13,** 81–90.
Munford, R. S., Sheppard, P. O., and O'Hara, P. J. (1995). Saposin-like proteins (SAPLIP) carry out diverse functions on a common backbone structure. *J. Lipid Res.* **36,** 1653–1663.
Munro, S. (2004). Organelle identity and the organization of membrane traffic. *Nat. Cell Biol.* **6,** 469–472.
Myers, E. W., Sutton, G. G., Delcher, A. L., Dew, I. M., Fasulo, D. P., Flanigan, M. J., Kravitz, S. A., Mobarry, C. M., Reinert, K. H., Remington, K. A., Anson, E. L., Bolanos, R. A., *et al.* (2000). A whole-genome assembly of *Drosophila*. *Science* **287,** 2196–2204.
Nakada-Tsukui, K., Saito-Nakano, Y., Ali, V., and Nozaki, T. (2005). A retromerlike complex is a novel Rab7 effector that is involved in the transport of the virulence factor cysteine protease in the enteric protozoan parasite *Entamoeba histolytica*. *Mol. Biol. Cell* **16,** 5294–5303.
Nakamura, H. (2005). Thioredoxin and its related molecules: Update 2005. *Antioxid. Redox. Signal.* **7,** 823–828.
Nakamura, M., Yamada, M., Hirota, Y., Sugimoto, K., Oka, A., and Takanami, M. (1981). Nucleotide sequence of the asnA gene coding for asparagine synthetase of *E. coli* K-12. *Nucleic Acids Res.* **9,** 4669–4676.

Nakatsu, F., and Ohno, H. (2003). Adaptor protein complexes as the key regulators of protein sorting in the post-Golgi network. *Cell Struct. Funct.* **28,** 419–429.
Nebreda, A. R. (2006). CDK activation by non-cyclin proteins. *Curr. Opin. Cell Biol.* **18,** 192–198.
Nesbo, C. L., L'Haridon, S., Stetter, K. O., and Doolittle, W. F. (2001). Phylogenetic analyses of two ''archaeal'' genes in thermotoga maritima reveal multiple transfers between archaea and bacteria. *Mol. Biol. Evol.* **18,** 362–375.
Nickel, R., Jacobs, T., Urban, B., Scholze, H., Bruhn, H., and Leippe, M. (2000). Two novel calcium-binding proteins from cytoplasmic granules of the protozoan parasite *Entamoeba histolytica. FEBS Lett.* **486,** 112–116.
Nixon, J. E. J., Wang, A., Field, J., Morrison, H. G., McArthur, A. G., Sogin, M. L., Loftus, B. J., and Samuelson, J. (2002). Evidence for lateral transfer of genes encoding ferredoxins, nitroreductases, NADH oxidase, and alcohol dehydrogenase 3 from anaerobic prokaryotes to *Giardia lamblia* and *Entamoeba histolytica. Eukaryot. Cell* **1,** 181–190.
Novick, P., and Zerial, M. (1997). The diversity of Rab proteins in vesicle transport. *Curr. Opin. Cell Biol.* **9,** 496–504.
Nozaki, T., Arase, T., Shigeta, Y., Asai, T., Leustek, T., and Takeuchi, T. (1998a). Cloning and bacterial expression of adenosine-5′-triphosphate sulfurylase from the enteric protozoan parasite *Entamoeba histolytica. Biochim. Biophys. Acta* **1429,** 284–291.
Nozaki, T., Asai, T., Kobayashi, S., Ikegami, F., Noji, M., Saito, K., and Takeuchi, T. (1998b). Molecular cloning and characterization of the genes encoding two isoforms of cysteine synthase in the enteric protozoan parasite *Entamoeba histolytica. Mol. Biochem. Parasitol.* **97,** 33–44.
Nozaki, T., Asai, T., Sanchez, L. B., Kobayashi, S., Nakazawa, M., and Takeuchi, T. (1999). Characterization of the gene encoding serine acetyltransferase, a regulated enzyme of cysteine biosynthesis from the protist parasites *Entamoeba histolytica* and *Entamoeba dispar*. Regulation and possible function of the cysteine biosynthetic pathway in Entamoeba. *J. Biol. Chem.* **274,** 32445–32452.
Nozaki, T., Ali, V., and Tokoro, M. (2005). Sulfur-containing amino acid metabolism in parasitic protozoa. *Adv. Parasitol.* **60,** 1–99.
Ochman, H., Lawrence, J. G., and Groisman, E. A. (2000). Lateral gene transfer and the nature of bacterial innovation. *Nature* **405,** 299–304.
Okada, M., Huston, C. D., Mann, B. J., Petri, W. A., Jr., Kita, K., and Nozaki, T. (2005). Proteomic analysis of phagocytosis in the enteric protozoan parasite *Entamoeba histolytica. Eukaryot. Cell* **4,** 827–831.
Ondarza, R. N., Tamayo, E. M., Hurtado, G., Hernandez, E., and Iturbe, A. (1997). Isolation and purification of glutathionyl-spermidine and trypanothione from *Entamoeba histolytica. Arch. Med. Res.* **28,** 73–75.
Ondarza, R. N., Hurtado, G., Iturbe, A., Hernandez, E., Tamayo, E., and Woolery, M. (2005). Identification of trypanothione from the human pathogen *Entamoeba histolytica* by mass spectrometry and chemical analysis. *Biotechnol. Appl. Biochem.* **42,** 175–181.
Ooi, C. E., Dell'Angelica, E. C., and Bonifacino, J. S. (1998). ADP-Ribosylation factor 1 (ARF1) regulates recruitment of the AP-3 adaptor complex to membranes. *J. Cell Biol.* **142,** 391–402.
Opperdoes, F. R., and Michels, P. A. (2001). Enzymes of carbohydrate metabolism as potential drug targets. *Int. J. Parasitol.* **31,** 482–490.
Ortner, S., Plaimauer, B., Binder, M., Scheiner, O., Wiedermann, G., and Duchêne, M. (1995). Molecular analysis of two hexokinase isoenzymes from *Entamoeba histolytica. Mol. Biochem. Parasitol.* **73,** 189–198.
Ortner, S., Binder, M., Scheiner, O., Wiedermann, G., and Duchêne, M. (1997a). Molecular and biochemical characterization of phosphoglucomutases from *Entamoeba histolytica* and *Entamoeba dispar. Mol. Biochem. Parasitol.* **90,** 121–129.
Ortner, S., Clark, C. G., Binder, M., Scheiner, O., Wiedermann, G., and Duchêne, M. (1997b). Molecular biology of the hexokinase isoenzyme pattern that distinguishes pathogenic

Entamoeba histolytica from nonpathogenic *Entamoeba dispar*. *Mol. Biochem. Parasitol.* **86,** 85–94.

Ostoa-Saloma, P., Carrero, J. C., Petrossian, P., Herion, P., Landa, A., and Laclette, J. P. (2000). Cloning, characterization and functional expression of a cyclophilin of *Entamoeba histolytica*. *Mol. Biochem. Parasitol.* **107,** 219–225.

Otomo, T., Tomchick, D. R., Otomo, C., Panchal, S. C., Machius, M., and Rosen, M. K. (2005). Structural basis of actin filament nucleation and processive capping by a formin homology 2 domain. *Nature* **433,** 488–494.

Parsons, M. (2004). Glycosomes: Parasites and the divergence of peroxisomal purpose. *Mol. Microbiol.* **53,** 717–724.

Pasqualato, S., Renault, L., and Cherfils, J. (2002). Arf, Arl, Arp and Sar proteins: A family of GTP-binding proteins with a structural device for 'front-back' communication. *EMBO Rep.* **3,** 1035–1041.

Patil, C., and Walter, P. (2001). Intracellular signaling from the endoplasmic reticulum to the nucleus: The unfolded protein response in yeast and mammals. *Curr. Opin. Cell Biol.* **13,** 349–355.

Peng, R., De Antoni, A., and Gallwitz, D. (2000). Evidence for overlapping and distinct functions in protein transport of coat protein Sec24p family members. *J. Biol. Chem.* **275,** 11521–11528.

Pereira-Leal, J. B., and Seabra, M. C. (2001). Evolution of the Rab family of small GTP-binding proteins. *J. Mol. Biol.* **313,** 889–901.

Petoukhov, M. V., Svergun, D. I., Konarev, P. V., Ravasio, S., van den Heuvel, R. H., Curti, B., and Vanoni, M. A. (2003). Quaternary structure of Azospirillum brasilense NADPH-dependent glutamate synthase in solution as revealed by synchrotron radiation x-ray scattering. *J. Biol. Chem.* **278,** 29933–29939.

Petri, W. A., Haque, R., and Mann, B. J. (2002). The bittersweet interface of parasite and host: Lectin-carbohydrate interactions during human invasion by the parasite *Entamoeba histolytica*. *Annu. Rev. Microbiol.* **56,** 39–64.

Petri, W. A., Jr., Smith, R. D., Schlesinger, P. H., Murphy, C. F., and Ravdin, J. I. (1987). Isolation of the galactose-binding lectin that mediates the in vitro adherence of *Entamoeba histolytica*. *J. Clin. Invest.* **80,** 1238–1244.

Petri, W. A., Jr., Chapman, M. D., Snodgrass, T., Mann, B. J., Broman, J., and Ravdin, J. I. (1989). Subunit structure of the galactose and *N*-acetyl-D-galactosamine-inhibitable adherence lectin of *Entamoeba histolytica*. *J. Biol. Chem.* **264,** 3007–3012.

Petri, W. A. J. (2002). Pathogenesis of amebiasis. *Curr. Opin. Microbiol.* **5,** 443–447.

Pillai, D. R., Britten, D., Ackers, J. P., Ravdin, J. I., and Kain, K. C. (1997). A gene homologous to *hgl2* of *Entamoeba histolytica* is present and expressed in *Entamoeba dispar*. *Mol. Biochem. Parasitol.* **87,** 101–105.

Pillutla, R. C., Yue, Z., Maldonado, E., and Shatkin, A. J. (1998). Recombinant human mRNA cap methyltransferase binds capping enzyme/RNA polymerase IIo complexes. *J. Biol. Chem.* **273,** 21443–21446.

Poirier, M. A., Xiao, W., Macosko, J. C., Chan, C., Shin, Y. K., and Bennett, M. K. (1998). The synaptic SNARE complex is a parallel four-stranded helical bundle. *Nat. Struct. Biol.* **5,** 765–769.

Poole, L. B., Chae, H. Z., Flores, B. M., Reed, S. L., Rhee, S. G., and Torian, B. E. (1997). Peroxidase activity of a TSA-like antioxidant protein from a pathogenic amoeba. *Free Radic. Biol. Med.* **23,** 955–959.

Poon, R. Y., Lew, J., and Hunter, T. (1997). Identification of functional domains in the neuronal Cdk5 activator protein. *J. Biol. Chem.* **272,** 5703–5708.

Pritham, E. J., Feschotte, C., and Wessler, S. R. (2005). Unexpected diversity and differential success of DNA transposons in four species of *Entamoeba* protozoans. *Mol. Biol. Evol.* **22,** 1751–1763.

Purdy, J. E., Pho, L. T., Mann, B. J., and Petri, W. A., Jr. (1996). Upstream regulatory elements controlling expression of the *Entamoeba histolytica* lectin. *Mol. Biochem. Parasitol.* **78,** 91–103.

Que, X., and Reed, S. L. (2000). Cysteine proteinases and the pathogenesis of amebiasis. *Clin. Microbiol. Rev.* **13,** 196–206.

Quevillon, E., Spielmann, T., Brahimi, K., Chattopadhyay, D., Yeramian, E., and Langsley, G. (2003). The *Plasmodium falciparum* family of Rab GTPases. *Gene* **306,** 13–25.

Quon, D. V., Delgadillo, M. G., and Johnson, P. J. (1996). Transcription in the early diverging eukaryote *Trichomonas vaginalis*: An unusual RNA polymerase II and alpha-amanitin-resistant transcription of protein-coding genes. *J. Mol. Evol.* **43,** 253–262.

Radhakrishna, H., and Donaldson, J. G. (1997). ADP-ribosylation factor 6 regulates a novel plasma membrane recycling pathway. *J. Cell Biol.* **139,** 49–61.

Ramakrishnan, G., Lee, S., Mann, B. J., and Petri, W. A., Jr. (2000). *Entamoeba histolytica*: Deletion of the GPI anchor signal sequence on the Gal/GalNAc lectin light subunit prevents Its assembly into the lectin heterodimer. *Exp. Parasitol.* **96,** 57–60.

Ramakrishnan, G., Gilchrist, C. A., Musa, H., Torok, M. S., Grant, P. A., Mann, B. J., and Petri, W. A., Jr. (2004). Histone acetyltransferases and deacetylase in *Entamoeba histolytica. Mol. Biochem. Parasitol.* **138,** 205–216.

Ramos, F., and Wiame, J. M. (1982). Occurrence of a catabolic L-serine (L-threonine) deaminase in *Saccharomyces cerevisiae. Eur. J. Biochem.* **123,** 571–576.

Ramos, M. A., Mercado, G. C., Salgado, L. M., Sanchez-Lopez, R., Stock, R. P., Lizardi, P. M., and Alagón, A. (1997). *Entamoeba histolytica* contains a gene encoding a homologue to the 54 kDa subunit of the signal recognition particle. *Mol. Biochem. Parasitol.* **88,** 225–235.

Ramponi, G., and Stefani, M. (1997). Structure and function of the low Mr phosphotyrosine protein phosphatases. *Biochim. Biophys. Acta* **1341,** 137–156.

Randazzo, P. A., Terui, T., Sturch, S., Fales, H. M., Ferrige, A. G., and Kahn, R. A. (1995). The myristoylated amino terminus of ADP-ribosylation factor 1 is a phospholipid- and GTP-sensitive switch. *J. Biol. Chem.* **270,** 14809–14815.

Ravdin, J. I., and Guerrant, R. L. (1981). Role of adherence in cytopathogenic mechanisms of *Entamoeba histolytica.* Study with mammalian tissue culture cells and human erythrocytes. *J. Clin. Invest.* **68,** 1305–1313.

Ravdin, J. I., and Guerrant, R. L. (1982). Separation of adherence, cytolytic, and phagocytic events in the cytopathogenic mechanisms of *Entamoeba histolytica. Arch. Invest. Méd.* **13,** 123–128.

Ravdin, J. I., Murphy, C. F., Guerrant, R. L., and Long-Krug, S. A. (1985). Effect of antagonists of calcium and phospholipase A on the cytopathogenicity of *Entamoeba histolytica. J. Infect. Dis.* **152,** 542–549.

Reed, S., Bouvier, J., Sikes Pollack, A., Engel, J. C., Brown, M., Hirata, K., Que, X., Eakin, A., Hagblom, P., Gillin, F., and McKerrow, J. H. (1993). Cloning of a virulence factor of *Entamoeba histolytica.* Pathogenic strains possess a unique cysteine proteinase gene. *J. Clin. Invest.* **91,** 1532–1540.

Reed, S. I. (1992). The role of p34 kinases in the G1 to S-phase transition. *Annu. Rev. Cell Biol.* **8,** 529–561.

Reed, S. L., Keene, W. E., and McKerrow, J. H. (1989). Thiol proteinase expression and pathogenicity of *Entamoeba histolytica. J. Clin. Microbiol.* **27,** 2772–2777.

Reeves, R. E. (1968). A new enzyme with the glycolytic function of pyruvate kinase. *J. Biol. Chem.* **243,** 3202–3204.

Reeves, R. E. (1970). Phosphopyruvate carboxylase from *Entamoeba histolytica. Biochim. Biophys. Acta* **220,** 346–349.

Reeves, R. E., Montalvo, F. E., and Lushbaugh, T. S. (1971). Nicotinamide-adenine dinucleotide phosphate-dependent alcohol dehydrogenase. Enzyme from *Entamoeba histolytica* and some enzyme inhibitors. *Int. J. Biochem.* **2,** 55–64.

Reeves, R. E., and South, D. J. (1974). Phosphoglycerate kinase (GTP). An enzyme from *Entamoeba histolytica* selective for guanine nucleotides. *Biochem. Biophys. Res. Commun.* **58,** 1053–1057.

Reeves, R. E., Serrano, R., and South, D. J. (1976). 6-phosphofructokinase (pyrophosphate). Properties of the enzyme from *Entamoeba histolytica* and its reaction mechanism. *J. Biol. Chem.* **251,** 2958–2962.

Reeves, R. E., Warren, L. G., Susskind, B., and Lo, H. S. (1977). An energy-conserving pyruvate-to-acetate pathway in *Entamoeba histolytica*. Pyruvate synthase and a new acetate thiokinase. *J. Biol. Chem.* **252,** 726–731.

Reeves, R. E. (1984). Metabolism of *Entamoeba histolytica* Schaudinn, 1903. *Adv. Parasitol.* **23,** 105–142.

Reid, M. F., and Fewson, C. A. (1994). Molecular characterization of microbial alcohol dehydrogenases. *Crit. Rev. Microbiol.* **20,** 13–56.

Rhee, S. G., Chae, H. Z., and Kim, K. (2005). Peroxiredoxins: A historical overview and speculative preview of novel mechanisms and emerging concepts in cell signaling. *Free Radic. Biol. Med.* **38,** 1543–1552.

Richards, T. A., Hirt, R. P., Williams, B. A., and Embley, T. M. (2003). Horizontal gene transfer and the evolution of parasitic protozoa. *Protist* **154,** 17–32.

Roberg, K. J., Crotwell, M., Espenshade, P., Gimeno, R., and Kaiser, C. A. (1999). LST1 is a SEC24 homologue used for selective export of the plasma membrane ATPase from the endoplasmic reticulum. *J. Cell Biol.* **145,** 659–672.

Rodriguez-Romero, A., Hernandez-Santoyo, A., del Pozo Yauner, L., Kornhauser, A., and Fernandez-Velasco, D. A. (2002). Structure and inactivation of triosephosphate isomerase from *Entamoeba histolytica*. *J. Mol. Biol.* **322,** 669–675.

Rodriguez, M. A., Baez-Camargo, M., Delgadillo, D. M., and Orozco, E. (1996). Cloning and expression of an *Entamoeba histolytica* $NADP^+$-dependent alcohol dehydrogenase gene. *Biochim. Biophys. Acta* **1306,** 23–26.

Rohmer, M., Knani, M., Simonin, P., Sutter, B., and Sahm, H. (1993). Isoprenoid biosynthesis in bacteria: A novel pathway for the early steps leading to isopentenyl diphosphate. *Biochem. J.* **295,** 517–524.

Rosenbaum, R. M., and Wittner, M. (1970). Ultrastructure of bacterized and axenic trophozoites of *Entamoeba histolytica* with particular reference to helical bodies. *J. Cell Biol.* **45,** 367–382.

Rotte, C., Stejskal, F., Zhu, G., Keithly, J. S., and Martin, W. (2001). Pyruvate: $NADP^+$ oxidoreductase from the mitochondrion of *Euglena gracilis* and from the apicomplexan *Cryptosporidium parvum*: A biochemical relic linking pyruvate metabolism in mitochondriate and amitochondriate protists. *Mol. Biol. Evol.* **18,** 710–720.

Rowe, J., Calegari, F., Taverna, E., Longhi, R., and Rosa, P. (2001). Syntaxin 1A is delivered to the apical and basolateral domains of epithelial cells: The role of munc-18 proteins. *J. Cell Sci.* **114,** 3323–3332.

Roy, S. W., and Gilbert, W. (2005). The pattern of intron loss. *Proc. Natl. Acad. Sci. USA* **102,** 713–718.

Rusnak, F., and Mertz, P. (2000). Calcineurin: Form and function. *Physiol. Rev.* **80,** 1483–1521.

Saavedra, E., Olivos, A., Encalada, R., and Moreno-Sanchez, R. (2004). *Entamoeba histolytica*: Kinetic and molecular evidence of a previously unidentified pyruvate kinase. *Exp. Parasitol.* **106,** 11–21.

Saavedra, E., Encalada, R., Pineda, E., Jasso-Chavez, R., and Moreno-Sanchez, R. (2005). Glycolysis in *Entamoeba histolytica*. Biochemical characterization of recombinant glycolytic enzymes and flux control analysis. *FEBS J.* **272,** 1767–1783.

Saavedra Lira, E., Robinson, O., and Pérez Montfort, R. (1992). Partial nucleotide sequence of the enzyme pyruvate, orthophosphate dikinase of *Entamoeba histolytica* HM-1:IMSS. *Arch. Med. Res.* **23,** 39–40.

Sahoo, N., Labruyere, E., Bhattacharya, S., Sen, P., Guillen, N., and Bhattacharya, A. (2004). Calcium binding protein 1 of the protozoan parasite *Entamoeba histolytica* interacts with actin and is involved in cytoskeleton dynamics. *J. Cell Sci.* **117,** 3625–3634.

Said-Fernandez, S., and Lopez-Revilla, R. (1988). Free fatty acids released from phospholipids are the major heat-stable hemolytic factor of *Entamoeba histolytica* trophozoites. *Infect. Immun.* **56,** 874–879.

Saito-Nakano, Y., Nakazawa, M., Shigeta, Y., Takeuchi, T., and Nozaki, T. (2001). Identification and characterization of genes encoding novel Rab proteins from *Entamoeba histolytica*. *Mol. Biochem. Parasitol.* **116,** 219–222.

Saito-Nakano, Y., Yasuda, T., Nakada-Tsukui, K., Leippe, M., and Nozaki, T. (2004). Rab5-associated vacuoles play a unique role in phagocytosis of the enteric protozoan parasite *Entamoeba histolytica*. *J. Biol. Chem.* **279,** 49497–49507.

Saito-Nakano, Y., Loftus, B. J., Hall, N., and Nozaki, T. (2005). The diversity of Rab GTPases in *Entamoeba histolytica*. *Exp. Parasitol.* **110,** 244–252.

Salzberg, S. L., White, O., Peterson, J., and Eisen, J. A. (2001). Microbial genes in the human genome: Lateral transfer or gene loss? *Science* **292,** 1903–1906.

Samarawickrema, N. A., Brown, D. M., Upcroft, J. A., Thammapalerd, N., and Upcroft, P. (1997). Involvement of superoxide dismutase and pyruvate:ferredoxin oxidoreductase in mechanisms of metronidazole resistance in *Entamoeba histolytica*. *J. Antimicrob. Chemother.* **40,** 833–840.

Samuelson, J., Banerjee, S., Magnelli, P., Cui, J., Kelleher, D. J., Gilmore, R., and Robbins, P. W. (2005). The diversity of dolichol-linked precursors to Asn-linked glycans likely results from secondary loss of sets of glycosyltransferases. *Proc. Natl. Acad. Sci. USA* **102,** 1548–1553.

Sanchez, L., Horner, D., Moore, D., Henze, K., Embley, T., and Müller, M. (2002). Fructose-1,6-bisphosphate aldolases in amitochondriate protists constitute a single protein subfamily with eubacterial relationships. *Gene* **295,** 51–59.

Sanderfoot, A. A., Assaad, F. F., and Raikhel, N. V. (2000). The *Arabidopsis* genome. An abundance of soluble *N*-ethylmaleimide-sensitive factor adaptor protein receptors. *Plant Physiol.* **124,** 1558–1569.

Sargeaunt, P. G., Williams, J. E., and Grene, J. D. (1978). The differentiation of invasive and non-invasive *Entamoeba histolytica* by isoenzyme electrophoresis. *Trans. R. Soc. Trop. Med. Hyg.* **72,** 519–521.

Sargeaunt, P. G. (1987). The reliability of *Entamoeba histolytica* zymodemes in clinical diagnosis. *Parasitol. Today* **3,** 40–43.

Sarti, P., Fiori, P. L., Forte, E., Rappelli, P., Teixeira, M., Mastronicola, D., Sanciu, G., Giuffre, A., and Brunori, M. (2004). *Trichomonas vaginalis* degrades nitric oxide and expresses a flavorubredoxin-like protein: A new pathogenic mechanism? *Cell. Mol. Life Sci.* **61,** 618–623.

Sata, M., Donaldson, J. G., Moss, J., and Vaughan, M. (1998). Brefeldin A-inhibited guanine nucleotide-exchange activity of Sec7 domain from yeast Sec7 with yeast and mammalian ADP ribosylation factors. *Proc. Natl. Acad. Sci. USA* **95,** 4204–4208.

Satish, S., Bakre, A. A., Bhattacharya, S., and Bhattacharya, A. (2003). Stress-dependent expression of a polymorphic, charged antigen in the protozoan parasite *Entamoeba histolytica*. *Infect. Immun.* **71,** 4472–4486.

Sawyer, M. K., Bischoff, J. M., Guidry, M. A., and Reeves, R. E. (1967). Lipids from *Entamoeba histolytica*. *Exp. Parasitol.* **20,** 295–302.

Schimmoller, F., Singer-Kruger, B., Schroder, S., Kruger, U., Barlowe, C., and Riezman, H. (1995). The absence of Emp24p, a component of ER-derived COPII-coated vesicles, causes a defect in transport of selected proteins to the Golgi. *EMBO J.* **14,** 1329–1339.

Schlessinger, J. (2000). Cell signaling by receptor tyrosine kinases. *Cell* **103,** 211–225.

Schofield, P. J., and Edwards, M. R. (1994). Biochemistry—is *Giardia* opportunistic in its use of substrates? *In* "Giardia: From Molecules to Disease" (R. C. A. Thompson, J. A. Reynoldson, and A. J. Lymbery, eds.), pp. 171–183. CAB International, Wallingford, UK.

Scholze, H., and Schulte, W. (1988). On the specificity of a cysteine proteinase from *Entamoeba histolytica. Biomed. Biochim. Acta* **47,** 115–123.

Scholze, H., Lohden-Bendinger, U., Müller, G., and Bakker-Grunwald, T. (1992). Subcellular distribution of amebapain, the major cysteine proteinase of *Entamoeba histolytica. Arch. Med. Res.* **23,** 105–108.

Schroeder, S. C., Schwer, B., Shuman, S., and Bentley, D. (2000). Dynamic association of capping enzymes with transcribing RNA polymerase II. *Genes Dev.* **14,** 2435–2440.

Schroepfer, G. J., Jr. (1981). Sterol biosynthesis. *Annu. Rev. Biochem.* **50,** 585–621.

Schulte, W., and Scholze, H. (1989). Action of the major protease from *Entamoeba histolytica* on proteins of the extracellular matrix. *J. Protozool.* **36,** 538–543.

Seabra, M. C., Mules, E. H., and Hume, A. N. (2002). Rab GTPases, intracellular traffic and disease. *Trends Mol. Med.* **8,** 23–30.

Shah, P. H., MacFarlane, R. C., Bhattacharya, D., Matese, J. C., Demeter, J., Stroup, S. E., and Singh, U. (2005). Comparative genomic hybridizations of *Entamoeba* strains reveal unique genetic fingerprints that correlate with virulence. *Eukaryot. Cell* **4,** 504–515.

Sharma, R., Azam, A., Bhattacharya, S., and Bhattacharya, A. (1999). Identification of novel genes of non-pathogenic *Entamoeba dispar* by expressed sequence tag analysis. *Mol. Biochem. Parasitol.* **99,** 279–285.

Shaywitz, D. A., Espenshade, P. J., Gimeno, R. E., and Kaiser, C. A. (1997). COPII subunit interactions in the assembly of the vesicle coat. *J. Biol. Chem.* **272,** 25413–25416.

Shen, Z., and Jacobs-Lorena, M. (1999). Evolution of chitin-binding proteins in invertebrates. *J. Mol. Evol.* **48,** 341–347.

Shire, A. M., and Ackers, J. P. (2007). SINE elements of *Entamoeba dispar. Mol. Biochem. Parasitol.* **152,** 47–52.

Shiu, S. H., and Bleecker, A. B. (2001). Receptor-like kinases from *Arabidopsis* form a monophyletic gene family related to animal receptor kinases. *Proc. Natl. Acad. Sci. USA* **98,** 10763–10768.

Shoemaker, N. B., Vlakamis, H., and Slyers, A. A. (2001). Evidence for extensive resistance gene transfer among *Bacteroides* spp., and among *Bacteroides* and other genera in the human colon. *Appl. Environ. Microbiol.* **67,** 561–568.

Sies, H. (1999). Glutathione and its role in cellular functions. *Free Radic. Biol. Med.* **27,** 916–921.

Silva, P. P., Martinez-Palomo, A., and Gonzalez-Robles, A. (1975). Membrane structure and surface coat of *Entamoeba histolytica.* Topochemistry and dynamics of the cell surface: Cap formation and microexudate. *J. Cell Biol.* **64,** 538–550.

Simanis, V., and Nurse, P. (1986). The cell cycle control gene cdc2^{+}of fission yeast encodes a protein kinase potentially regulated by phosphorylation. *Cell* **45,** 261–268.

Singh, U., Purdy, J., Mann, B. J., and Petri, W. A., Jr. (1997). Three conserved cis-acting sequences in the core promoter control gene expression in the protozoan parasite *Entamoeba histolytica. Arch. Med. Res.* **28**(Spec No), 41–42.

Singh, U., and Rogers, J. B. (1998). The novel core promoter element GAAC in the hgl5 gene of *Entamoeba histolytica* is able to direct a transcription start site independent of TATA or initiator regions. *J. Biol. Chem.* **273,** 21663–21668.

Singh, U., Gilchrist, C. A., Schaenman, J. M., Rogers, J. B., Hockensmith, J. W., Mann, B. J., and Petri, W. A. (2002). Context-dependent roles of the *Entamoeba histolytica* core promoter element GAAC in transcriptional activation and protein complex assembly. *Mol. Biochem. Parasitol.* **120,** 107–116.

Snell, K. (1986). The duality of pathways for serine biosynthesis is a fallacy. *Trends Biochem. Sci.* **11,** 241–243.

Springer, S., Chen, E., Duden, R., Marzioch, M., Rowley, A., Hamamoto, S., Merchant, S., and Schekman, R. (2000). The p24 proteins are not essential for vesicular transport in *Saccharomyces cerevisiae*. *Proc. Natl. Acad. Sci. USA* **97,** 4034–4039.
Srivastava, G., Anand, M. T., Bhattacharya, S., and Bhattacharya, A. (1995). Lipophosphoglycan is present in distinctly different form in different *Entamoeba histolytica* strains and absent in *Entamoeba moshkovskii* and *Entamoeba invadens*. *J. Eukaryot. Microbiol.* **42,** 617–622.
Stamnes, M. A., and Rothman, J. E. (1993). The binding of AP-1 clathrin adaptor particles to Golgi membranes requires ADP-ribosylation factor, a small GTP-binding protein. *Cell* **73,** 999–1005.
Stanley, S. L., Jr., Zhang, T., Rubin, D., and Li, E. (1995). Role of the *Entamoeba histolytica* cysteine proteinase in amebic liver abscess formation in severe combined immunodeficient mice. *Infect. Immun.* **63,** 1587–1590.
Stark, M. J. (1996). Yeast protein serine/threonine phosphatases: Multiple roles and diverse regulation. *Yeast* **12,** 1647–1675.
Strauss, E., Kinsland, C., Ge, Y., McLafferty, F. W., and Begley, T. P. (2001). Phosphopantothenoylcysteine synthetase from *Escherichia coli*. Identification and characterization of the last unidentified coenzyme A biosynthetic enzyme in bacteria. *J. Biol. Chem.* **276,** 13513–13516.
Surana, U., Robitsch, H., Price, C., Schuster, T., Fitch, I., Futcher, A. B., and Nasmyth, K. (1991). The role of CDC28 and cyclins during mitosis in the budding yeast *S. cerevisiae*. *Cell* **65,** 145–161.
Susskind, B. M., Warren, L. G., and Reeves, R. E. (1982). A pathway for the interconversion of hexose and pentose in the parasitic amoeba *Entamoeba histolytica*. *Biochem. J.* **204,** 191–196.
Swain, A. L., Jaskolski, M., Housset, D., Rao, J. K., and Wlodawer, A. (1993). Crystal structure of *Escherichia coli* L-asparaginase, an enzyme used in cancer therapy. *Proc. Natl. Acad. Sci. USA* **90,** 1474–1478.
Sykes, D. E., and Band, R. N. (1977). Superoxide dismutase and peroxide activity of *Acanthamoeba* and two anaerobic *Entamoeba* species. *J. Cell Biol.* **75,** 85a.
Takeuchi, T., Weinbach, E. C., and Diamond, L. S. (1977). *Entamoeba histolytica*: Localization and characterization of phosphorylase and particulate glycogen. *Exp. Parasitol.* **43,** 107–114.
Takeuchi, T., Weinbach, E. C., Gottlieb, M., and Diamond, L. S. (1979). Mechanism of L-serine oxidation in *Entamoeba histolytica*. *Comp. Biochem. Physiol. B Comp. Biochem.* **62,** 281–285.
Talamas-Rohana, P., Aguirre-Garcia, M. M., Anaya-Ruiz, M., and Rosales-Encina, J. L. (1999). *Entamoeba dispar* contains but does not secrete acid phosphatase as does *Entamoeba histolytica*. *Exp. Parasitol.* **92,** 219–222.
Tannich, E., Bruchhaus, I., Walter, R. D., and Horstmann, R. D. (1991a). Pathogenic and nonpathogenic *Entamoeba histolytica*: Identification and molecular cloning of an iron-containing superoxide dismutase. *Mol. Biochem. Parasitol.* **49,** 61–71.
Tannich, E., Ebert, F., and Horstmann, R. D. (1991b). Primary structure of the 170-kDa surface lectin of pathogenic *Entamoeba histolytica*. *Proc. Natl. Acad. Sci. USA* **88,** 1849–1853.
Tannich, E., Scholze, H., Nickel, R., and Horstmann, R. D. (1991c). Homologous cysteine proteinases of pathogenic and nonpathogenic *Entamoeba histolytica*. Differences in structure and expression. *J. Biol. Chem.* **266,** 4798–4803.
Tannich, E., Nickel, R., Buss, H., and Horstmann, R. D. (1992). Mapping and partial sequencing of the genes coding for two different cysteine proteinases in pathogenic *Entamoeba histolytica*. *Mol. Biochem. Parasitol.* **54,** 109–111.
Temesvari, L. A., Harris, E. N., Stanley, J. S. L., and Cardellia, J. A. (1999). Early and late endosomal compartments of *Entamoeba histolytica* are enriched in cysteine proteases, acid phosphatase and several Ras-related Rab GTPases. *Mol. Biochem. Parasitol.* **103,** 225–241.
Todd, J., Post-Beittenmiller, D., and Jaworski, J. G. (1999). KCS1 encodes a fatty acid elongase 3-ketoacyl-CoA synthase affecting wax biosynthesis in *Arabidopsis thaliana*. *Plant J.* **17,** 119–130.

Tokoro, M., Asai, T., Kobayashi, S., Takeuchi, T., and Nozaki, T. (2003). Identification and characterization of two isoenzymes of methionine γ-lyase from *Entamoeba histolytica*: A key enzyme of sulfur-amino acid degradation in an anaerobic parasitic protist that lacks forward and reverse trans-sulfuration pathways. *J. Biol. Chem.* **278,** 42717–42727.

Toonen, R. F., and Verhage, M. (2003). Vesicle trafficking: Pleasure and pain from SM genes. *Trends Cell Biol.* **13,** 177–186.

Torii, S., Takeuchi, T., Nagamatsu, S., and Izumi, T. (2004). Rab27 effector granuphilin promotes the plasma membrane targeting of insulin granules via interaction with syntaxin 1a. *J. Biol. Chem.* **279,** 22532–22538.

Touz, M. C., Nores, M. J., Slavin, I., Carmona, C., Conrad, J. T., Mowatt, M. R., Nash, T. E., Coronel, C. E., and Lujan, H. D. (2002). The activity of a developmentally regulated cysteine proteinase is required for cyst wall formation in the primitive eukaryote *Giardia lamblia*. *J. Biol. Chem.* **277,** 8474–8481.

Tovar, J., Fischer, A., and Clark, C. G. (1999). The mitosome, a novel organelle related to mitochondria in the amitochondriate parasite *Entamoeba histolytica*. *Mol. Microbiol.* **32,** 1013–1021.

Trilla, J. A., Duran, A., and Roncero, C. (1999). Chs7p, a new protein involved in the control of protein export from the endoplasmic reticulum that is specifically engaged in the regulation of chitin synthesis in *Saccharomyces cerevisiae*. *J. Cell Biol.* **145,** 1153–1163.

Trombetta, E. S., and Parodi, A. J. (2003). Quality control and protein folding in the secretory pathway. *Annu. Rev. Cell Dev. Biol.* **19,** 649–676.

Tye, B. K. (1999). MCM proteins in DNA replication. *Annu. Rev. Biochem.* **68,** 649–686.

Uemura, T., Ueda, T., Ohniwa, R. L., Nakano, A., Takeyasu, K., and Sato, M. H. (2004). Systematic analysis of SNARE molecules in *Arabidopsis*: Dissection of the post-Golgi network in plant cells. *Cell Struct. Funct.* **29,** 49–65.

Uwanogho, D. A., Hardcastle, Z., Balogh, P., Mirza, G., Thornburg, K. L., Ragoussis, J., and Sharpe, P. T. (1999). Molecular cloning, chromosomal mapping, and developmental expression of a novel protein tyrosine phosphatase-like gene. *Genomics* **62,** 406–416.

Van Dellen, K., Field, J., Wang, Z., Loftus, B., and Samuelson, J. (2002a). LINEs and SINE-like elements of the protist *Entamoeba histolytica*. *Gene* **297,** 229–239.

Van Dellen, K., Ghosh, S. K., Robbins, P. W., Loftus, B., and Samuelson, J. (2002b). *Entamoeba histolytica* lectins contain unique 6-Cys or 8-Cys chitin-binding domains. *Infect. Immun.* **70,** 3259–3263.

Van Dellen, K. L., Bulik, D. A., Specht, C. A., Robbins, P. W., and Samuelson, J. C. (2006a). Heterologous expression of an *Entamoeba histolytica* chitin synthase in *Saccharomyces cerevisiae*. *Eukaryot. Cell* **5,** 203–206.

Van Dellen, K. L., Chatterjee, A., Ratner, D. M., Magnelli, P. E., Cipollo, J. F., Steffen, M., Robbins, P. W., and Samuelson, J. (2006b). Unique posttranslational modifications of chitin-binding lectins of *Entamoeba invadens* cyst walls. *Eukaryot. Cell* **5,** 836–848.

van der Giezen, M., Cox, S., and Tovar, J. (2004). The iron–sulfur cluster assembly genes iscS and iscU of *Entamoeba histolytica* were acquired by horizontal gene transfer. *BMC Evol. Biol.* **4,** 7.

Van Mullem, V., Wery, M., Werner, M., Vandenhaute, J., and Thuriaux, P. (2002). The Rpb9 subunit of RNA polymerase II binds transcription factor TFIIE and interferes with the SAGA and elongator histone acetyltransferases. *J. Biol. Chem.* **277,** 10220–10225.

Vanacova, S., Liston, D. R., Tachezy, J., and Johnson, P. J. (2003). Molecular biology of the amitochondriate parasites, *Giardia intestinalis*, *Entamoeba histolytica* and *Trichomonas vaginalis*. *Int. J. Parasitol.* **33,** 235–255.

Vandepoele, K., Raes, J., De Veylder, L., Rouze, P., Rombauts, S., and Inze, D. (2002). Genome-wide analysis of core cell cycle genes in *Arabidopsis*. *Plant Cell* **14,** 903–916.

Varela-Gomez, M., Moreno-Sanchez, R., Pardo, J. P., and Perez-Montfort, R. (2004). Kinetic mechanism and metabolic role of pyruvate phosphate dikinase from *Entamoeba histolytica*. *J. Biol. Chem.* **279,** 54124–54130.

Vargas-Villarreal, J., Olvera-Rodriguez, A., Mata-Cardenas, B. D., Martinez-Rogriguez, H. G., Said-Fernandez, S., and Alagon-Cano, A. (1998). Isolation of an *Entamoeba histolytica* intracellular alkaline phospholipase A2. *Parasitol. Res.* **84,** 310–314.
Vargas, M., Sansonetti, P., and Guillen, N. (1996). Identification and cellular localization of the actin-binding protein ABP-120 from *Entamoeba histolytica*. *Mol. Microbiol.* **22,** 849–857.
Vartiainen, M. K., and Machesky, L. M. (2004). The WASP-Arp2/3 pathway: Genetic insights. *Curr. Opin. Cell Biol.* **16,** 174–181.
Vats, D., Vishwakarma, R. A., Bhattacharya, S., and Bhattacharya, A. (2005). Reduction of cell surface glycosylphosphatidylinositol conjugates in *Entamoeba histolytica* by antisense blocking of *E. histolytica* GlcNAc-phosphatidylinositol deacetylase expression: Effect on cell proliferation, endocytosis, and adhesion to target cells. *Infect. Immun.* **73,** 8381–8392.
Vayssie, L., Vargas, M., Weber, C., and Guillen, N. (2004). Double-stranded RNA mediates homology-dependent gene silencing of gamma-tubulin in the human parasite *Entamoeba histolytica*. *Mol. Biochem. Parasitol.* **138,** 21–28.
Villalobo, E., Wender, N., and Mirelman, D. (2005). *Entamoeba histolytica*: Molecular characterization of an aldose 1-epimerase (mutarotase). *Exp. Parasitol.* **110,** 298–302.
Vishwakarma, R. A., Anand, M. T., Arya, R., Vats, D., and Bhattacharya, A. (2006). Glycosylated inositol phospholipid from *Entamoeba histolytica*: Identification and structural characterization. *Mol. Biochem. Parasitol.* **145,** 121–124.
Volpicelli-Daley, L. A., Li, Y., Zhang, C. J., and Kahn, R. A. (2005). Isoform-selective effects of the depletion of ADP-ribosylation factors 1–5 on membrane traffic. *Mol. Biol. Cell* **16,** 4495–4508.
Vowels, J. J., and Payne, G. S. (1998). A dileucine-like sorting signal directs transport into an AP-3-dependent, clathrin-independent pathway to the yeast vacuole. *EMBO J.* **17,** 2482–2493.
Walenta, J. H., Didier, A. J., Liu, X., and Kramer, H. (2001). The Golgi-associated hook3 protein is a member of a novel family of microtubule-binding proteins. *J. Cell Biol.* **152,** 923–934.
Wang, C. C., and Cheng, H. W. (1984). The deoxyribonucleoside phosphotransferase of *Trichomonas vaginalis*. A potential target for anti-trichomonial chemotherapy. *J. Exp. Med.* **160,** 987–1000.
Wang, Z., Samuelson, J., Clark, C. G., Eichinger, D., Paul, J., Van Dellen, K., Hall, N., Anderson, I., and Loftus, B. (2003). Gene discovery in the *Entamoeba invadens* genome. *Mol. Biochem. Parasitol.* **129,** 23–31.
Wassmann, C., Hellberg, A., Tannich, E., and Bruchhaus, I. (1999). Metronidazole resistance in the protozoan parasite *Entamoeba histolytica* is associated with increased expression of iron-containing superoxide dismutase and peroxiredoxin and decreased expression of ferredoxin 1 and flavin reductase. *J. Biol. Chem.* **274,** 26051–26056.
Watson, W. H., Yang, X., Choi, Y. E., Jones, D. P., and Kehrer, J. P. (2004). Thioredoxin and its role in toxicology. *Toxicol. Sci.* **78,** 3–14.
Weber, C., Guigon, G., Bouchier, C., Frangeul, L., Moreira, S., Sismeiro, O., Gouyette, C., Mirelman, D., Coppee, J. Y., and Guillen, N. (2006). Stress by heat shock induces massive down regulation of genes and allows differential allelic expression of the Gal/GalNAc lectin in *Entamoeba histolytica*. *Eukaryot. Cell* **5,** 871–875.
Wegmann, D., Hess, P., Baier, C., Wieland, F. T., and Reinhard, C. (2004). Novel isotypic gamma/zeta subunits reveal three coatomer complexes in mammals. *Mol. Cell. Biol.* **24,** 1070–1080.
Weinbach, E. C., and Diamond, L. S. (1974). *Entamoeba histolytica*. I. Aerobic metabolism. *Exp. Parasitol.* **35,** 232–243.
Weinberg, M. V., Jenney, F. E., Jr., Cui, X., and Adams, M. W. (2004). Rubrerythrin from the hyperthermophilic archaeon Pyrococcus furiosus is a rubredoxin-dependent, iron-containing peroxidase. *J. Bacteriol.* **186,** 7888–7895.

Wennerberg, K., Rossman, K. L., and Der, C. J. (2005). The Ras superfamily at a glance. *J. Cell Sci.* **118,** 843–846.
Wera, S., and Hemmings, B. A. (1995). Serine/threonine protein phosphatases. *Biochem. J.* **311,** 17–29.
Werries, E., and Thurn, I. (1989). Breakdown of glucopolysaccharides in *Entamoeba histolytica* by phosphorylase. *J. Protozool.* **36,** 607–612.
Werries, E., Franz, A., and Geisemeyer, S. (1990). Detection of glycogen-debranching system in trophozoites of *Entamoeba histolytica*. *J. Protozool.* **37,** 576–580.
West, C. M. (2003). Comparative analysis of spore coat formation, structure, and function in *Dictyostelium*. *Int. Rev. Cytol.* **222,** 237–293.
West, C. M., van der Wel, H., Coutinho, P. M., and Henrissat, B. (2005). Glycosyltransferase genomics in *Dictyostelium*. *In* "Dictyostelium Genomics" (W. F. Loomis and A. Kuspa, eds.), pp. 235–264. Horizon Bioscience, Norfolk, UK.
Whitney, J. A., Gomez, M., Sheff, D., Kreis, T. E., and Mellman, I. (1995). Cytoplasmic coat proteins involved in endosome function. *Cell* **83,** 703–713.
WHO (1998). "The world health report 1998. Life in the 21st century: A vision for all." World Health Organization, Geneva, Switzerland.
Willhoeft, U., Buss, H., and Tannich, E. (1999). Analysis of cDNA expressed sequence tags from *Entamoeba histolytica*: Identification of two highly abundant polyadenylated transcripts with no overt open reading frames. *Protist* **150,** 61–70.
Willhoeft, U., and Tannich, E. (1999). The electrophoretic karyotype of *Entamoeba histolytica*. *Mol. Biochem. Parasitol.* **99,** 41–53.
Willhoeft, U., Buss, H., and Tannich, E. (2002). The abundant polyadenylated transcript 2 DNA sequence of the pathogenic protozoan parasite *Entamoeba histolytica* represents a nonautonomous non-long-terminal-repeat retrotransposon-like element which is absent in the closely related nonpathogenic species *Entamoeba dispar*. *Infect. Immun.* **70,** 6798–6804.
Winder, S. J., and Ayscough, K. R. (2005). Actin-binding proteins. *J. Cell Sci.* **118,** 651–654.
Winkelmann, J., Leippe, M., and Bruhn, H. (2006). A novel saposin-like protein of *Entamoeba histolytica* with membrane-fusogenic activity. *Mol. Biochem. Parasitol.* **147,** 85–94.
Wittenberg, C., Sugimoto, K., and Reed, S. I. (1990). G1-specific cyclins of *S. cerevisiae*: Cell cycle periodicity, regulation by mating pheromone, and association with the p34CDC28 protein kinase. *Cell* **62,** 225–237.
Wohlschlegel, J. A., Dhar, S. K., Prokhorova, T. A., Dutta, A., and Walter, J. C. (2002). Xenopus Mcm10 binds to origins of DNA replication after Mcm2–7 and stimulates origin binding of Cdc45. *Mol. Cell* **9,** 233–240.
Wood, Z. A., Poole, L. B., and Karplus, P. A. (2003a). Peroxiredoxin evolution and the regulation of hydrogen peroxide signaling. *Science* **300,** 650–653.
Wood, Z. A., Schroder, E., Robin Harris, J., and Poole, L. B. (2003b). Structure, mechanism and regulation of peroxiredoxins. *Trends Biochem. Sci.* **28,** 32–40.
Woods, S. A., Schwartzbach, S. D., and Guest, J. R. (1988). Two biochemically distinct classes of fumarase in *Escherichia coli*. *Biochim. Biophys. Acta* **954,** 14–26.
Wostmann, C., Tannich, E., and Bakker-Grunwald, T. (1992). Ubiquitin of *Entamoeba histolytica* deviates in 6 amino acid residues from the consensus of all other known ubiquitins. *FEBS Lett.* **308,** 54–58.
Wu, G., and Müller, M. (2003). Glycogen phosphorylase sequences from the amitochondriate protists, *Trichomonas vaginalis, Mastigamoeba balamuthi, Entamoeba histolytica* and *Giardia intestinalis*. *J. Eukaryot. Microbiol.* **50,** 366–372.
Xu, Y., Moseley, J. B., Sagot, I., Poy, F., Pellman, D., Goode, B. L., and Eck, M. J. (2004). Crystal structures of a Formin Homology-2 domain reveal a tethered dimer architecture. *Cell* **116,** 711–723.

Yadava, N., Chandok, M. R., Prasad, J., Bhattacharya, S., Sopory, S. K., and Bhattacharya, A. (1997). Characterization of EhCaBP, a calcium binding protein of *Entamoeba histolytica* and its binding proteins. *Mol. Biochem. Parasitol.* **84,** 69–82.

Yamawaki-Kataoka, Y., Tamaoki, T., Choe, H. R., Tanaka, H., and Kataoka, T. (1989). Adenylate cyclases in yeast: A comparison of the genes from *Schizosaccharomyces pombe* and *Saccharomyces cerevisiae*. *Proc. Natl. Acad. Sci. USA* **86,** 5693–5697.

Yang, B., Steegmaier, M., Gonzalez, L. C., Jr., and Scheller, R. H. (2000). nSec1 binds a closed conformation of syntaxin1A. *J. Cell Biol.* **148,** 247–252.

Yang, W., Li, E., Kairong, T., and Stanley, S. L., Jr. (1994). *Entamoeba histolytica* has an alcohol dehydrogenase homologous to the multifunctional *adhE* gene product of *Escherichia coli*. *Mol. Biochem. Parasitol.* **64,** 253–260.

Yasui, K., Ishiguro, A., and Ishihama, A. (1998). Location of subunit-subunit contact sites on RNA polymerase II subunit 3 from the fission yeast *Schizosaccharomyces pombe*. *Biochemistry* **37,** 5542–5548.

Yeo, M., Lin, P. S., Dahmus, M. E., and Gill, G. N. (2003). A novel RNA polymerase II C-terminal domain phosphatase that preferentially dephosphorylates serine 5. *J. Biol. Chem.* **278,** 26078–26085.

Yin, Q. Y., de Groot, P. W., Dekker, H. L., de Jong, L., Klis, F. M., and de Koster, C. G. (2005). Comprehensive proteomic analysis of *Saccharomyces cerevisiae* cell walls: Identification of proteins covalently attached via glycosylphosphatidylinositol remnants or mild alkali-sensitive linkages. *J. Biol. Chem.* **280,** 20894–20901.

Yoshihisa, T., Barlowe, C., and Schekman, R. (1993). Requirement for a GTPase-activating protein in vesicle budding from the endoplasmic reticulum. *Science* **259,** 1466–1468.

Young, R. A. (1991). RNA polymerase II. *Annu. Rev. Biochem.* **60,** 689–715.

Yu, X., and Warner, J. R. (2001). Expression of a micro-protein. *J. Biol. Chem.* **276,** 33821–33825.

Zerial, M., and McBride, H. (2001). Rab proteins as membrane organizers. *Nat. Rev. Mol. Cell Biol.* **2,** 107–117.

Zhang, J. (2000). Protein-length distributions for the three domains of life. *Trends Genet.* **16,** 107–109.

Zhang, W. W., Shen, P. S., Descoteaux, S., and Samuelson, J. (1994). Cloning and expression of the gene for an NADP$^+$-dependent aldehyde dehydrogenase of *Entamoeba histolytica*. *Mol. Biochem. Parasitol.* **63,** 157–161.

Zhang, Z., Wang, L., Seydel, K. B., Li, E., Ankri, S., Mirelman, D., and Stanley, S. L., Jr. (2000). *Entamoeba histolytica* cysteine proteinases with interleukin-1 beta converting enzyme (ICE) activity cause intestinal inflammation and tissue damage in amoebiasis. *Mol. Microbiol.* **37,** 542–548.

Zhang, Z., Duchene, M., and Stanley, S. L., Jr. (2002). A monoclonal antibody to the amebic lipophosphoglycan-proteophosphoglycan antigens can prevent disease in human intestinal xenografts infected with *Entamoeba histolytica*. *Infect. Immun.* **70,** 5873–5876.

Zhang, Z. Y., and VanEtten, R. L. (1991). Pre-steady-state and steady-state kinetic analysis of the low molecular weight phosphotyrosyl protein phosphatase from bovine heart. *J. Biol. Chem.* **266,** 1516–1525.

Zuo, X., and Coombs, G. H. (1995). Amino acid consumption by the parasitic, amoeboid protists *Entamoeba histolytica* and *E. invadens*. *FEMS Microbiol. Lett.* **130,** 253–258.

CHAPTER 3

Epidemiological Modelling for Monitoring and Evaluation of Lymphatic Filariasis Control

Edwin Michael,* **Mwele N. Malecela-Lazaro,**† and **James W. Kazura**‡

Contents

* Department of Infectious Disease Epidemiology, Imperial College London, Norfolk Place, London W2 1PG, United Kingdom
† National Institute for Medical Research, Dar es Salaam, Tanzania
‡ Center for Global Health & Diseases, Case Western Reserve University, Cleveland, Ohio 44106–7286, USA

Advances in Parasitology, Volume 65
ISSN 0065-308X DOI: 10.1016/S0065-308X(07)65003-9

Abstract

Monitoring and evaluation are critically important to the scientific management of any parasite control programme. If a management action is prescribed, monitoring plays a pivotal role in assessing the effectiveness of implemented actions, identifying necessary adaptations for management and determining when management objectives are achieved. Here, we focus on the control of the vector-borne parasitic disease, lymphatic filariasis, to show how mathematical models of parasite transmission can provide a scientific framework for supporting the optimal design of parasite control monitoring programmes by their ability to (1) enable the estimation of endpoint targets, (2) provide information on expected trends in infection due to interventions to allow rational evaluation of intervention effectiveness and calculation of the frequency of monitoring, (3) support the selection of indicators that permit reliable and statistically powerful measurement of the effects of interventions while minimizing costs and sampling intensity and (4) aid the interpretation of monitoring data for improving programme management and knowledge of the population dynamics of parasite control. The results also highlight how the use of a model-based monitoring framework will be vitally enhanced by adopting an adaptive management approach that acknowledges uncertainty and facilitates the use of monitoring data to learn about effective control, and which allows future decisions to be modified as we learn by doing. We conclude by emphasizing a pressing need to incorporate mathematical models coupled with changes to existing management systems in ongoing efforts to design and implement rational monitoring plans for evaluating filariasis and other parasitic disease control programmes.

1. INTRODUCTION

The current renewed global interest in achieving the control or eradication of selected human helminths of public health significance has focussed attention on how best to plan, implement and manage large-scale community-based parasite control programmes (Albonico *et al.*, 2006; Burkot *et al.*, 2006; Fenwick *et al.*, 2006; Habbema *et al.*, 1992; Michael *et al.*, 2004, 2006a,b; Molyneux and Nantulya, 2004; Molyneux *et al.*, 2005; Ottesen, 2006; Ottesen *et al.*, 1997; Savioli *et al.*, 2004; Stolk *et al.*, 2003). Recent work in this area has highlighted the need, rationale, strategies and available tools, as well as issues surrounding the implementation of proposed interventions that aim to control or eliminate these parasites from human communities in the developing world. These efforts also underscore the crucial importance and need to engage and forge novel public–private, volunteer group and regional–global partnerships for developing political will, policy and operational frameworks required to initiate and maintain programmes concerned with the control of the major human helminthiases (Molyneux and Zagaria, 2002). Yet, while these developments have undoubtedly led to significant progress in our understanding and capacity to plan the control or even elimination of these parasites, a conspicuous gap is the lack of a firm quantitative framework regarding the design of a key element of a successful parasite control programme, namely how best to undertake and achieve the scientific monitoring and evaluation of community-based parasite control programmes in order to ensure that management objectives are attained (Brooker *et al.*, 2004; Gyorkos, 2003; Habicht *et al.*, 1999; Michael *et al.*, 2006b). This gap is despite acknowledgement that monitoring and evaluation are important to the rational management of any mass parasite control programme, and the need to continue development of more accurate diagnostic tools for assessing the effects of interventions on parasitic infection in hosts and their vectors (Albonico *et al.*, 2006; Doenhoff *et al.*, 2004; Nicolas, 1997; Weil *et al.*, 1997; Williams *et al.*, 2002).

From a programme management perspective, monitoring and evaluation activities underpin effective parasite control because they are integral to informed decision making via their ability to facilitate (Binns and Nyrop, 1992; Binns *et al.*, 2000; Elzinga *et al.*, 2001; Maddox *et al.*, 1999; Michael *et al.*, 2006b; Mulder *et al.*, 1999):

- The periodic assessment of the effectiveness of an implemented programme as intervention proceeds in order to ascertain whether satisfactory progress is being made towards meeting set endpoint targets

- The identification and specification of necessary adaptations to ensure the meeting of programme objectives if deficiencies in expected infection trends are encountered during the intervention period
- The determination of when objectives are achieved

More recently, attention has focussed on how monitoring and periodic evaluation of programme progress are also key to providing relevant data for validating the assumptions underlying the expected outcomes (e.g., magnitude of change in infection expected over time for an endemicity level) of a given intervention strategy, which implies that implementing such actions is fundamental to improving knowledge regarding the population impacts of proposed interventions, especially when understanding concerning such impacts is hampered by uncertainty prior to intervention (Maddox *et al.*, 1999; Parma and NCEAS working group on population management, 1998; Shea *et al.*, 2002; Walters, 1986; Walters and Hilborn, 1978).

These considerations indicate that monitoring for the purpose of decision making in parasite control ultimately should reflect the impact of intervention decisions on the underlying population dynamics of parasite transmission and infection (Dent, 2000; Kendall, 2001; Maddox *et al.*, 1999; Michael *et al.*, 2006b; Parma and NCEAS working group on population management, 1998; Shea *et al.*, 2002; Walters, 1986). In other words, monitoring of parasite interventions at its core is to provide information to evaluate the effects of perturbations on a dynamical ecological system. This is because the norm against which the current state of the parasite population size is compared to in order to determine whether satisfactory progress is being made towards attaining a set endpoint and values of the endpoint targets signifying desired end states of the parasite population themselves, either the controlled state or elimination of transmission, are each properties of the dynamics of parasite transmission, persistence and response to interventions (Anderson and May, 1991; Michael *et al.*, 2006a; Woolhouse, 1996). This implies that the development of a monitoring programme for evaluating parasite intervention progress or success cannot simply be considered as synonymous with generating infection indicators and the measurement of temporal changes in these indicators without evidence of causation (Boyle *et al.*, 2001; Maddox *et al.*, 1999; Mulder *et al.*, 1999), that is without an explicit cause–effect link to the underlying parasite population dynamics.

Our purpose here is to extend recent work on the use of mathematical models of parasite transmission to show how, by articulating critical relationships and dynamic interactions between parasite population infection processes, indicators of these processes and intervention effects, such tools can provide the scientific template for the development of defensible monitoring and evaluation plans aimed at the control of parasitic infection (Maddox *et al.*, 1999; Michael *et al.*, 2006b). In particular, we demonstrate how these models can guide the successful resolution of the three major components of an effective monitoring and evaluation programme, namely (1) the optimal design of programmes for monitoring effectiveness, including the setting of monitoring targets and provision of information on expected trends in infection due to an intervention, (2) the determination of monitoring indicators and sampling requirements and (3) the interpretation and use of monitoring data for supporting management decision making. An important secondary objective is also to portray how the application of a model-based monitoring and evaluation programme will improve understanding (decrease gaps in knowledge and uncertainty in assumptions) of the population dynamics of control, and hence refine current knowledge of management options by using the monitoring and evaluation results to 'learn' about the parasite system and its response to control perturbations (Maddox *et al.*, 1999; Parma and NCEAS working group on population management, 1998; Shea *et al.*, 2002; Tolle *et al.*, 1999; Walters and Hilborn, 1978). Although we illustrate here the use of this dynamic modelling approach for monitoring the control and/or elimination of the vector-borne helminthiasis, lymphatic filariasis, the fundamental principles and framework described are equally applicable to the control of the other major human helminths currently adopted for global control (Albonico *et al.*, 2006; Fenwick *et al.*, 2006).

2. WHY MONITOR AND EVALUATE FILARIASIS CONTROL PROGRAMMES

Three reasons may be considered to underlie the rational and need for monitoring and evaluation of lymphatic filariasis and indeed any other parasitic disease control programme. First, these activities are required to fulfil mandatory obligations set by public health agencies, governments and donors that implemented programmes are in conformance to and attain prescribed standards and targets, for example a prescribed drug coverage or infection threshold level (Michael *et al.*, 2004, 2006a). Compliance with these standards and targets will be critical to fulfilling both national and global requirements for certification of parasite elimination in an endemic area (World Health Organization, 1998a,b). Second, monitoring of programme effectiveness during the course of an intervention (see Table 3.1 for the different types of monitoring that can be carried out for assessing parasite control) will be important to determine if an intervention is on track to meet a set target or if observed changes due to the intervention have deviated significantly from that expected, thereby warranting remedial action to correct for such deviations (Binns *et al.*, 2000; Mulder *et al.*, 1999). Finally, an even more important reason for carrying

TABLE 3.1 Types of monitoring in parasite control programmes

Implementation monitoring	This type of monitoring is used to ensure that strategies and treatments are implemented in accord with stated management standards and guidelines. It is used to determine if the basic management directives (such as maintaining pre-defined optimal coverages) are followed.
Effectiveness monitoring	This monitoring aims to assess whether the implemented strategy is having the predicted effect on the targeted attribute. It is also used to determine whether changes or corrections to strategies are needed during the course of an intervention or management due to deficiencies in either strategy implementation or some other external conditions.
Validation monitoring	This monitoring determines whether key assumptions made in the implemented programme, such as treatment effectiveness or expectations provided by the transmission model, are valid.

out monitoring and evaluation activities is that information from these actions can improve understanding of process, model and intervention uncertainties such that we may be able to iteratively or adaptively revise, update and alter our intervention strategies, including reassessing initially set targets and thresholds, to successfully facilitate the attainment of set programme objectives (Parma and NCEAS working group on population management, 1998; Shea *et al.*, 2002; Walters, 1986; Williams *et al.*, 2001).

3. MATHEMATICAL MODELS AND THE DESIGN OF MONITORING AND EVALUATION PLANS

Monitoring facilitates the successful evaluation of a parasite intervention programme by providing information on changes in indicators of the underlying parasite system, whereas *evaluation* is the process by which monitoring data is used to assess the effectiveness of an implemented programme as well as to guide and improve management to achieve desired outcomes (Elzinga *et al.*, 2001; Gibbs *et al.*, 1999; Habicht *et al.*, 1999; Maddox *et al.*, 1999; Mulder *et al.*, 1999). These definitions support the hypothesis that monitoring and evaluation are most useful when both are purpose oriented (i.e., have clear targets) and when objectives of a parasite intervention programme, and its monitoring indicators, are closely related to the underlying infectious disease transmission process they are intended to mirror (Maddox *et al.*, 1999; Tolle *et al.*, 1999).

3.1. Three roles of mathematical models in parasite monitoring programmes

Informed decision making in any parasite control programme consists of four key elements: (1) setting of objectives, (2) specification of management options or actions, (3) understanding of the structure and dynamics of the parasite system to be managed, that is how the parasite population responds dynamically to perturbations, and (4) periodic monitoring of the results of management that informs and possibly adjusts subsequent management decisions (Kendall, 2001; Michael *et al.*, 2006b). Figure 3.1 summarizes how parasite transmission models can be used to address

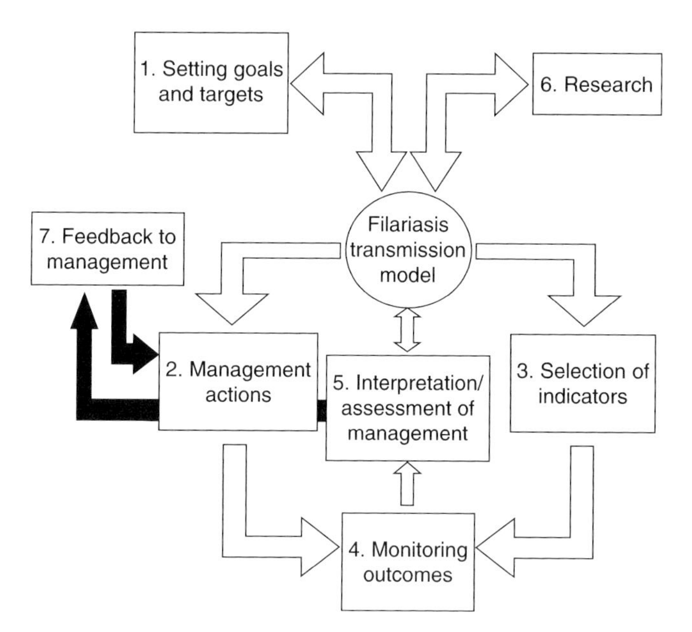

FIGURE 3.1 The central role of the parasite transmission model in integrating management actions and monitoring data for providing policy- and management-relevant support to the design and monitoring of parasite control programmes. Models can help define endpoint targets for intervention goals, predict the effects of management actions, aide the selection of outcomes and indicators for monitoring and guide interpretation of monitoring data. Feedback from model-based evaluation of monitoring outcomes will be vital for adapting and refining both management actions and the model (via revealing topics for research).

these elements and hence provide a framework for supporting the design of parasite control monitoring and evaluation plans. Specifically, these models can aide the development of such plans via their ability to unveil information in the following ways: (1) determination of intervention endpoint targets to meet control programme objectives (see Table 3.2) and provision of predictions regarding the expected magnitude of changes in the states of the parasite system under various interventions (Michael *et al.*, 2004, 2006a); (2) determination and exploration of the roles of the major indicators for monitoring effectiveness and (3) supporting interpretation of monitoring results to improve programme management for accomplishing successful parasite control (Maddox *et al.*, 1999; Tolle *et al.*, 1999). In addition, by providing feedback to both management and research needs, we may also use monitoring results in conjunction with model refinement to revise and update our understanding of the population dynamics of parasite transmission and control as well as for assessing possible alternative or recently discovered control strategies (Fig. 3.1).

4. MODELS AND QUANTIFYING INTERVENTION ENDPOINT TARGETS

4.1. Parasite elimination endpoints: Vector biting thresholds and worm break points

The articulation of clear and precise programme objectives is a fundamental requirement of any parasite control programme since by unambiguously describing the desired endpoint or state of parasitic infection, for example whether elimination, eradication or control of infection or disease is desired, they not only provide the benchmarks against which the progress or success of the programme will be judged but also contribute to the conceptual features of the monitoring and evaluation plan [specifically, what ought to be measured and where, the frequency and duration of such measurements, what the optimal sample size for undertaking reliable monitoring ought to be and when success could be declared to have been attained (Table 3.2)]. As noted above, the first contribution of mathematical models of parasite transmission to this question is that they facilitate the analysis and estimation of thresholds signifying desired endpoints (Anderson and May, 1991; Michael *et al.*, 2006a; Woolhouse, 1992), and thus to the determination of scientifically defensible and measurable indicators of these thresholds. This role of transmission models in aiding the construction of the objective function of the programme (Kendall, 2001; Michael *et al.*, 2004) has been highlighted by the results presented in Michael *et al.* (2006a). In particular, these show how

TABLE 3.2 Potential objectives of interventions against lymphatic filariasis, monitoring targets and the design of monitoring plan (after Michael *et al.*, 2006b)

Objective	Definition	Monitoring targets	Design of monitoring programme
Disease control	Reduction of morbidity to a locally acceptable level	Infection threshold (prevalence or intensity) at which disease is negligible	Repeated monitoring required until and following target attainment to maintain infection below target levels. Frequency of monitoring related to both initial endemicity and the rate of reinfection following the attainment of disease control target.
Elimination/ Eradication	Reduction of the incidence of infection to zero	Infection break point prevalence/intensity in human or vector hosts or vector biting threshold	Repeated monitoring required until target attainment, and following this to ensure certification requirements. For elimination, monitoring needs to continue to prevent re-establishment of infection from untreated regions. Note that monitoring will cease only if coordinated interventions at the regional or global levels occur to ensure parasite eradication. Frequency of monitoring is related to initial endemicity.

mathematical analyses of models of lymphatic filariasis transmission enable detection of the two elimination thresholds for use as targets in those programmes aiming to eliminate infection—one related to the infection transmission process from vectors and the other to worm infection levels in the human host (Anderson and May, 1991; Woolhouse, 1992)—and allow estimation of the values of potential indicators of these thresholds, namely the threshold annual vector biting density from the mosquito population (Fig. 3.2A) and the break point microfilaraemia prevalence in the human host (Fig. 3.2B). These analyses highlight that the derivation of such 'infection thresholds' essentially requires an analysis of the dynamic properties of the parasite transmission system (Anderson and May, 1991), as these variables represent stability or persistence components of (such) dynamical systems (Deredec and Courchamp, 2003; Pugliese and Tonetto, 2004; Pugliese *et al.*, 1998). Note also that an important insight afforded by this modelling analysis, otherwise obscured, is that the specific indicator values of such thresholds (i.e., the break point microfilaria prevalence and threshold vector biting densities) will be strongly dependent on particulars of the density—dependent mechanisms governing parasite infection in the human and local vector hosts and the degree of aggregation of infection in different endemic communities (Anderson and May, 1991; Duerr *et al.*, 2005; Michael *et al.*, 2006a,b). This suggests that elimination targets in the case of filariasis are very likely to be sensitive to local variations in both these processes. The import of this result for filariasis elimination is that the estimation and setting of locally varying endpoint targets will be necessary in different endemic areas. Nonetheless, as highlighted previously (Duerr *et al.*, 2005; Michael *et al.*, 2006a; Pichon, 2002), at least in the case of the annual vector transmission biting threshold density, the present modelling-derived figures [833 and 2375 bites per person per year estimated for culicine and anopheline vectors respectively (Fig. 3.2A)] could be taken to represent the lowest benchmark values, which may be required to be set by the global programme for securing the worldwide eradication of filariasis by vector control. This is in contrast to the situation in the case of worm break point values, whereby modelling work conducted thus far suggests that it might be more prudent to specify and use break point microfilaraemia prevalence or intensity figures that more closely reflect average infection aggregation measures observed in a locality (Michael *et al.*, 2006a). This will make these threshold values higher or lower than the global estimate of 0.5% microfilaraemia prevalence depicted in Fig. 3.2B for a particular region.

The identification and quantification of the major density-dependent processes regulating infection in the vector and host populations, improved understanding of factors underlying parasite aggregation and

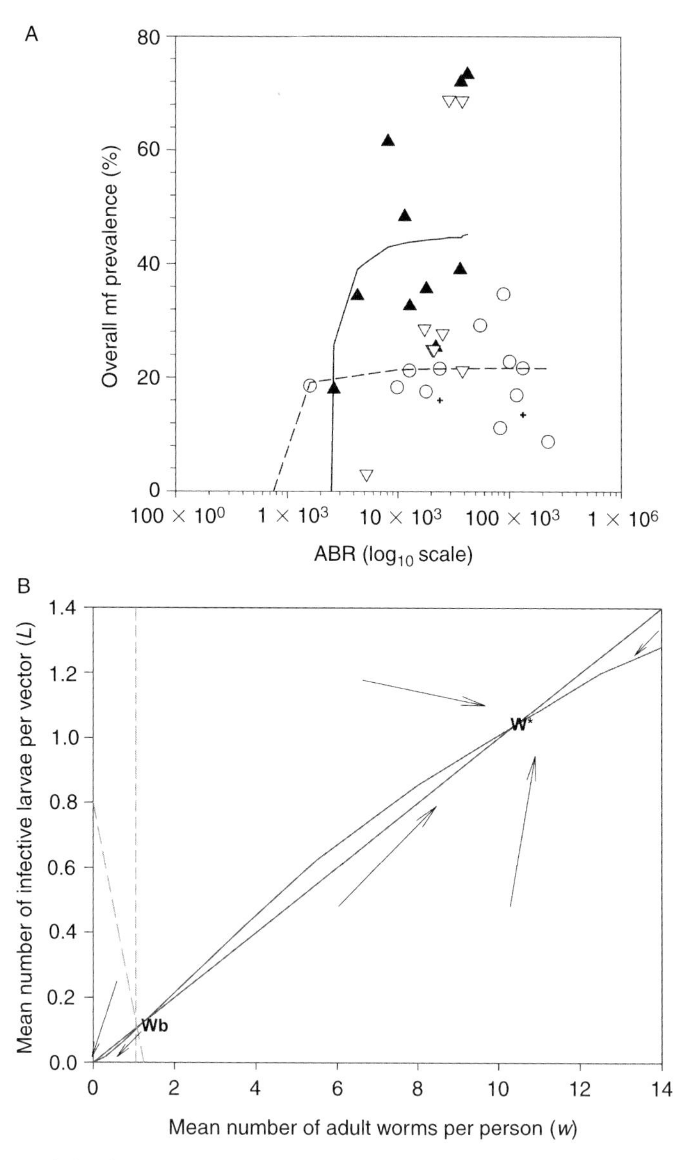

FIGURE 3.2 (A) Relationship depicting the overall community *Wuchereria bancrofti* microfilarial (mf) prevalence (%) as a function of the annual biting rate (ABR) on a $\log_{10}$ scale. Symbols denote published data from single communities in areas with culicine (circles), *Aedes* (crosses), anopheline (filled triangles) and mixed culicine–anopheline (inverted open triangles) parasite transmitting mosquito vectors. Lines portray the best-fitting population dynamic models relating host community infection ABR to the observed data from the culicine–*Aedes* (dashed line) and anopheline-mixed anopheline/culicine (solid line) transmission sites, respectively (see Michael *et al.*, 2006a for details

the derivation of transmission models incorporating such processes, represent urgent current research priority areas in filariasis epidemiology if more pertinent regional values of these thresholds are to be estimated and applied realistically in control programmes. Until such data and modelling results become available, current estimates of infection break points may be considered as first approximations to be used in a precautionary manner (Gollier and Treich, 2003).

4.2. Vector infection thresholds

Although the above thresholds represent the major parasite elimination thresholds in vector-borne infections (Anderson and May, 1991), there has been increasing recent interest in determining if larval infection thresholds exist in the vector population that could be used as monitoring targets for assessing the elimination of lymphatic filariasis. There are two principal reasons for this interest. First, larval infection thresholds rather than vector biting densities represent the parasite elimination target for drug intervention programmes. Second, xenomonitoring tools, such as the parasite DNA-based PCR poolscreen method (Fischer *et al.*, 2003;

---

of data, model structure and fitting procedures). The fits indicate the occurrence of a threshold biting value of 833 bites for the culicine–*Aedes* data set, and a corresponding value of 2,375 for the anopheline-mixed vector transmission dataset. (B) Phase-plane analysis of a basic transmission model for lymphatic filariasis indicating the existence of a breakpoint worm load in the human host below which transmission is interrupted. The depicted equilibrium numerical solutions are for the basic model assuming that filarial worms are polygamous and are distributed in a negative binomial manner among hosts. The straight line represents values for the mean number of infective larvae, L, and the corresponding mean number of filarial worms per person, W, for which $dW/dt = 0$, while the curved line represents values of W and L for which $dL/dt = 0$. Points of intersection represent equilibrium solutions for both W and L. The arrows indicate the directions of changes in W and L starting from initial non-equilibrium values. These depicted dynamical trajectories indicate that the equilibrium solution at the origin is stable, as is the endemic equilibrium, W^*. In between these points lies an unstable equilibrium, the breakpoint, W_B. The dashed line passing through the unstable breakpoint indicates the boundary between the two stable states, and governs which equilibrium values of W and L are reached for given initial values of these variables. Using reasonable values for density-dependent factors regulating infection in both the human and vector hosts together with a mean mid-point value obtained from published data for the parameter K of the negative binomial distribution model describing parasite distribution in hosts and a function relating mean worm intensity to prevalence, model analyses suggest that worm break points might be as low as 0.5% mf prevalence for culicine-mediated filariasis (details in Michael *et al.*, 2006a).

Rao *et al.*, 2006; Williams *et al.*, 2002), that feature high throughput and sensitivity may offer a more reliable and cost-effective monitoring tool for determining the interruption of parasite transmission compared to methods based on measurement of infection in humans.

Theoretically, larval infection (presence of microfilariae to L3 larvae in the mosquito) thresholds may arise in vector populations as a result of (1) the dynamic interrelationship between the prevalence of infectiousness, vector life expectancy and the latent or development period of a parasite in the vector (Fig. 3.3A), and (2) the functional forms of the processes governing the development of L3 infection from ingested microfilariae (Snow and Michael, 2002; Snow *et al.*, 2006). Unfortunately, at the present time, field mosquito infection data sufficient for both validating these predictions and estimating the infection thresholds below which infective (L3) infections do not develop are scarce, but our preliminary analysis of the available data depicted in Fig. 3.3B indicates the possibility that such thresholds could occur in vector populations for filariasis under natural field settings. Again, more data providing parallel information on vector filarial infection versus infective rates from areas differing in the primary transmitting mosquito species are required to estimate this threshold, if parasite elimination endpoints based on quantifying infection prevalence in vector populations by tools, such as the PCR-dependent poolscreen method, are to be successfully applied in filariasis elimination monitoring programmes.

4.3. Disease control targets

Given that the ultimate health dividend of parasite eradication could be large (Ramaiah and Das, 2004; Ramaiah *et al.*, 2000), this objective represents the socially optimal intervention policy for lymphatic filariasis. Nevertheless, this optimality can be tempered by (1) the marginal cost (often likely to be much larger than the average cost) of targeting the last few individuals to meet any set elimination or eradication criteria, (2) the technical feasibility of maintaining adequate population coverage with Mass Drug Administration (MDA) necessary to achieve interruption of transmission within the mooted period of 4–to 6 years and (3) whether the population ecology of infection permits eradication in finite time (Barrett and Hoel, 2004; Gersovitz and Hammer, 2003, 2004; Gyldmark and Alban, 1998). These considerations suggest that in certain cases, especially where lymphatic filariasis endemicity is high and budgetary constraints apply, it may be desirable to adopt intervention strategies aimed at reducing infection to levels sufficiently low to prevent the occurrence of disease in the community, even though in the absence of interventions a state with a low positive level of infection is sustained.

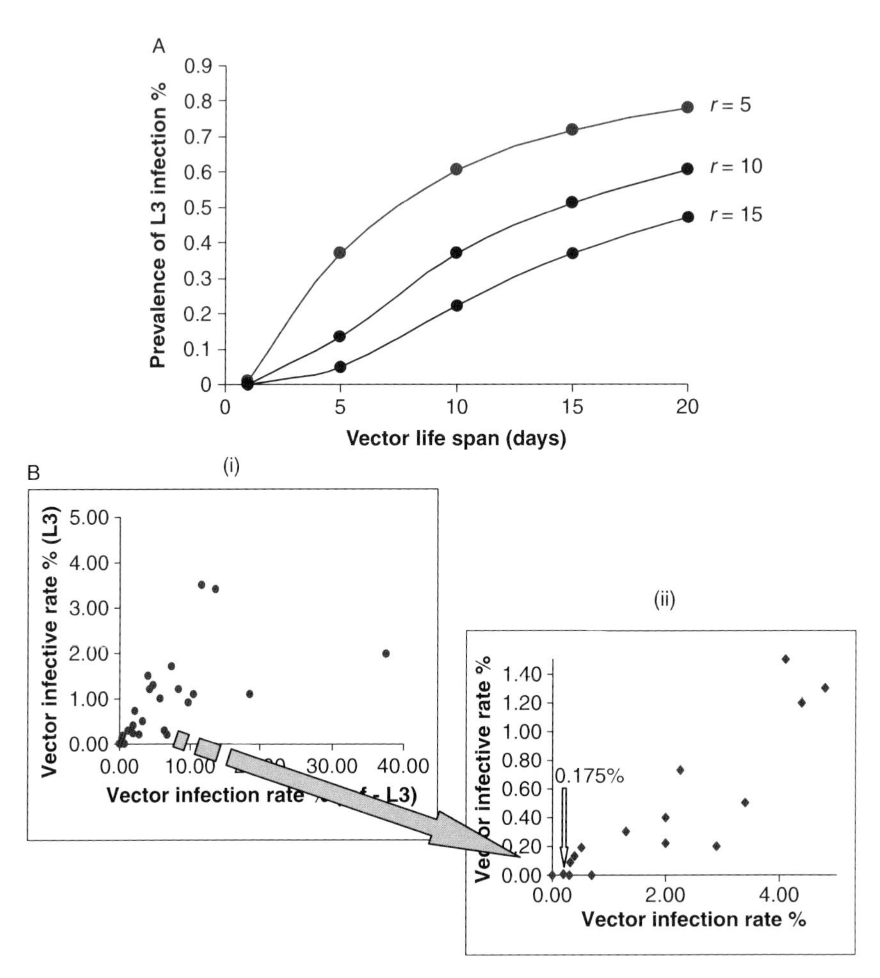

FIGURE 3.3 (A) The relationship between the prevalence of infectious vectors (positive for L3 larvae) and vector life expectancy ($1/b$) for varying infection latent periods, r. Prevalence of infectious vectors (y') is related to vector infection (y) by the model: $y' = [y \exp(-br)]/(b + y)$, where $1/b$ denotes vector life expectancy and r represents the infection latent period. Lines are model predictions assuming y, the vector infection prevalence, to be 5.0% and denote that short vector life expectancy and long latent periods can result in very low prevalences of infectious vectors even when infection prevalence is high (Anderson and May, 1991). (B) Observed relationship between vector L3 infective and infection prevalences for *W. bancrofti* infections in mosquito vectors. Data are from the published literature assessing these infection variables in wild caught mosquito samples (Aslamkhan and Wolfe, 1972; Basu *et al.*, 1965; Bockarie *et al.*, 1998; Bushrod, 1979; Chand *et al.*, 1961a,b; Dzodzomenyo *et al.*, 1999; Gubler and Battacharya, 1974; Hati *et al.*, 1989; Michael *et al.*, 2001; Nair *et al.*, 1960; Pawar and Mittal, 1968; Rajagopalan *et al.*, 1977; Ramaiah *et al.*, 1989; Rao *et al.*, 1976, 1981; Rozeboom *et al.*, 1968)

Carrying out an epidemiological analysis of observed data on the relationship between the overall prevalence of chronic filarial disease (combined lymphoedema and hydrocele cases) and microfilarial infection in communities to estimate the threshold infection level below which chronic disease is negligible, in conjunction with mathematical model predictions of filariasis reinfection, can allow the examination of the value of employing infection control as an intervention option in the control of lymphatic filariasis (E.M., M.N.M.-L. and J.W.K., unpublished observations). Figure 3.4A shows the observed association of prevalence of microfilarial infection (standardized to reflect the sampling of 1 ml of blood in order to maximize the sensitivity of infection detection) and prevalence of filariasis-related chronic disease, based on data complied from published field studies from the filarial endemic regions that yielded a sufficiently large number of these studies, namely sub-Saharan Africa, India and other Asia (apart from India and China). Although portraying some degree of between-region variation, the data indicate the occurrence globally of an overall positive but non-linear association between chronic disease and filarial infection at the community level, with chronic disease apparently constant up to a threshold microfilaria prevalence level and then increasing positively with microfilaria prevalence (Fig. 3.4A). Such a disease-infection pattern suggests the operation of an infection dose-response function in the development of filarial disease, whereby in communities with low-infection prevalence the observed chronic disease may be due to causes other than filariasis while disease is attributable to filarial infection only above a specific microfilaria threshold (van der Werf *et al.*, 2002). The microfilaria prevalence threshold, below which the risk of filariasis-associated disease is zero, can be empirically estimated from the data depicted in Fig. 3.4A via the fit of a logistic regression-based dose-response model incorporating a threshold parameter (Hunt and Rai, 2005; Ulm, 1991; Whitehead *et al.*, 2002). The curve in the figure shows the predictions of the best-fitting of these models, and indicates that a preliminary threshold mf prevalence value that could be used as a target for disease control in filariasis may lie in the region of 3.50% at the scale of using a 1-ml mf sampling blood volume.

Figure 3.4B shows how using predictions of a mathematical model of filariasis transmission (Chan *et al.*, 1998; Michael *et al.*, 2004; Norman *et al.*, 2000) on the annual effects of the two major mass-treatment regimens

from different endemic regions and mix of vector species. Graph ii is to provide a better resolution of the relationship at the lower range of infection values. Vertical arrow indicates the infection region where the infection threshold for development of infectiousness may occur in these data. Given the limited nature of the available data, this is estimated by eye here and so provides only preliminary evidence for the occurence of a threshold.

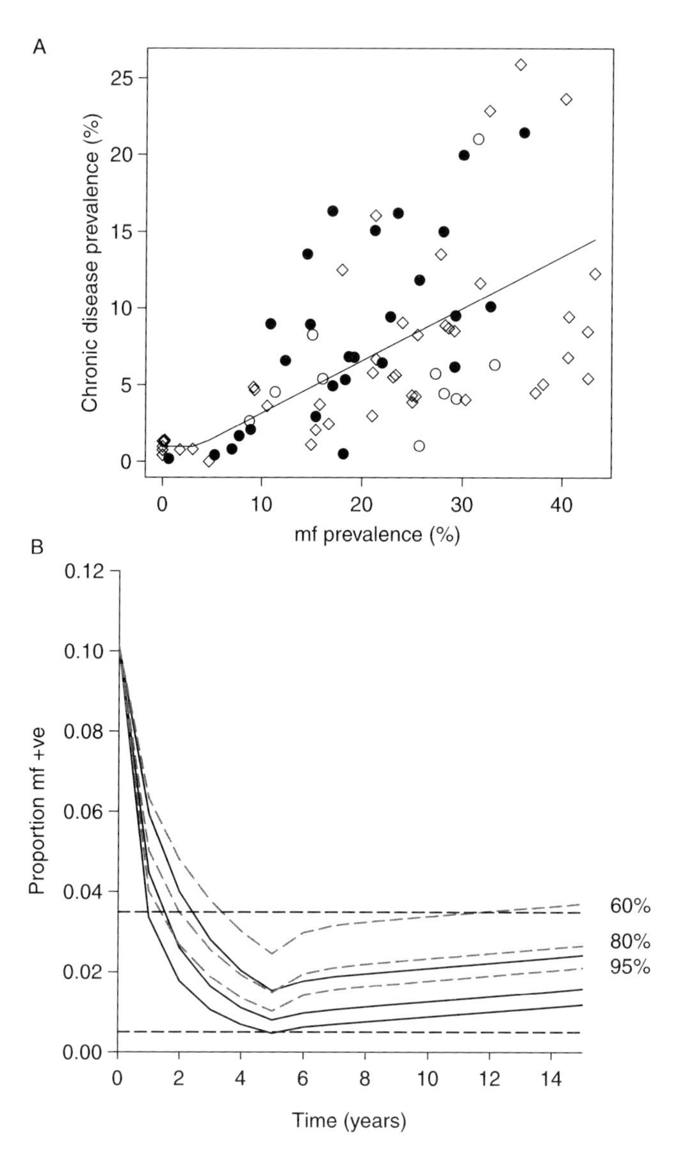

FIGURE 3.4 (A) The association between prevalence of *W. bancrofti* microfilarial (mf) infection and prevalence of combined lymphoedema and hydrocele disease in filarial endemic communities from sub-Saharan Africa (◇), Asia other than India (○) and India (•). Field study ($n = 81$) references are given in E.M., M.N.M.-L. and J.W.K. (unpublished observations). All mf prevalence values have been standardized to reflect sampling of 1-ml blood volumes using a transformation factor of 1.150 and 1.375 for values originally estimated using 100- or 20-μl blood volumes, respectively (factors derived using

[combined diethylcarbamazine and albendazole (DEC/ALB) and combined ivermectin and albendazole (IVM/ALB)] currently proposed for filariasis control on mean community microfilaria prevalence following a 5-year annual mass intervention programme in relation to microfilaria threshold targets for either signifying parasite elimination [set to be around 0.5% prevalence here (Michael *et al.*, 2006a)] or disease control (i.e., 3.50% microfilaria prevalence) can allow an examination of the usefulness of implementing the latter strategy as an intervention option against filariasis. The simulations in the figure are all based on a moderately high overall community pre-control microfilaria prevalence of 10%, and for each treatment regimen (drug efficacy parameters are given in the legend to the figure), are illustrated for MDA coverage values of 65%, 80% and 95%, respectively. The results show that while both these regimens, for all the three treatment coverages considered, failed to

---

data from Dreyer *et al.* (1996), McMahon *et al.* (1979), Moulia-Pelat *et al.* (1992) and Sabry (1991)]. The association of infection with disease is described by the following logistic regression model incorporating a threshold parameter (Bender, 1999; Ulm, 1991):

$$\log\left(\frac{p}{1-p}\right) = \begin{Bmatrix} a \\ a+\beta(x-\tau) \end{Bmatrix} \text{ for } \begin{matrix} X \leq \tau \\ X > \tau \end{matrix}$$

where p is the probability of the occurrence of combined disease in the community, τ is the threshold value of the risk factor, X (i.e., mf prevalence), a defines the baseline morbidity prevalence due to non-filarial causes and β describes the degree of association between infection and disease. This model assumes that the risk of an event (chronic disease here) is constant below the threshold τ and increases according to the logistic equation above τ. The model was fitted to the data using maximum likelihood estimation methods assuming a binomial distribution. The curve shows the best-fit model for the data with estimated values for a and τ of 1.07% (95% confidence limits: 0.21–1.93%) and 3.49% (95% confidence limits: 1.36–5.62%), respectively. (B) Simulation results showing annual changes in overall community mf prevalence predicted by the EPIFIL model for filariasis transmission (Chan *et al.*, 1998; Michael, 2002; Michael *et al.*, 2004; Norman *et al.*, 2000) following a 5-year annual mass intervention programme with either the DEC/ALB (solid lines) or the IVM/ALB (dashed lines) drug regimen. Initial community mf prevalence set at 10%, and for each regimen predictions are shown for treatment coverages of 60%, 80% and 95%, respectively (portrayed by curves going from top to bottom, respectively, for each regimen). Upper horizontal dashed line shows the disease control mf threshold of 3.49%, while the lower line depicts the parasite elimination threshold of 0.5% mf prevalence. All figures are given at the scale of 1-ml blood sampling volume. Drug efficacy values: DEC/ALB—55% worm kill, 95% mf cured and 6 months mf suppression; IVM/ALB—35% worm kill, 99% mf cured and 9 months mf suppression (Michael *et al.*, 2004).

achieve the parasite elimination target, each (even in the case of the least effective IVM/ALB regimen given at a relatively low coverage of 60%) achieved the disease control target of 3.50% microfilaria prevalence before 5 years. An additional feature is that following the end of the treatment programme, the rates of rebound of infection (given that transmission has not been interrupted) are also predicted to be slow. For the IVM/ALB regimen with only 60% population coverage, infection levels reduced to ~2.35% are predicted to remain under the disease control threshold for at least 7 years after cessation of the 5-year intervention programme. For the more effective DEC/ALB regimen even at moderate coverage rates, infection levels remained below the threshold level for more than 10 years (Fig. 3.4B). This result is largely a function of the relatively long life span of the adult filarial worm in the human host (Michael, 2000, 2002; Michael *et al.*, 2004) [set in these simulations to 8 years (Chan *et al.*, 1998)], and indicates that a strategy aimed at disease control by reducing infection prevalence to below the disease threshold is also likely to be highly stable at least for those communities afflicted with moderate to moderately high pre-control infection levels. These slow parasite reinfection rates will reduce the need for frequent retreatment and monitoring activities (Section 5.1), and would also retard the potential evolution of drug resistance (Michael *et al.*, 2004). These results highlight a current need not only to reappraise intervention objectives against filariasis but also to evaluate where and when disease control may play a contributory role towards achieving parasite elimination. Given that monitoring designs are intrinsically linked to the selected endpoint targets (Table 3.2), carrying out such an analysis will also be key to the development and implementation of the relevant optimal monitoring strategy for assessing the progress and success of an anti-parasite intervention in a particular endemic situation.

5. MONITORING CHANGES IN INFECTION LEVELS DUE TO INTERVENTIONS FOR AIDING MANAGEMENT DECISION MAKING

Apart from aiding assessments of whether an implemented intervention has successfully attained desired objectives, an equally important goal of monitoring parasite control is to facilitate timely evaluations to determine if an ongoing intervention is on target to attain set endpoint objectives or else requires the introduction of mitigating measures to rectify observed deviations from expected declines in infection (Elzinga *et al.*, 2001; Maddox *et al.*, 1999; Michael *et al.*, 2006b; Mulder *et al.*, 1999). Two major strategic uses of model predictions of the dynamic impact of an intervention for aiding the undertaking of such monitoring are emphasized here.

5.1. Assessing intervention progress and specification of the frequency of monitoring

Given that the major impetus of infection trend monitoring during a parasite control programme is to assess intervention progress towards a set goal, the first utility of models in this type of monitoring is that they can generate expected values or expected time trends of indicator values of a parasite population between an observation time, t, and its state projected at time, $t + \Delta$, given some management action. This contributes to assessment of progress because only by comparing observed with such expected values or trends can a determination be made about whether the outcomes or measured effectiveness of an intervention programme are in line with expectations, and hence whether the programme is on track to achieve set targets, for example whether as exemplified in Fig. 3.5A an MDA programme will meet the target of 0.5% mf within 6 years given initial endemicity level and drug coverage patterns. A key model-based quantity to aid this decision making is the specification of management triggers or *action thresholds* (*at*s), below which the observed parasite population sample means/prevalence or trajectory following episodes of an intervention is deemed to be acceptable [in terms of enabling the intervention programme to achieve a set target threshold (Binns and Nyrop, 1992; Binns *et al.*, 2000)]. This decision boundary, ideally based on the accuracy and variability of chosen indicators, may lead to the setting of an upper uncertainty band around model predictions, such as the 10% error band shown around the modelled microfilaria prevalence trajectory in Fig. 3.5A, to serve as the *at*s for assisting decision making on this component of control programme assessment (see also Section 6.3.1 regarding the statistics underlying the use of thresholds for aiding decision making in parasite control management).

Second, as also shown in Fig. 3.5A, this combination of model predictions of the impact of an intervention, observed data and the specification of an *at* can also allow the estimation of the optimal periodicity or frequency of monitoring required for the timely correction of any observed deficiencies in programme effectiveness. This is based on the premise that if the observed values of an indicator (say prevalence p) are greater than *at* at any monitoring time during the period of intervention, the implemented intervention is unlikely to achieve the corresponding endpoint target, and hence would require the triggering of a remedial management action (Binns and Nyrop, 1992; Binns *et al.*, 2000; Dent, 2000; Walters, 1986).

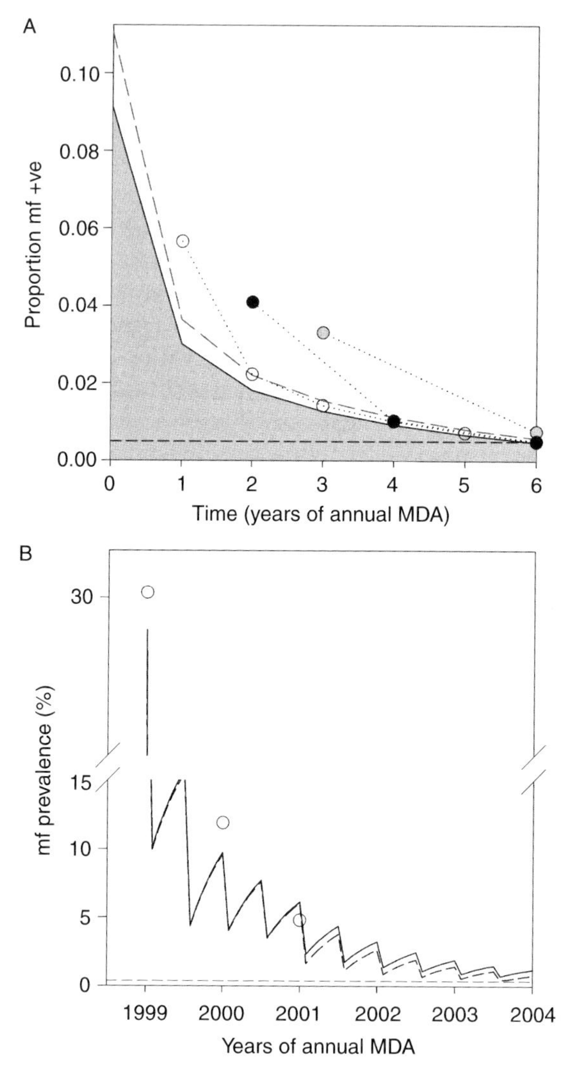

FIGURE 3.5 (A) Illustrative EPIFIL model predictions of the rate of decline of microfilaria (mf) prevalence due to annual Mass Drug Administration (MDA) with DEC/ALB at 80% coverage from a baseline pre-control prevalence of 10% to demonstrate how predicted effects may be used to examine and optimize monitoring of an intervention. The solid curve in the graph denotes the mean expected population trend, while the dashed curve portrays the upper limit of a 10% error band around the mean trajectory, which in this example also serves as an *action threshold* for initiating remedial action. Any observed parasite prevalence above this threshold during the intervention period would be deemed to be unacceptable and hence would trigger the taking of remedial action (see text). The circles joined by dotted lines show simulation results of the impacts of three different monitoring protocols for assessing the decline in filarial infection following annual MDA by the DEC/ALB regimen. Effects on infection are shown for an initial drug coverage of 50% per year until the first monitoring bout at year 1 for the annual (open circles), year 2 for the biennial (closed circles) and year 3 for the triennial protocols (greyed circles), as well as for

When this model-based triggering threshold is combined with parasite control dynamics (e.g., depletion and subsequent reinfection or recrudescence of infection following each annual treatment) and various monitoring protocols describing different periodicities of resampling (e.g., once a year, once every two years, once in three years), joint simulations of these monitoring plans and changes in parasite population size allow an assessment of the required frequency of monitoring, as illustrated by the results shown in Fig. 3.5A. Essentially, this is based on the fact that the frequency of monitoring necessary to assess whether an intervention will attain its set target is a function of how quickly infection rebounds in a community before elimination occurs following intervention episodes (Michael *et al.*, 2006b). Although obviously dependent on the initial levels of falloff from the optimal coverage value of 80%, the results in Fig. 3.5A indicate that for filariasis, the optimal monitoring protocol may be one based on either an annual or a biennial but not on a triennial monitoring scheme. This is further illustrated by the results depicted in Fig. 3.5B that portray a real-world application of using model predictions and an annual monitoring protocol to track progress and determine if remedial management actions are required in order to ensure filariasis elimination in a community administered with a mass DEC programme in Tanzania (Michael *et al.*, 2004).

5.2. Role of spatial distribution of infection for monitoring and evaluation

The spatial distribution of a parasite may be considered to have two principal implications for monitoring large-scale parasite control programmes. First, parasite spatial distributions play a fundamental role in guiding the efficient and scientifically valid sampling of representative sites for monitoring and evaluating the effects of control (Brooker and Michael, 2000; Brooker *et al.*, 2004; Rogers and Williams, 1993). This includes the selection of appropriate sentinel sites for conducting follow-up monitoring of the effects of an intervention within treatment

---

the subsequent increases in drug coverage to 85% (after year 1), 90% (after year 2) and 95% (after year 3) for the remainder of the 6-year programme in the case of the annual, biennial and triennial monitoring protocols, respectively. The results suggest that the optimal monitoring protocol for the successful management of filarial MDA programmes may be one in which monitoring is carried out biennially (see text). (B) Using model predictions of the effect of an intervention to support decision making in filariasis elimination. Open circles denote the observed change in mf prevalence following a 6-month mass DEC treatment administered for 2 years to an endemic community in Tanzania with an initial mf prevalence of $\sim$30% (Simonsen *et al.*, 2004). The solid line shows the prediction of the EPIFIL model incorporating the observed coverages (which ranged from 67.9% to 86.5%) up to the end of year 2 and the average of these coverages (76.8%) thereafter, while the dashed line indicates the predicted decline in mf prevalence when the coverage is increased to 85% after year 2. The horizontal dashed line in the graph denotes the 0.5% mf threshold.

administrative units, which has been mandated by the World Health Organization to form the major basis of evaluating the effectiveness of filariasis intervention programmes in endemic countries (World Health Organization, 1998a,b). Such model-based approaches do not rely on random sampling, but on the selection of sites representative of different classes or strata of parasite transmission dynamics, within which both the design of interventions to interrupt transmission and the response of the parasite system to such interventions are comparable. The concept of spatial patterns in parasite distribution is central to planning such monitoring schemes since by taking account of the spatial distribution of parasitic infection it facilitates the selection of the appropriately stratified sentinel sites or communities for undertaking monitoring (Binns *et al.*, 2000; Venette *et al.*, 2002; Yoccoz *et al.*, 2001).

Figure 3.6 (which is plate 3.6 in the separate Colour Plate Section) illustrates this impact of spatial heterogeneity by showing how the mapping of pretreatment baseline microfilaria prevalence data from communities in conjunction with geostatistical modelling carried out in a geographic information system (GIS) can provide a powerful empirical tool for undertaking the selection of these sites for monitoring and evaluating the effectiveness of filariasis control at the district level in an endemic country. Succinctly, this method entails first the mapping of observed mean microfilaria prevalence data by community spatial location, followed by geostatistical analysis of the mapped data to derive predictive maps defining and stratifying areas with similar ecological and autocorrelated risk factors (Fig. 3.6). The overlaying of such risk maps on village location maps will then facilitate the stratified (on different parasite transmission risks) random sampling (which may or may not be based on equal probability sampling) of sites to carry out sentinel monitoring activities (Fig. 3.6). Note, however, that this method relies on the availability of spatial data on pretreatment infection prevalence as well as the existence of up-to-date community map and census information. If pretreatment infection data are not available, a generalized stratified sampling design could still be employed based on using other ecologically meaningful categories for stratification [e.g., via the mapping of ecological/climatic risk factors associated with filariasis infection prevalence (Lindsay and Thomas, 2000)]. Once geographic stratification has been achieved (Fig. 3.6), the number of sentinel sites that need to be monitored for evaluating programme effectiveness per stratum will depend on the expected within- and between-stratum variation in the effects of interventions on infection and the magnitude of 'noise' in the parasite system and in its control dynamics (Binns *et al.*, 2000; Yoccoz *et al.*, 2001). This suggests that a more systematic investigation of such variations will therefore also be an important prerequisite to the reliable application of the proposed sentinel site approach to carrying out the monitoring of filariasis control in the field. The impact of spatial heterogeneity on estimating optimal sample sizes for reliably determining infection levels or changes in these levels at the community level is discussed further in Section 6.3.3 below.

The second implication of parasite spatial distribution is less well recognized. As depicted in Fig. 3.7, this influence of spatial heterogeneity in infection arises as a result of the positive relationships between variations in the pre-control prevalence of infection and (1) the duration of an intervention required to meet a set target (such as shown for the annual DEC/ALB MDA programme for lymphatic filariasis in Fig. 3.7A) and (2) the optimal population coverage of an intervention needed to achieve a particular programmatic endpoint (Fig. 3.7B). The required time frame of an intervention is linked to the design of an effective monitoring programme principally because, as noted in Section 5.1, it can influence the frequency of monitoring episodes that may be required to be performed for both tracking the progress of the intervention and providing sufficiently early warning of any deficiencies (Binns *et al.*, 2000). This is perceived by comparing the results in Fig 3.5A versus Fig. 3.7C. Thus, the simulations depicted in Fig. 3.7C show that although a longer period of annual MDA using the DEC/ALB regimen is required to meet the set 0.5% microfilaria prevalence elimination threshold when the pre-control prevalence is comparatively high, paradoxically the optimal monitoring frequency need only be once every three years compared to the higher frequency (once very two years) required in the case of the shorter period of intervention predicted for the same MDA regimen in situations when initial infection prevalences are lower (Fig. 3.5A). This finding underlines a basic outcome for monitoring infection trends between control programmes with variable time lengths: in general, the frequency or intensity of monitoring will be inversely related to the duration of a parasite intervention programme. Combining this model-based monitoring design with filariasis distribution maps to derive more precise locally varying frequencies of monitoring bouts can thus potentially reduce the overall costs of monitoring a national-level control programme.

The positive (though non-linear) association between the optimal drug coverage and the level of pre-control mf infection, which occurs for an anti-filarial annual MDA programme aiming to meet a set target within a given time frame portrayed in Fig. 3.7B, indicates that an estimation of the spatial distribution of filariasis will also be key to undertaking both implementation and effectiveness monitoring (Table 3.1) of control programmes in different parts of a country. This is based on the observation that spatial variations in pre-control infection prevalence will require the need for implementing different localized optimal coverage patterns with corresponding monitoring plans. The impact of parasite spatial distributions on control programming and monitoring, despite long

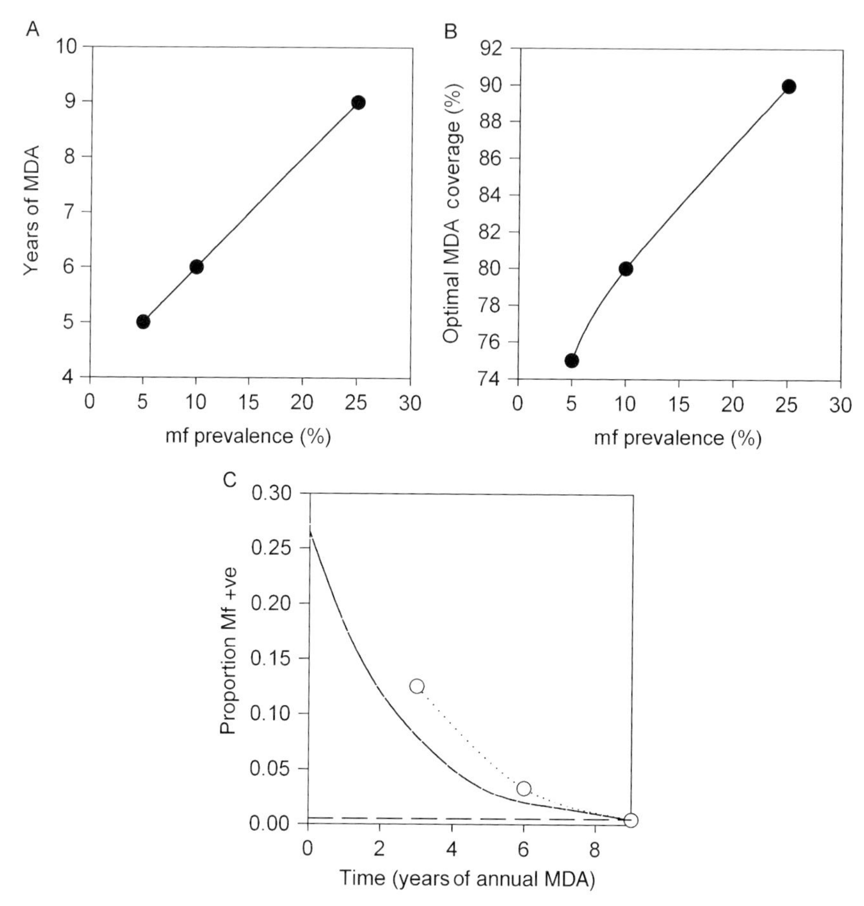

FIGURE 3.7 (A) Predictions of the EPIFIL model of the number of years of intervention required by a DEC/ALB annual Mass Drug Administration (MDA) regimen at 80% coverage to achieve a target parasite elimination threshold of 0.5% microfilaria (mf) prevalence for a range of pre-treatment community endemicity (mf %) levels. Drug efficacy values: 55% worm kill, 95% mf cured and 6 months mf suppression following each treatment episode. (B) Model simulations of the optimal drug coverages needed by the annual mass DEC/ALB regimen to achieve the 0.5% mf prevalence threshold in relation to variations in pre-treatment community mf prevalences. (C) Similar to Fig. 5(A) except that the solid curve shows EPIFIL model predictions of the mf prevalence trajectory in response to annual mass MDA with the DEC/ALB regimen given at 80% coverage to a community with an initial pre-control mf prevalence of ~25%. The open circles joined by dotted lines show that carrying out a triennial monitoring protocol would be sufficient for assessing the decline in filarial infection until the target of 0.5% mf prevalence is reached in this case for an initial drug coverage of 50% per year until the first monitoring bout at year 3 and subsequent increases in drug coverage to 90% (after year 3) for the remainder of the intervention programme.

being acknowledged as important (Brooker and Michael, 2000), has yet to be fully realized in the management of parasite control.

6. MATHEMATICAL MODELS AND THE SELECTION OF MONITORING INDICATORS

The vital importance of choosing the right infection indicators for monitoring the effects of a parasite control programme is patently obvious since for most parasitic diseases, because of the difficulties in observing adult parasite numbers directly, intervention impacts can only be evaluated in the context of induced changes in one or more indicators of the underlying variations in the parasite population size. The high value of this one step for the ultimate success or failure of any parasite monitoring programme is underlined by fact that the use of wrong indicators could indicate programme failure even when a programme has been successfully implemented and vice versa (Mulder *et al.*, 1999). Table 3.3 summarizes and lists the essential properties required by indicators for aiding the effective monitoring of parasite control programmes. Here, we describe how mathematical models of parasite transmission can play an important role in resolving these criteria, and hence facilitate the informed selection of the most appropriate markers for monitoring lymphatic filariasis control.

6.1. Models and the impact of diagnostic accuracy of indicators for monitoring filariasis control

A fundamental feature of using indicators of infection for monitoring the effect of an intervention on parasite prevalence is that the estimated prevalence data will almost always represent the apparent infection prevalence simply because of the imperfect accuracy of available diagnostic tests

TABLE 3.3 Essential properties of indicators for monitoring parasite control (after Michael *et al.*, 2006b)

Their dynamics should parallel that of the underlying parasite population dynamics
They can be accurately and precisely measured [i.e., have high diagnostic accuracy (high sensitivity and specificity)]
The likelihood (or statistical power) for detecting a change in their magnitude is high given the rate of change in the state of the parasite population being monitored
Demonstrate low natural variability
The costs of their measurement are not prohibitive

(Pepe, 2003; Zhou *et al.*, 2002). The major outcome of this effect for programme evaluation is highlighted by the simulated impacts of an annual DEC/ALB MDA regimen on the prevalence of filarial infection presented in Fig. 3.8A, which show how the sensitivity of a diagnostic test for measuring infection can systematically generate underestimates of the true prevalence of infection to the extent that, when the sensitivity of the test is insufficiently low, it can suggest the successful attainment of parasite eradication thresholds when the true prevalence indicates that this has not been achieved. This result is a consequence of the detection of false negatives by diagnostic tests with less than perfect sensitivities even when specificities are very high or perfect as in the case of the present simulations. The lower dashed grey line in Fig. 3.8A furthermore indicates that if the sensitivity of a test changes as intervention progresses (e.g., sensitivities for detecting positive individuals may decline as infection intensity is sharply reduced due to intervention), this will significantly increase the error of declaring programme success leading to the incorrect decision to stop further interventions. Note also that the present results assume perfect specificity in filariasis diagnostic tests: any deficiencies in this parameter will decrease the positive predictive value (PPV) of a test and hence increase detection of false positives as infection prevalence declines due to interventions, further complicating the interpretation of test results for evaluating programme success. This inherent diagnostic dynamics of a parasite monitoring programme can also influence the optimal interval of any control monitoring plan through their effects on apparent durations of interventions required to meet set targets under different endemic conditions.

These results thus indicate that estimates of filariasis prevalence must take the inherent performance (sensitivity and specificity) of a diagnostic test and possibly the change in community infection prevalence into account when they are used for programme evaluation. The first step here evidently relates to the crucial need for more reliably estimating the performance accuracy of currently proposed diagnostic tools for detecting filarial infection (Fischer *et al.*, 2003; Lammie *et al.*, 2004; Rao *et al.*, 2006; Simonsen and Dunyo, 1999; Tobian *et al.*, 2003; Williams *et al.*, 2002). However, as pointed out before (de Vlas *et al.*, 1993), empirical analytical methods normally also treat the performance properties of a diagnostic test as intrinsic, fixed characteristics of the test, and thus do not take account of the intervention dynamics of the underlying worm population. This underscores another potential application of transmission models in filariasis monitoring, namely, that using model predictions of changes in true filariasis prevalence in conjunction with observed prevalences generated by a given diagnostic test may offer an effective way forward to estimate the performance characteristics of currently proposed tools for diagnosing filariasis infection (de Vlas *et al.*, 1993). For example, if specificity is not a problem, then test sensitivity can be easily estimated

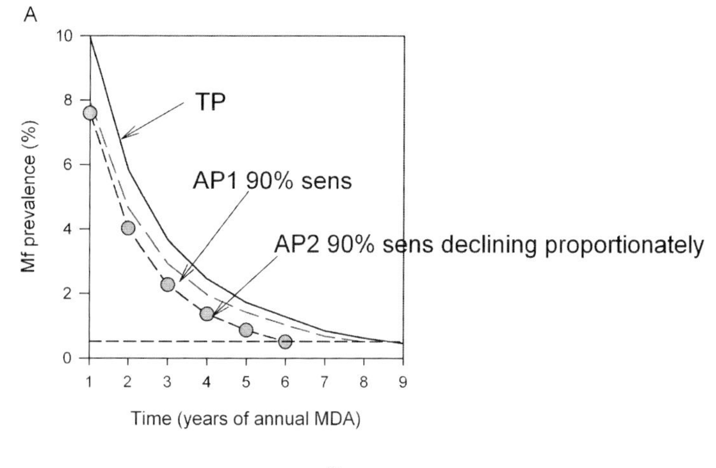

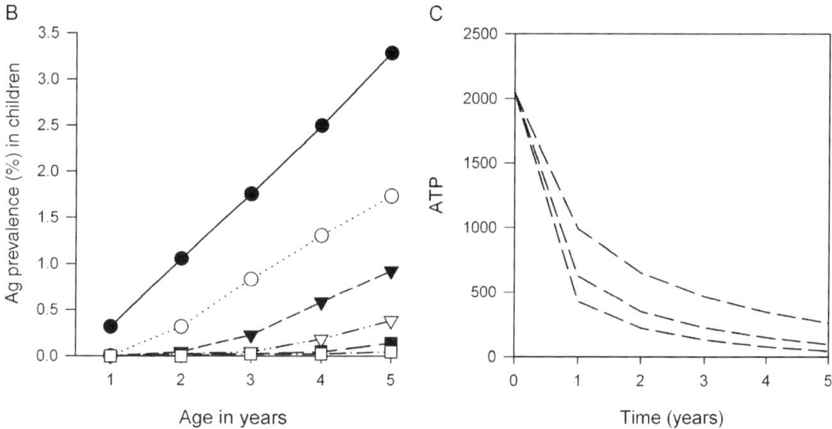

FIGURE 3.8 The impact of indicators and performance of diagnostic tools for monitoring filariasis control. (A) The sensitivity of a diagnostic test for quantifying change in infection prevalence over time. The solid line in the graph denotes EPIFIL model predictions of the rate of decline in the true mean microfilaria (mf) prevalence (see text) due to an annual DEC/ALB Mass Drug Administration (MDA) therapy given at 70% coverage to a community with an initial mf prevalence of 20%, while the dashed lines show the apparent declines in mf prevalence as a result of applying diagnostic tests for quantifying mf prevalence either with a sensitivity of 0.90% (upper dashed line) or as sensitivity declines progressively from 90% (lower dashed grey line). Apparent prevalences decline faster and appear to reach the endpoint target of 0.5% mf prevalence compared to the true prevalence as a result of the imperfect sensitivity of the diagnostic test. The apparent rate of decline in mf prevalence occurs even faster and the target is achieved more earlier when the test sensitivity is simulated to decrease by 10% each year of monitoring from 0.90% initially (redrawn from Michael *et al.*, 2006b). Panels (B) and (C) show how by facilitating linkages between different parasite transmission components, mathematical models can also provide quantitative expectations

using the equation: $se = \mathrm{AP/TP}$, where AP is the apparent prevalence as estimated by the diagnostic test, TP is the true modelled prevalence [where this is modelled as presently as a function of both the underlying mean worm burden (M) and the distribution of worm loads between individuals (Chan *et al.*, 1998; Norman *et al.*, 2000)] and *se* denotes the sensitivity of the test (de Vlas *et al.*, 1993). This finding suggests that developing frameworks integrating transmission models with assessments of diagnostic tool performances will provide a useful tool not only for evaluating diagnostic tests but also for interpreting the results of applying such tests to the monitoring of parasite control.

6.2. Selecting appropriate monitoring indicators

Because models can describe dynamic relationships between various parasite system components and processes, thresholds and expectations of change (the effect size) due to interventions may also be defined on the bases of any currently proposed indicator of filarial infection and transmission (e.g., parasite antigenemia, antibody responses, vector infection rates), once linkages between these variables can be quantified. This role of models is illustrated in Fig. 3.8, which shows how changes in microfilaria prevalence due to the effects of a 5-year annual DEC/ALB MDA programme (at 80% coverage) is reflected by corresponding predictions of (1) changes in the age prevalence of circulating filarial antigen (CFA) in children up to 5 years of age following each annual mass treatment (Fig. 3.8B), and (2) yearly changes in mean L3 infection in vector populations (Fig. 3.8C). Thus, by combining this ability of models to predict interlinked changes or provide information on relative magnitudes of effect sizes in each specific parasite system component due to an intervention with estimates of natural variation in values of indicators of such changes, their sensitivities and costs of usage, it will be possible to facilitate the informed selection of the most appropriate indicators for monitoring the effects of control. This is achieved by assessing which of the available measures provide the highest precision and lowest variability, the greatest statistical power for detecting a significant change in the underlying parasite population size (which is linked to the magnitude of the effect size) and the least cost (see Section 6.4 for a more comprehensive data theoretical method that facilitates the selection of cost-effective indicators for parasite control management decision making by addressing these factors simultaneously).

---

of change in infection/transmission (the effect size) due to intervention based on different measures of infection, such as antigenemia age-prevalence in children up to 5 years of age (redrawn from Michael *et al.*, 2006b) (B) and vector infection indices (C).

A further significant but little appreciated outcome of using models in this regard is illustrated by the results in Fig. 3.9. These show how mathematical models can guide the planning and execution of the second major type of monitoring involved in evaluating programme success, namely programme implementation or process monitoring (Table 3.1). While this type of monitoring normally addresses how well a programme has met planned activities to achieve an objective (Gyorkos, 2003; Habicht *et al.*, 1999; Margoluis and Salafsky, 1998), the modelling results relating the primary parasite control programme activity of treatment or intervention coverage to effectiveness (Fig. 3.9) suggest that models also allow the use of observed coverage values as indicators to indirectly evaluate the effectiveness of an intervention programme. In particular, the model-based estimation of optimal coverages required to achieve programme objectives for different strategies and endemic conditions (Michael *et al.*, 2004) means that yardstick coverage values can be derived against which observed coverages at different time points may be compared to determine if acceptable progress is being made. The fact that coverage values can normally be calculated at the end of each annual MDA for each treated community indicates that annual and timely monitoring of both process and effectiveness of treatment programmes can be achieved in a cost-effective manner using this method. Validation of these findings, particularly how using coverage values compares with or adds to using

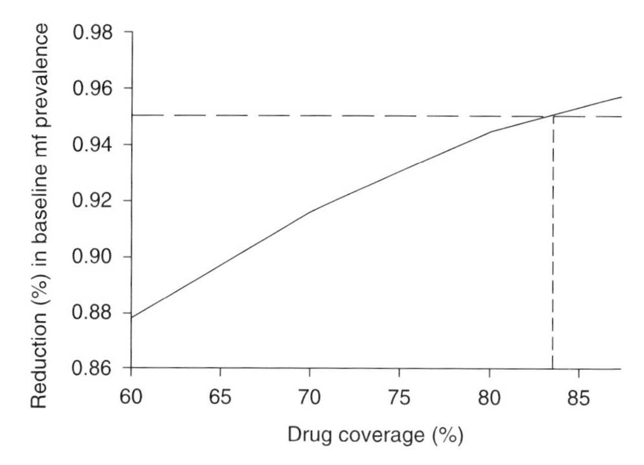

FIGURE 3.9 The relationship between population Mass Drug Administration (MDA) coverage with DEDC/ALB and the rate of decline in microfilaria (mf) prevalence from a baseline prevalence of 10% as predicted by the EPIFIL model. The horizontal dashed line represents the elimination threshold of 0.5% mf prevalence proportional to the baseline prevalence. The vertical dashed line denotes the optimal annual coverage required over 5 years to achieve the elimination threshold; any drop in coverage below this optimum will result in failure to meet the elimination target in 5 years (solid curve).

other measures of programme effectiveness, is yet to be investigated and is suggested to be a research priority in lymphatic filariasis epidemiology and model-based programme evaluation studies.

6.3. Models and sampling for parasite monitoring

Since monitoring the status or trend of an indicator essentially represents a problem in estimating the value of an unknown parameter (e.g., the state of the underlying parasite population), it is apparent that collection of data in monitoring programmes should be founded soundly on statistical sampling theory to ensure that adequate effort is being made to obtain sufficiently precise population estimates or to detect important changes. Sampling methodology and experimental design are huge topics, which cannot be adequately reviewed here. Several excellent papers in the context of ecological monitoring are noteworthy, including Binns and Nyrop (1992), Nyrop *et al.* (1999), Madden and Hughes (1999) and Venette *et al.* (2002). Several key features of sampling relevant to parasite monitoring are discussed below.

6.3.1. Estimation or hypothesis testing

Making a distinction between these two statistical processes is important in a monitoring programme because it is linked to the purpose for which the sample data are collected, and will also determine the required sampling intensity or number of samples to be collected (Binns and Nyrop, 1992; Madden and Hughes, 1999; Nyrop *et al.*, 1999). In brief, estimation procedures are driven by the following questions: how many samples are needed so that an estimate (of infection prevalence or density) is within some range of a true value, or how many samples are required so that an estimate has a specified level of precision? However, as noted in Section 5.1 , information on parasite prevalence or abundance should also be used to direct management actions in parasite control programmes, for example whether to either accept that an intervention is progressing as expected or if not to institute a remedial action, or whether an intervention has reduced infection (or not) below a specified endpoint threshold. This represents the use of monitoring data for the purpose of guiding programme decision making in parasite control (Madden and Hughes, 1999; Nyrop *et al.*, 1999). Thus, compared to the goal of estimation, it is more informative for this purpose to ask how many samples are required so that the likelihoods of incorrect management decisions are acceptable (Nyrop *et al.*, 1999). This is a hypothesis testing or classification problem as one is interested in testing the null hypothesis H_0: $p \leq p_c$ versus the alternative hypothesis H_1: $p > p_c$, in which p_c is some critical value [e.g., the model-derived *at* in Section 5.1 or some endpoint threshold value (Sections 4.1 and 4.2)]. Because as with any hypothesis test, the probability of accepting the null hypothesis given any true value of p is governed by balancing the making of two types of errors: (1) rejecting the null hypothesis when it is true (falsely deciding that $p > p_c$), called a Type I error; and (2) accepting the null hypothesis when it is false (falsely deciding that $p \leq p_c$), called a Type II error; sampling for classification will produce its own optimal sample size values that will turn out to be radically different from those for estimation (Binns *et al.*, 2000). This conclusion has thus two major implications for developing effective parasite monitoring programmes. First, it indicates that sample size calculations must be linked to the question being asked by monitoring, whether this is for reliable estimation of parasite population size or for making decisions based on such estimations. Note that a clear understanding of this difference is of particular relevance to parasite monitoring as in general sample sizes required for classification problems will be feasibly smaller than those required for estimation problems (Fig. 3.10). Second, it can guide the selection of monitoring indicators based on the contributions they make to the statistical power of a monitoring programme to detect meaningful changes in infection values (Maddox *et al.*, 1999; Mulder *et al.*, 1999), as described in Section 6.3.2.

6.3.2. The statistical power of parasite monitoring programmes

The power of a monitoring programme to detect an expected rate of change in infection is defined as its ability to detect the change when it occurs, that is power is statistically specified as $1 - \beta$, where β is the probability of committing a Type II error [falsely concluding that a difference or trend did not occur when it in fact did (see above)]. In general, the statistical power of a test is a function of the following: the sample size, N, variance of the estimated measures of infection, s^2, which in turn is a combination of errors due to the inherent sampling variation in the parasite component being measured plus the detection error of the diagnostic tool used, effect size (the rate of change in infection that is to be detected) and α (the maximum rate of Type I error tolerated, concluding falsely that a difference or trend has occurred when it in fact has not), with power usually increasing with increasing N and effect size and decreasing with increasing s^2 and α (Anderson, 1998; Elzinga *et al.*, 2001; Mulder *et al.*, 1999). These considerations indicate that the key, and indeed one of the most difficult, challenges in developing a parasite control monitoring programme with sufficient detection power is the determination of the effect size, that is the value of an indicator, or the magnitude of change in its value over some interval, that indicates a biologically significant impact of an intervention on parasite infection (Anderson, 1998; Gibbs *et al.*, 1999; Mulder *et al.*, 1999). Mathematical model predictions of the impact of interventions can manifestly help resolve this question in lymphatic filariasis via their ability to provide estimates of such effect sizes (and their variances) for various interventions (Michael *et al.*, 2004, 2006a). Indicators

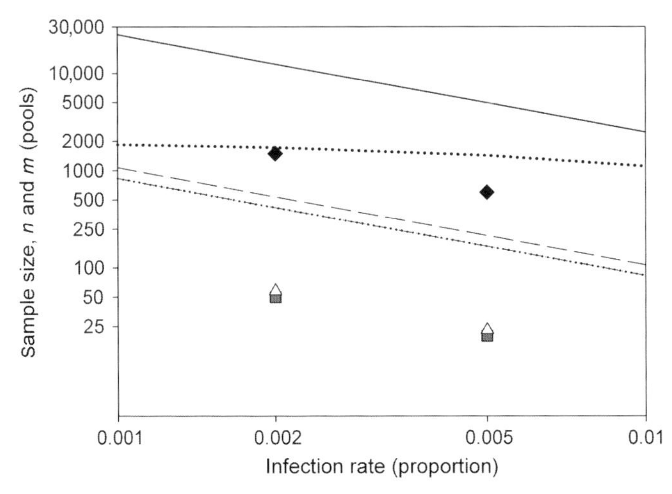

FIGURE 3.10 Effective sizes of individual samples (*n*) and pools of samples (*m*) required to either estimate (curves) or detect (symbols) various infection proportions. Solid curve denotes the optimal individual sampling units (human or vector hosts) required to estimate varying infection proportions for a coefficient of variation of 20% around the mean proportion assuming a binomial distribution model. Dotted curve shows the corresponding individual sampling units required following the application of a finite population correction factor (FPC) to estimates portrayed by the solid curve. The dot-dashed and dashed curves are the optimal number of pools of samples (*m*) of size 30 individual units each (e.g., mosquitoes) required to estimate varying infection proportions for the same coefficient of variation of 20% assuming a simple binomial distribution of infection (dot-dashed curve) or a clustered distribution of infection (dashed curve) in the community. A value of 0.01 was used as the intracluster correlation coefficient to derive the estimates for the latter in this example. Infection aggregation will always raise the number of sample pools required to maintain a desired precision of infection proportion or prevalence estimation. The filled diamonds represent the effective individual sampling units required to detect either a larval infection prevalence of 0.2% [approximating the likely vector infection threshold (Fig. 3B)] or a microfilaria prevalence of 0.5% [denoting the likely worm break point prevalence (see text)] at a desired probability of 0.95% assuming a binomially distributed infection. The corresponding number of pools (of size 30 sampling units) to detect these proportions are portrayed by the closed squares (assuming that infection is distributed according to the binomial distribution model) and the open triangles (assuming clustered infection). Again aggregation of infection increases the number of sample pools required to detect infection at a desired probability of parasite detection (here illustrated for 95%). Sample size formulae are as given in Elzinga *et al.* (2001), Hughes *et al.* (1995), Madden and Hughes (1999) and Venette *et al.* (2002).

related to components of the parasite system that undergo the biggest or fastest rates of change as a result of an intervention, for example induced rates of change may be larger in the case of infection prevalence in humans than in mosquito populations; see Fig. 3.8A versus C, or change in abundance is more rapid and larger than change in infection prevalence (Anderson and May, 1991), should therefore represent the choice markers for monitoring as they would not only improve monitoring power but could also be used to reduce the required sampling effort.

6.3.3. The impacts of target population size and spatial infection patterns

A clear understanding of the ratio of the size of the sampled population to the target population about which an inference is to be made is another important factor in devising optimal sampling plans given its direct bearing on both sample size determinations and in adjusting test statistics (Elzinga *et al.*, 2001). For monitoring of filariasis control, primary inferences regarding infection status and change in this status are normally made for the total human or mosquito hosts in a community or village based on representative samples drawn at this level (although these are then secondarily aggregated to the district level by summarizing findings from a representative sample of these villages). The importance of sample population size (n) is that if this population is large relative to the target population (N) (and the rule of thumb is that the ratio n/N should not be more than 0.05, i.e. not more than 5% of the target population), then there is a need to apply a finite population correction factor that reduces the size of the standard error (Elzinga *et al.*, 2001). This will enhance the precision of an estimate and thus reduce the sample size required for both estimation and hypothesis testing problems. The practical significance of this result is that sample sizes for estimating very low threshold infection prevalences, such as those concerned with the infection threshold in the mosquito vector population (0.175% from Fig. 3.3B), need not be as large as those determined by binomial models, which assume infinite populations (Fig. 3.10). This is an important insight given that such large sample sizes as determined routinely but inappropriately using the latter model will be limiting if the target host population size are likely to be small in some areas (as has been suggested for some vector species that transmit lymphatic filariasis).

The intensity of sampling is also influenced by the spatial distribution of parasites in host populations (Hughes *et al.*, 1995; Madden and Hughes, 1999). This is also illustrated in Fig. 3.10, which shows how the number of samples required to maintain a desired precision in parasite detection, for example 20%, will increase as the proportion of samples infected falls or as the degree of infection aggregation rises. This situation may exist even when assessment of infection is based on aggregates of populations, such as mosquito pools used in tools like the poolscreen PCR-tool for determining filarial infection in vector populations.

The issues described above indicate that sampling plans must not only account for the objectives of the monitoring programme (estimation or classification), indicator variability and diagnostic performance, but should also consider the ecology of infection (host population size, parasite spatial distribution and expected or modelled effect sizes) if they are to aid effective parasite monitoring. Indeed, as can be appreciated from Fig. 3.10, when sampling size estimates are based on probability models that incorporate such information, an assessment can be made about whether a particular infection detection method (e.g., blood sampling or dissection of mosquitoes) may be feasibly used to estimate a required infection level (e.g., the various infection thresholds discussed here (Figs. 3.2A,B and 3.3B)]. In some cases, it may be expected that certain infection levels in the host population, for example the infection threshold in the vector population depicted in Fig. 3.3B, cannot be feasibly quantified by available and widely used diagnostic tools, for example mosquito dissection, with any reasonable precision due to the inordinately large sample sizes that would be necessary to achieve their estimation.

6.4. Applying a decision-theoretical approach to selecting cost-effective monitoring tools for assessing filariasis control

It is essential from the perspectives of feasibility and sustainability that selected monitoring tools are cost-effective (Table 3.3). However, when assessing the cost-effectiveness of a monitoring tool for aiding the management of parasite control, it is important to not only consider the cost and performance of the tool, as is traditionally done, but also the costs of taking a management decision based on the data obtained with such a tool. Here, we show how model predictions of the effect of an intervention in conjunction with a decision-theoretical framework that sets the objective of minimizing the overall cost of monitoring and making a management decision can offer a useful method to achieve these dual goals and hence find the most cost-effective monitoring tool for evaluating filariasis control (Field *et al.*, 2004; Mapstone, 1995; Williams *et al.*, 2001). This approach is based on the expectation that failing to detect an abnormal impact of intervention on infection (a Type II error) during a monitoring bout will result in a serious failure in achieving a set programme target, while on the other hand, mistakenly concluding an aberrant effect of treatment (a Type I error) will usually cause a relatively smaller economic impact for an intervention programme (Fig. 3.11). The method thus involves the development of a cost function for carrying out a management decision based on three main components: (1) an estimate of the probability that an aberrant change in infection (compared to the expected effect size due to intervention) has occurred; (2) the probability that analysis of the monitoring data from using a particular diagnostic tool will correctly indicate whether that change has occurred and (3) the

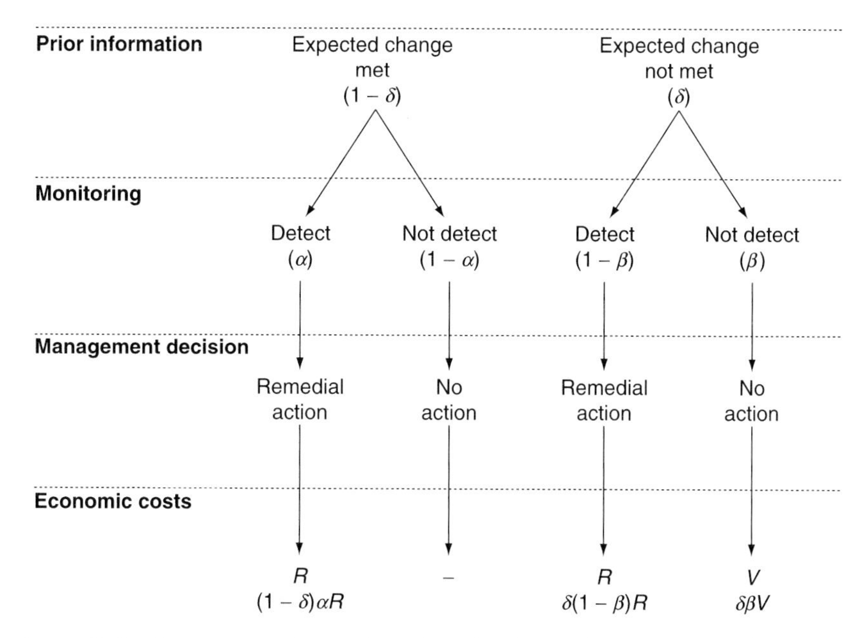

FIGURE 3.11 Decision tree diagram for calculating the total expected cost of a parasite control programme management decision based on monitoring data. δ, probability that a deviant response has occurred; α, Type I error rate; β, Type II error rate; *R*, cost of remedial action; *V*, economic cost of an undetected aberrant response.

monetary costs of actions triggered by the conclusions of the analysis (also referred to as 'utility'). By constructing a decision tree that incorporates these components (Fig. 3.11), multiplying down the branches and summing across their termini, an expression for the overall expected cost of monitoring and management may be easily derived as (Field *et al.*, 2004):

$$E[C] = (1 - \delta)\alpha R + \delta[(1 - \beta)R + \beta V] + M \quad (1)$$

where $E[C]$ is the expected total cost of monitoring and management, δ is the prior expectation of an aberrant change in infection occurring given pre-control endemicity levels, programme implementation statistics and the dynamics of indicator data used (note here that the more rapid the parasite attribute measured by an indicator declines due to intervention, the greater the power of detecting it), α is the Type I error rate (probability of falsely detecting an aberrant response), β is the Type II error rate (probability of missing a real aberrant outcome), R is the cost of remedial action, V is the economic loss associated with an undetected aberrant response (i.e., in the present case, the costs of extra years of treatment required to ensure attainment of target or the economic costs associated with reinfection when elimination targets are missed) and M is the economic cost of carrying out monitoring. The monitoring tool giving rise to the least total overall cost of monitoring and management using expression (1) will thus represent the most cost-effective among available diagnostic tools for undertaking filariasis monitoring.

7. USING MONITORING DATA FOR PROGRAMME MANAGEMENT

The arguments above indicate that monitoring programmes do not end merely with the collection of data, its analysis or even synthesis and summary. Instead, they imply that the results of parasite intervention monitoring programmes are of value to the extent that they provide information for making informed decisions regarding effective management of a control programme, particularly with respect to deciding if an intervention is on track to meet set programme objectives or if early adjustments can be triggered to mitigate against deficient outcomes as the intervention progresses. This conclusion indicates that monitoring results need to be effectively linked to management decision making if they are to be useful for undertaking successful parasite control (Kendall, 2001; Mulder *et al.*, 1999; Williams *et al.*, 2001). A core problem, however, relates to using the monitoring data for decision making in the context of uncertainty and incomplete information regarding parasite population dynamics and intervention effectiveness (Michael *et al.*, 2004, 2006a). This implies that a decision aid procedure that links monitoring data with uncertainty in current knowledge and management decisions will be required to facilitate the effective use of such data for guiding parasite control management actions. Here, we show how one formalization using the principles of statistical decision theory (Lindley, 1985) may offer a way forward to addressing this question. Briefly, statistical decision theory can account for uncertainty in guiding optimal decision making in the management of parasite control by: determining the probability of observing alternative infection outcomes as estimated using the monitoring data, describing the management decisions in response to these outcomes and estimating the 'utilities' associated with each possible combination of decision and intervention outcome (i.e., the costs of wrong decisions and misinterpretation of the monitoring signal) (Table 3.4).

As illustrated in Fig. 3.12, once these elements are quantified, the management decision is chosen that maximizes the expected (average) utility (Lindley, 1985; Mulder *et al.*, 1999; Williams *et al.*, 2001). The main feature of the method, which deals with uncertainty, is that the 'true' status of the parasite population is explicitly acknowledged as unknowable; instead, all we can do is to estimate the likelihood of the occurrence

TABLE 3.4 A sequential list of steps to follow while using decision theory to make an optimal decision in parasite control in the context of uncertainty and incomplete information (Mulder *et al.*, 1999)

Steps	Decisions
1	Set the bounds of the management decision space
2	Define a range of possible management actions in response to monitoring data
3	Estimate the probabilities associated with each possible interpretation of the monitoring data
4	Estimate the utilities associated with each possible combination of decision and monitoring data (i.e., the costs of wrong decisions and misinterpretation of the monitoring signal)
5	Determine the decision that maximizes utility

of different status categories based on the best available data taking into account diagnostic test accuracy and the expected effect size and its likely variation (which together is used to ascertain the probabilities of whether the monitored parasite population is on target to meet programme objectives given particular values of α and β) (Fig. 3.12). Another equally critical feature is that the method lends itself to revision and updating as new information becomes available during the intervention period leading to recomputation of the probabilities associated with the possible states of the parasite system. The decision process is then revisited to determine if a different decision [or a new management strategy, e.g., adding vector control to MDA (Michael *et al.*, 2004, 2006a)] now maximizes overall utility. This example shows that using monitoring data for decision making in parasite control will require the institution of a process, such as the application of statistical decision theory, to connect the questions of decision makers to the analysis and summary of the monitored data. It also highlights the critical role that mathematical models can play in the development and implementation of such decision tools.

8. UNCERTAINTY, MONITORING AND ADAPTIVE MANAGEMENT

The foregoing sections show that monitoring of parasite control is ultimately intended to track the progress of an intervention towards meeting a management objective. A difficulty is that it implies that current knowledge regarding the population dynamics of parasite control is sufficient to set monitoring targets and thresholds that allow achievement of these intervention goals. For filariasis, as noted in previous

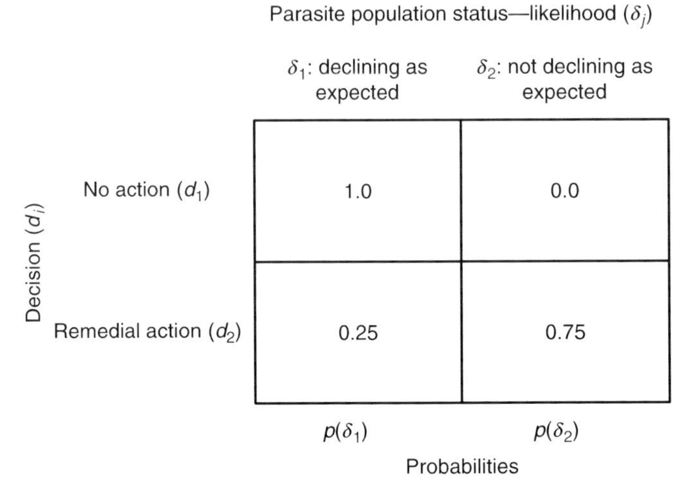

FIGURE 3.12 Hypothetical utility table illustrating the likelihood of different parasite population states given monitoring data, the possible management decisions, and the utilities associated with the combinations of states and decisions. Monitoring data is first used in conjunction with model predictions to estimate the probability that observed parasite population trend is declining as expected ($p(\delta_1)$), or not declining as expected ($p(\delta_2)$). The alternative decisions (here either no action or take remedial action) for each of these states are then presented in a 2-way decision table as in the figure. The next task in using this approach is to assign values or 'utilities', $u(d_i,\delta_j)$ associated with the various outcomes. Utilities are scaled to the unit interval, with $u = 1$ 'best' and $u = 0$ 'worst'. Thus, in the figure, if the parasite population is accepted to be declining as expected, the decision of not taking an action is best. However, taking a remedial action when none is needed in this case may be assigned a lower utility because of the cost of the action and to hedge against the fact that our acceptance of expected decline is based on statistical probability and not certainty. Once the elements of the table are complete (d_i, δ_j, and u_{ij} (d_i)), the management decision (d_i) is chosen that maximizes the expected (average) utility ($u(d_i)$ (Burgman, 2005; Mulder *et al.*, 1999).

sections, major uncertainties exist in almost every area of our understanding of parasite population dynamics and control, including (1) model and process uncertainty that gives rise to approximate thresholds and predictions of the effects of intervention, (2) observation uncertainty related to diagnostic tool accuracy and the roles of spatial and population ecologies on the sampling process and (3) intervention uncertainty in which there is a little reliable information available regarding the effectiveness of the various proposed interventions to reduce or interrupt parasite transmission (Hughes *et al.*, 1995; Michael *et al.*, 2006a,b; Venette *et al.*, 2002). This situation implies that currently running or soon to be implemented control programmes must proceed in the face of all these sources of uncertainty.

Recent work in fisheries, forestry and wildlife management, however, has shown how combining monitoring results with a management method that forces us to acknowledge uncertainty can provide a process to help resolve these uncertainties and hence revise and improve our understanding of parasite control. Termed adaptive management, this process involves using current interventions and responses of a system to management actions to 'learn' about the behaviour of that system, which is then used to modify or adapt future management actions (Holling, 1978; Parma and NCEAS working group on population management, 1998; Shea *et al.*, 2002; Walters, 1986; Walters and Hilborn, 1978). Briefly, it may be thought of as managing ecological interventions with a plan for learning about the system. Its most effective form—*active adaptive management*—uses management programmes that are designed to experimentally compare selected policies or practices in order to evaluate alternate hypotheses about the system being managed. Thus, when used in this way, monitoring results may either support or conflict with current understanding, thereby inspiring a development of knowledge that includes resolving gaps in system or transmission model dynamics and reinforcing or improving well-understood relationships. Finally, this approach can help prioritize future research according to such gaps.

The above considerations suggest that there is an imperative need to modify existing management models for filariasis control in order to incorporate adaptive management principles. Specifically, we need to develop an adaptive management plan that (1) acknowledges uncertainty, (2) contemplates using monitoring system responses to interventions to learn about the system and (3) anticipates that future management interventions will be modified as we gather information and learn more about system behaviour (Fig. 3.13). Such a management approach will also require the close linking of decision makers and mangers with research groups, which in turn will allow not only the tailoring of research to answering management problems in parasite control but also the rapid transfer of research outcomes to the field (Dent, 2000). Of course, developing and instituting an adaptive parasite management plan will have a cost. However, we suggest that this short-term cost is well worth paying if our ignorance about how nature works in reality leads to an even bigger loss from present investments in parasite control, for example when choices made in a myopic way, which ignore future learning, prevents the attainment of intervention goals.

9. CONCLUSIONS

This chapter highlights three major conclusions regarding the role of epidemiological modelling for the development of effective monitoring plans for aiding the evaluation of lymphatic filariasis elimination. First, successful

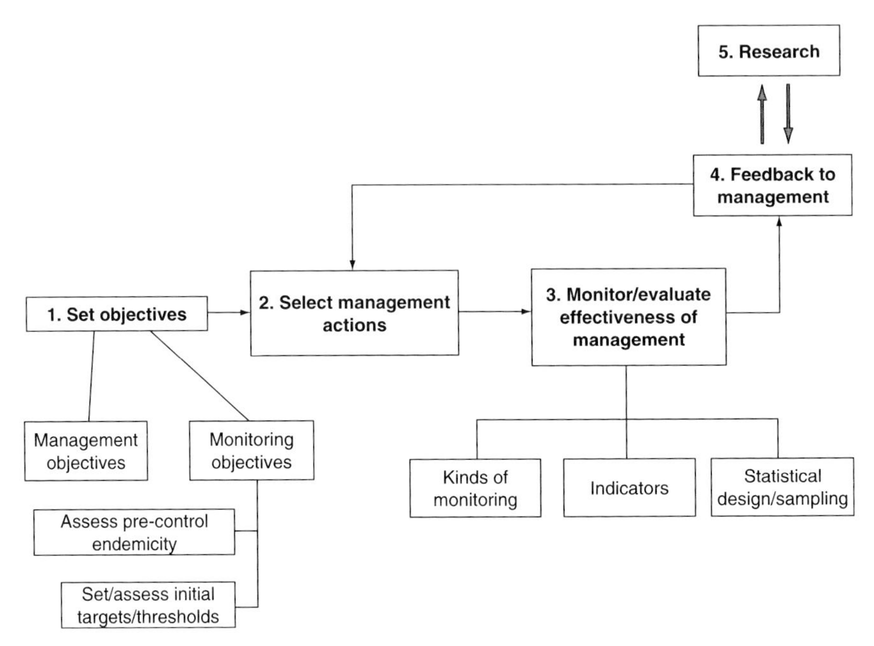

FIGURE 3.13 The main steps and components in developing a sound adaptive management process for parasite control programmes. The process begins by the setting up of both management and monitoring objectives, followed by the selection of the appropriate management action, its monitoring and evaluation, and finally using feedback from the evaluation to adaptively alter management actions as well as guide required research. Mathematical models can clearly guide each of these steps.

monitoring of parasite control strategies is intricately related to the underlying parasite transmission dynamics since these govern the target thresholds that signify various management goals related to either elimination or parasite control and provide the expected magnitudes of change needed to determine if interventions are proceeding as expected. It is also the dynamics of the parasite population response to intervention episodes that govern the optimal frequency of monitoring, warning of deficiencies in effectiveness and suggestion of remedial actions to get the intervention back on track to meet set endpoint targets. Parasite population dynamics also inform the selection of appropriate indicators for monitoring the effects of interventions via providing comparative information on their expected effect sizes. We suggest that a purely empirical approach that does not consider how parasite transmission processes and components interact dynamically with a perturbation is unlikely to fulfil these functions.

These considerations in turn imply that developing effective parasite control monitoring plans will be enhanced if their designs, including selection of indicators, are based on the use of mathematical models that organize, depict and integrate best existing scientific knowledge on the processes and components underlying the population dynamics of parasite transmission and control. This chapter has aimed to show how such models can provide the scientific framework for the development and implementation of sound filariasis monitoring plans by synthesizing current understanding of parasite system dynamics, identifying important processes and illustrating dynamic linkages between indicators and parasite population states or processes. Yet, surprisingly, existing models so far have been little used in the design of monitoring programmes. We suggest that rectifying this situation be a priority in the management of currently running or soon to be initiated parasite control programmes.

The second major contribution of this chapter is to clarify that the essential purpose of parasite control monitoring is to allow decision making regarding whether an intervention is on track to meet a set goal. This implies that development of monitoring plans be based on clear articulation and selection of intervention objectives, available management actions, sensitive indicators of parasite population change related to an implemented intervention and statistically valid methods for collection of data using these indicators. From a practical perspective, these considerations suggest that monitoring should be conducted only if opportunities to allow change in management actions exist. If no alternative management options are available, expending resources in monitoring is futile as information from carrying out such an exercise is wasted. It also suggests that simply gathering monitoring information without designing decision aids that can be used by managers to determine the effects of management actions, such as that based on statistical decision theory discussed in this chapter (Section 7), is again not useful. We recommend that emphasis be given to the development of such decision-making aids, linked closely to mathematical models, in the development of parasite monitoring plans.

Finally, we have demonstrated that monitoring and evaluation activities must be viewed more as an adaptive management rather than a one-step process (Holling, 1978; Parma and NCEAS working group on population management, 1998; Shea *et al.*, 2002; Walters, 1986; Walters and Hilborn, 1978). This is because of the important need to acknowledge uncertainties in our understanding of parasite population dynamics and the effects of interventions, which hampers predictions on whether our interventions are likely to meet set management goals. By contrast, adaptive parasite control management by explicitly acknowledging the existence of uncertainty and the importance of the need to incorporate learning into management aims to institute a plan with the following goals: (1) manage to the best of current knowledge, (2) learn from management actions and (3) improve/modify management in the future.

Again, mathematical models can play a critical role in this process by facilitating the iterative use of monitoring data to better refine our understanding of both the structure of the underlying parasite population dynamics and intervention effectiveness, and hence improve predictions of the effects of management actions. We indicate an imperative need to reassess current management strategies at all levels to ensure that these adaptive principles of management are adopted as far as possible. The success of our ambitious goals for controlling or eliminating lymphatic filariasis, and indeed other major human helminthiases, may well depend on evolving and implementing such innovative management changes.

ACKNOWLEDGEMENTS

E.M. and M.N.M.-L. are grateful to the British Council Higher Education Link Programme for financial support of this work. Collaboration between E.M. and J.W.K. was made possible by a United States Public Health Service NIH ICIDR grant U19 AI33061 and a R01 grant AI69387. Contributions by J.W.K. were in part also supported by the United States Public Health Service NIH grant U19 AI065717.

REFERENCES

Albonico, M., Montressor, A., Crompton, D. W. T., and Savioli, L. (2006). Intervention for the control of soil-transmitted helminthiasis in the community. *Adv. Parasitol.* **61,** 311–348.

Anderson, J. L. (1998). Errors of inference. *In* "Statistical Methods for Adaptive Management Studies" (V. Sit and B. Taylor, eds.), pp. 69–87. B.C. Ministry of Forests, Victoria, BC, Canada.

Anderson, R. M., and May, R. M. (1991). "Infectious Diseases of Humans. Dynamics and Control." Oxford University Press, Oxford.

Aslamkhan, M., and Wolfe, M. S. (1972). Bancroftian filariasis in two villages in Dinajpur District, East Pakistan. II. Entomological investigations. *Am. J. Trop. Med. Hyg.* **21,** 30–37.

Barrett, S., and Hoel, M. (2004). Optimal disease eradication. *The Fondazione Eni Enrico Mattei Note di Lavoro Series.* No. 2004.50.

Basu, P. C., Rao, V. N., and Pattanayak, S. (1965). Filariasis in greater Bombay – results of a rapid survey conducted in June, 1965. *Bulletin of the National Society of India Malaria and Mosquito Diseases* **13,** 296–317.

Bender, R. (1999). Quantitative risk assessment in epidemiological studies investigating threshold effects. *Biometrical Journal* **41,** 305–319.

Binns, M. R., and Nyrop, J. P. (1992). Sampling insect populations for the purpose of IPM decision making. *Annu. Rev. Entomol.* **37,** 427–453.

Binns, M. R., Nyrop, J. P., and van der Werf, W. (2000). "Sampling and Monitoring in Crop Protection." CAB International, Wallingford, UK.

Bockarie, M. J., Alexander, N. D., Hyun, P., Dimber, Z., Bockarie, F., Ibam, E., Alpers, M. P., and Kazura, J. W. (1998). Randomised community-based trial of annual single-dose diethylcarbamazine with or without ivermectin against *Wuchereria bancrofti* infection in human beings and mosquitoes. *Lancet* **351,** 162–168.

Boyle, M., Kay, J., and Pond, B. (2001). Monitoring in support of policy: An adaptive ecosystem apprach. *In* "Encyclopedia of Global Environmental Change" (M. K. Tolba, ed.), pp. 116–137. John Wiley & Sons, New York, NY.

Brooker, S., and Michael, E. (2000). The potential of geographical information systems and remote sensing in the epidemiology and control of human helminth infections. *Adv. Parasitol.* **47,** 245–288.

Brooker, S., Whawell, S., Kabatereine, N. B., Fenwick, A., and Anderson, R. M. (2004). Evaluating the epidemiological impact of national control programmes for helminths. *Trends Parasitol.* **20,** 537–545.

Burgman, M. A. (2005). ''Risks and Decisions for Conservation and Environmental Management.'' Cambridge University Press, Cambridge, UK.

Burkot, T., Durrheim, D., Melrose, W., Speare, R., and Ichimori, K. (2006). The argument for integrating vector control with multiple drug administration compaigns to ensure elimination of lymphatic filariasis. *Filaria J.* **5,** 10.

Bushrod, F. M. (1979). Studies on filariasis transmission in Kwale, a Tanzanian coastal village, and the results of mosquito control measures. *Ann. Trop. Med. Parasitol.* **73,** 277–285.

Chan, M. S., Srividya, A., Norman, R. A., Pani, S. P., Ramaiah, K. D., Vanamail, P., Michael, E., Das, P. K., and Bundy, D. A. (1998). Epifil: A dynamic model of infection and disease in lymphatic filariasis. *Am. J. Trop. Med. Hyg.* **59,** 606–614.

Chand, D., Singh, M. V., Gupta, B. B., and Srivastava, R. N. (1961a). A note on filariasis in Gonda Town (Utter Pradesh). *Indian J. Malariol.* **15,** 39–47.

Chand, D., Singh, M. V., and Pathak, V. K. (1961b). Filariasis in the District of Ghazipur (Uttar Pradesh). *Indian J. Malariol.* **15,** 21–29.

Cressie, N. A. C. (1993). ''Statistics for Spatial Data.'' Wiley, New York, NY.

de Vlas, S. J., Nagelkerke, N. J. D., Habbema, J. D. F., and Van Oortmarssen, G. J. (1993). Statistical models for estimating prevalence and incidence of parasite diseases. *Stat. Methods Med. Res.* **2,** 3–21.

Dent, D. (2000). ''Insect Pest Management.'' CABI Publishing, Wallingford, Oxon.

Deredec, A., and Courchamp, F. (2003). Extinction thresholds in host-parasite dynamics. *Ann. Zool. Fennici* **40,** 115–130.

Doenhoff, M. J., Chiodini, P. L., and Hamilton, J. V. (2004). Specific and sensitive diagnosis of schistosome infection: Can it be done with antibodies. *Trends Parasitol.* **20,** 35–39.

Dreyer, G., Pimentael, A., Medeiros, Z., Beliz, F., Moura, I., Coutinho, A., De Andrade, L. D., Rocha, A., Da Silva, L. M., and Piessens, W. F. (1996). Studies on the periodicity and intravascular distribution of *Wuchereria bancrofti* microfilariae in paired samples of capillary and venous blood from Recife, Brazil. *Trop. Med. Int. Health* **1,** 264–272.

Duerr, H. P., Dietz, K., and Eichner, M. (2005). Determinants of the eradicability of filarial infections: A conceptual approach. *Trends Parasitol.* **21,** 88–96.

Dzodzomenyo, M., Dunyo, S. K., Ahorlu, C. K., Coker, W. Z., Appawu, M. A., Pedersen, E. M., and Simonsen, P. E. (1999). Bancroftian filariasis in an irrigation project community in southern Ghana. *Trop. Med. Int. Health* **4,** 13–18.

Elzinga, C. L., Salzer, D. W., Willoughby, J. W., and Gibbs, J. P. (2001). ''Monitoring Plant and Animal Populations.'' Blackwell Science Inc, Malden, MA, USA.

Fenwick, A., Rollinson, D., and Southgate, V. (2006). Implementation of human schistosomiasis control: Challenges and prospects. *Adv. Parasitol.* **61,** 567–622.

Field, S. A., Tyre, A. J., Jonzen, N., Rhodes, J. R., and Possingham, H. P. (2004). Minimizing the cost of environmental management decisions by optimizing statistical thresholds. *Ecol. Lett.* **7,** 669–675.

Fischer, P., Boakye, D., and Hamburger, J. (2003). Polymerase chain reaction-based detection of lymphatic filariasis. *Med. Microbiol. Immunol.* **192,** 3–7.

Gersovitz, M., and Hammer, J. S. (2003). Infectious diseases, public policy, and the marriage of economics and epidemiology. *World Bank Res. Obs.* **18,** 129–157.

Gersovitz, M., and Hammer, J. S. (2004). The economic control of infectious diseases. *Econ. J.* **114,** 1–27.

Gibbs, J. P., Snell, H. L., and Causton, C. E. (1999). Effective monitoring for adaptive wildlife management: Lessons from the Galapagos islands. *J. Wildl. Manage.* **63,** 1055–1065.
Gollier, C., and Treich, N. (2003). Decision-making under scientific uncertainty: The economics of the precautionary principle. *J. Risk Uncertain.* **27,** 77–103.
Gubler, D. J., and Bhattacharya, N. C. (1974). A quantitative approach to the study of Bancroftian filariasis. *Am. J. Trop. Med. Hyg.* **23,** 1027–1036.
Gyldmark, M., and Alban, A. (1998). Economic evaluation of programmes aiming at eradicating infectious diseases. *Health Policy* **45,** 69–79.
Gyorkos, T. W. (2003). Monitoring and evaluation of large scale helminth control programmes. *Acta Trop.* **86,** 275–282.
Habbema, J. D., Alley, E. S., Plaisier, A. P., Van Oortmarssen, G. J., and Remme, J. H. (1992). Epidemiological modelling for onchocerciasis control. *Parasitol. Today* **8,** 99–103.
Habicht, J. P., Victoria, C. G., and Vaughan, J. P. (1999). Evaluation designs for adequacy, plausibility and probability of public health programme performance and impact. *Int. J. Epidemiol.* **28,** 10–18.
Hati, A. K., Chandra, G., Bhattacharyya, A., Biswas, D., Chatterjee, K. K., and Dwibedi, H. N. (1989). Annual transmission potential of Bancroftian filariasis in an urban and a rural area of West Bengal, India. *Am. J. Trop. Med. Hyg.* **40,** 365–367.
Holling, C. S. (1978). ''Adaptive Environmental Assessment and Management.'' John Wiley & Sons, New York, NY.
Hughes, G., Madden, L. V., and Munkvold, G. P. (1995). Cluster sampling for disease incidence data. *Phytopathology* **86,** 132–137.
Hunt, D. L., and Rai, S. N. (2005). A new threshold dose-response model including random effects foe data from development toxicity studies. *J. Appl. Toxicol.* **25,** 435–439.
Kendall, W. L. (2001). Using models to facilitate complex decisions. *In* ''Modeling in Natural Resource Management'' (T. M. Shenk and A. B. Franklin, eds.), pp. 147–170. Island Press, Washington.
Lammie, P. J., Weil, G., Noordin, R., Kaliraj, P., Steel, C., Goodman, D., Lakshmikanthan, V. B., and Ottesen, E. (2004). Recombinant antigen-based antibody assays for the diagnosis and surveillance of lymphatic filariasis – a multicenter trial. *Filaria J.* **3,** 9.
Lindley, D. V. (1985). ''Making Decisions.'' Wiley, New York, USA.
Lindsay, S. W., and Thomas, C. J. (2000). Mapping and estimating the population at risk from lymphatic filariasis in Africa. *Trans. R. Soc. Trop. Med. Hyg.* **94,** 37–45.
Madden, L. V., and Hughes, G. (1999). Sampling for plant disease incidence. *Phytopathology* **89,** 1088–1103.
Maddox, D. E., Polani, K., and Unnasch, R. (1999). Evaluating management success: Using ecological models to ask the right monitoring questions. *In* ''Ecological Stewardship. A Common Reference for Ecosystem Management'' (N. C. Johnson, A. J. Malk, R. C. Szaro, and W. T. Sexton, eds.), pp. 563–584. Elsevier Science Ltd., Oxford.
Mapstone, B. (1995). Scalable decision rules for environmental impact studies: Effect size, type 1, and type 2 errors. *Ecol. Appl.* **5,** 401–410.
Margoluis, R., and Salafsky, N. (1998). ''Measures of Success: Designing, Managing, and Monitoring Conservation and Development Projects.'' Island Press, Washington, DC.
McMahon, J. E., Marshall, T. F., Vaughan, J. P., and Abaru, D. E. (1979). Bancroftian filariasis: A comparison of microfilariae counting techniques using counting chamber, standard slide and membrane (nuclepore) filtration. *Ann. Trop. Med. Parasitol.* **73,** 457–464.
Michael, E. (2000). The population dynamics and epidemiology of lymphatic filariasis. *In* ''Lymphatic Filariasis'' (T. B. Nutman, ed.), pp. 41–81. Imperial College Press, London.
Michael, E. (2002). The epidemiology of filariasis control. *In* ''The Filaria'' (T. R. Klei and T. V. Rajan, eds.), pp. 60–74. Kluwer Academic Publishers, Boston.
Michael, E., Simonsen, P. E., Malecela, M., Jaoko, W. G., Pedersen, E. M., Mukoko, D., Rwegoshora, R. T., and Meyrowitsch, D. W. (2001). Transmission intensity and the

immunoepidemiology of Bancroftian filariasis in East Africa. *Parasite Immunol.* **23,** 373–388.

Michael, E., Malecela-Lazaro, M. N., Simonsen, P. E., Pedersen, E. M., Barker, G., Kumar, A., and Kazura, J. W. (2004). Mathematical modelling and the control of lymphatic filariasis. *Lancet Infect. Dis.* **4,** 223–234.

Michael, E., Malecela-Lazaro, M. N., Kabali, C., Snow, L. C., and Kazura, J. W. (2006a). Mathematical models and lymphatic filariasis control: Endpoints and optimal interventions. *Trends Parasitol.* **22,** 226–233.

Michael, E., Malecela-Lazaro, M. N., Maegga, B. T. A., Fischer, P., and Kazura, J. W. (2006b). Mathematical models and lymphatic filariasis control: Monitoring and evaluating interventions. *Trends Parasitol.* **22,** 529–535.

Molyneux, D. H., and Nantulya, V. M. (2004). Linking disease control programmes in rural Africa: A pro-poor strategy to reach Abuja targets and millennium development goals. *Br. Med. J.* **328,** 1129–1132.

Molyneux, D. H., and Zagaria, N. (2002). Lymphatic filariasis elimination: Progress in global programme development. *Ann.Trop. Med. Parasitol.* **96**(Suppl. 2), S15–S40.

Molyneux, D. H., Hotez, P. J., and Fenwick, A. (2005). "Rapid-Impact Interventions": How a policy of integrated control for Africa's neglected tropical diseases could benefit the poor. *PLoS Med.* **2,** 1064–1070.

Moulia-Pelat, J. P., Glaziou, P., Nguyen-Ngoc, L., Cardines, D., Spiegel, A., and Cartel, J. L. (1992). A comparative study of detection methods for evaluation of microfilaremia in lymphatic filariasis control programmes. *Trop. Med. Parasitol.* **43,** 146–148.

Mulder, B. S., Noon, B. R., Spies, T. A., Raphael, M. G., Palmer, G. J., Olsen, A. R., Reeves, G. H., and Welsh, H. H. (1999). "The strategy and design of the effectiveness monitoring program for the Northwest Forest Plan." *General Technical Report.* Department of Agriculture, Forest Service, Pacific Northwest Research Station, Portland, OR, USA.

Nair, C. P., Radhagovinda, R., and Joseph, C. (1960). Filariasis in Kerala State. VI. Filaria survey of Fort Cochin municipality. *Indian J. Malario.* **14,** 223–231.

Nicolas, L. (1997). New tools for diagnosis and monitoring of Bancroftian filariasis parasitism: The polynesian experience. *Parasitol. Today* **13,** 370–375.

Norman, R. A., Chan, M. S., Srividya, A., Pani, S. P., Ramaiah, K. D., Vanamail, P., Michael, E., Das, P. K., and Bundy, D. A. (2000). EPIFIL: The development of an age-structured model for describing the transmission dynamics and control of lymphatic filariasis. *Epidemiol. Infect.* **124,** 529–541.

Nyrop, J. P., Binns, M. R., and van der Werf, W. (1999). Sampling for IPM decision making: Where should we invest time and resources? *Phytopathology* **89,** 1104–1111.

Ottesen, E. A. (2006). Lymphatic filariasis: Treatment, control and elimination. *Adv. Parasitol.* **61,** 395–441.

Ottesen, E. A., Duke, B. O., Karam, M., and Behbehani, K. (1997). Strategies and tools for the control/elimination of lymphatic filariasis. *Bull. World Health Organan.* **75,** 491–503.

Parma, A. M., and NCEAS working group on population management (1998). What can adaptive management do for our fish, forests, food, and biodiversity? *Integrative Biology* **1,** 16–26.

Pawar, R. G., and Mittal, M. C. (1968). Filariasis in Jamnagar. *Indian J. Med. Res.* **56,** 370–376.

Pepe, M. S. (2003). "The Statistical Evaluation of Medical Tests for Classification and Prediction." Oxford University Press, Oxford, UK.

Pichon, G. (2002). Limitation and facilitation in the vectors and other aspects of the dynamics of filarial transmission: The need for vector control against Anopheles-transmitted filariasis. *Ann. Trop. Med. Parasitol.* **96**(Suppl 2), S143–S152.

Pugliese, A., and Tonetto, L. (2004). Thresholds for macroparasite infections. *J. Math. Biol.* **49,** 83–110.

Pugliese, A., Rosa, R., and Damaggio, M. L. (1998). Analysis of a model for macroparasitic infection with variable aggregation and clumped infections. *J. Math. Biol.* **36,** 419–447.

Rajagopalan, P. K., Kazmi, S. J., and Mani, T. R. (1977). Some aspects of transmission of *Wuchereria bancrofti* and ecology of the vector Culex pipiens fatigans in Pondicherry. *Indian J. Med. Res.* **66,** 200–215.

Ramaiah, K. D., and Das, P. K. (2004). Mass drug administration to eliminate lymphatic filariasis in India. *Trends Parasitol.* **20,** 499–502.

Ramaiah, K. D., Pani, S. P., Balakrishnan, N., Sadanandane, C., Das, L. K., Mariappan, T., Rajavel, A. R., Vanamail, P., and Subramanian, S. (1989). Prevalence of Bancroftian filariasis & its control by single course of diethyl carbamazine in a rural area in Tamil Nadu. *Indian J. Med. Res.* **89,** 184–191.

Ramaiah, K. D., Das, P. K., Michael, E., and Guyatt, H. (2000). The economic burden of lymphatic filariasis in India. *Parasitol. Today* **16,** 251–253.

Rao, C. K., Sundaram, R. M., Krishna Rao, C., Rao, J. S., and Venkata Narayana, M. (1976). Trend of Bancroftian filariasis in two villages and long-term effects of mass diethylcarbamazine treatment. *J. Commun. Dis.* **8,** 28–34.

Rao, C. K., Sundaram, R. M., Venkatanarayana, M., Rao, J. S., and Chandrasekharan, A. (1981). Epidemiological studies on Bancroftian filariasis in East Godavari District (Andhra Pradesh): Entomological aspects. *J. Commun. Dis.* **13,** 81–91.

Rao, R. U., Atkinson, L. J., Ramzy, R. M., Helmy, H., Farid, H. A., Bockarie, M. J., Susapu, M., Laney, S. J., Williams, S. A., and Weil, G. J. (2006). A real-time PCR-based assay for detection of *Wuchereria bancrofti* DNA in blood and mosquitoes. *Am. J. Trop. Med. Hyg.* **74,** 826–832.

Rogers, D. J., and Williams, B. G. (1993). Monitoring trypanosomiasis in space and time. *Parasitology* **106,** S77–S92.

Rozeboom, L. E., Bhattacharya, N. C., and Gilotra, S. K. (1968). Observations on the transmission of filariasis in urban Calcutta. *Am. J. Trop. Med. Hyg.* **87,** 616–631.

Sabry, M. (1991). A quantitative approach to the relationship between *Wuchereria bancrofti* microfilaria counts by venous blood filtration and finger-prick blood films. *Trans. R. Soc. Trop. Med. Hyg.* **85,** 506–510.

Savioli, L., Albonico, M., Engels, D., and Montresor, A. (2004). Progress in the prevention and control of schistosomiasis and soil-transmitted helminthiasis. *Parasitol. Int.* **53,** 103–113.

Shea, K., Possingham, H. P., Murdoch, W. W., and Roush, R. (2002). Active adaptive management in insect pest and weed control: Intervention with a plan for learning. *Ecol. Appl.* **12,** 927–936.

Simonsen, P. E., and Dunyo, S. K. (1999). Comparative evaluation of three new tools for diagnosis of Bancroftian filariasis based on detection of specific circulating antigens. *Trans. R. Soc. Trop. Med. Hyg.* **93,** 278–282.

Simonsen, P. E., Meyrowitsch, D. W., Mukoko, D. A., Pedersen, E. M., Malecela-Lazaro, M. N., Rwegoshora, R. T., Ouma, J. H., Masese, N., Jaoko, W. G., and Michael, E. (2004). The effect of repeated half-yearly diethylcarbamazine mass treatment on *Wuchereria bancrofti* infection and transmission in two East African communities with different levels of endemicity. *Am. J. Trop. Med. Hyg.* **70,** 63–71.

Snow, L. C., and Michael, E. (2002). Transmission dynamics of lymphatic filariasis: Density-dependence in the uptake of *Wuchereria bancrofti* microfilariae by vector mosquitoes. *Med. Vet. Entomol.* **16,** 409–423.

Snow, L. C., Bockarie, M. J., and Michael, E. (2006). Transmission dynamics of lymphatic filariasis: Vector-specific density dependence in the development of *Wuchereria bancrofti* infective larvae in mosquitoes. *Med. Vet. Entomol.* **20,** 261–272.

Srividya, A., Michael, E., Palaniyandi, M., Pani, S. P., and Das, P. K. (2002). A geostatistical analysis of the geographic distribution of lymphatic filariasis prevalence in southern India. *Am. J. Trop. Med. Hyg.* **67,** 480–489.

Stolk, W. A., Swaminathan, S., Van Oortmarssen, G. J., Das, P. K., and Habbema, J. D. (2003). Prospects for elimination of Bancroftian filariasis by mass drug treatment in Pondicherry, India: A simulation study. *J. Infect. Dis.* **188,** 1371–1381.

Tobian, A. A., Tarongka, N., Baisor, M., Bockarie, M., Kazura, J. W., and King, C. L. (2003). Sensitivity and specificity of ultrasound detection and risk factors for filarial-associated hydroceles. *Am. J. Trop. Med. Hyg.* **68,** 638–642.

Tolle, T., Powell, D. S., Breckenridge, R., Cone, L., Keller, R., Kreshner, J., Smith, K. S., White, G. J., and Williams, G. L. (1999). Managing the monitoring and evaluation process. *In* "Ecological Stewardship. A Common Reference for Ecosystem Management" (N. C. Johnson, A. J. Malk, R. C. Szaro, and W. T. Sexton, eds.), pp. 585–601. Elsevier Science Ltd., Oxford.

Ulm, K. (1991). A statistical method for assessing a threshold in epidemiological studies. *Stat. Med.* **10,** 341–349.

van der Werf, M. J., De Vlas, S. J., Casper, W. N. L., Nagelkerke, N. J. D., Habbema, J. D. F., and Engels, D. (2002). Associating community prevalence of *Schistosoma mansoni* infection with prevalence of signs and symptoms. *Acta Trop.* **82,** 127–137.

Venette, R. C., Moon, R. D., and Hutchison, W. D. (2002). Strategies and statistics of sampling for rare individuals. *Annu. Rev. Entomol.* **47,** 143–174.

Walters, C. J. (1986). "Adaptive Management of Renewable Resources." McGraw-Hill, New York, NY.

Walters, C. J., and Hilborn, R. (1978). Ecological optimization and adaptive mangement. *Annu. Rev. Ecol. Syst.* **9,** 157–188.

Weil, G. J., Lammie, P. J., and Weiss, N. (1997). The ICT Filariasis Test: A rapid-format antigen test for diagnosis of Bancroftian filariasis. *Parasitol. Today* **13,** 401–404.

Whitehead, N., Hill, H. A., Brogan, D. J., and Blackmore-Prince, C. (2002). Exploration of threshold analysis in the relation between stressful life events and preterm delivery. *Am. J. Epidemiol.* **155,** 117–124.

Williams, B. K., Nichols, J. D., and Conroy, M. J. (2001). "Analysis and Management of Animal Populations. Modeling, Estimation, and Decision Making." Academic Press, San Diego, CA, USA.

Williams, S. A., Laney, S. J., Bierwert, L. A., Saunders, L. J., Boakye, D. A., Fischer, P., Goodman, D., Helmy, H., Hoti, S. L., Vasuki, V., Lammie, P. J., Plichart, C., *et al.* (2002). Development and standardization of a rapid, PCR-based method for the detection of *Wuchereria bancrofti* in mosquitoes, for xenomonitoring the human prevalence of Bancroftian filariasis. *Ann. Trop. Med. Parasitol.* **96**(Suppl 2), S41–S46.

Woolhouse, M. E. (1992). On the application of mathematical models of schistosome transmission dynamics. II. Control. *Acta Trop.* **50,** 189–204.

Woolhouse, M. E. (1996). Mathematical models of transmission dynamics and control of schistosomiasis. *Am. J. Trop. Med. Hyg.* **55,** 144–148.

World Health Organization (1998a). "Guidelines for Certifying Lymphatic Filariasis Elimination (including Discussion of Critical Issues and Rationale)." WHO/FIL/99/197, Geneva.

World Health Organization (1998b). "Report of a WHO Informal Consultation on Epidemiologic Approaches to Lymphatic Filariasis Elimination: Initial Assessment, Monitoring and Certification." WHO/FIL/99/195, Geneva.

Yoccoz, N. G., Nichols, J. D., and Boulinier, T. (2001). Monitoring of biological diversity in space and time. *Trends Ecol. Evol.* **16,** 446–453.

Zhou, X.-H., Obuchowski, N. A., and Mcclish, D. K. (2002). "Statistical Methods in Diagnostic Medicine." John Wiley & Sons, Inc, New York, USA.

CHAPTER 4

The Role of Helminth Infections in Carcinogenesis

David A. Mayer* and **Bernard Fried**[†]

Contents

Abstract

This review examines the significant literature on the role of helminth infections in carcinogenesis. Both parasitic infections and cancer have complex natural histories and long latent periods during which numerous exogenous and endogenous factors interact to obfuscate causality. Although only two helminths, *Schistosoma haematobium* and *Opisthorchis viverrini*, have been proven to be definitely carcinogenic to humans, others have been implicated in facilitating malignant transformation. The known mechanisms of helminth-induced cancer include chronic inflammation, modulation of the host immune system, inhibition of intracellular communication, disruption of proliferation–antiproliferation pathways, induction of genomic instability and stimulation of malignant

* Department of Surgery, New York Medical College, Valhalla, New York 10595
† Department of Biology, Lafayette College, Easton, Pennsylvania 10842

Advances in Parasitology, Volume 65
ISSN 0065-308X DOI: 10.1016/S0065-308X(07)65004-0

stem cell progeny. Approximately 16% of all cancer cases worldwide are attributable to pathogenic agents, including schistosomes and liver flukes. This equates to 1,375,000 preventable cancer deaths per year. Means to reduce the incidence of helminth-associated malignancies are discussed.

1. INTRODUCTION

The purpose of this chapter is to explore the role of helminth infections in carcinogenesis. Johanes Fibiger won the only Nobel Prize for helminthology in 1926 for the induction of gastric cancer in rats by feeding them cockroaches infected with *Spiroptera neoplastica* larvae (Campbell, 1997). The idea of parasitic worms causing cancer was not new. Fibiger had been aware of the causal relationship of *Schistosoma haematobium* and human urinary bladder cancer. His dissection of wild rats had yielded remarkable finding—stomach tumours containing worms. The problem of proving the role of parasites in cancer induction is difficult because of their complex natural histories and long asymptomatic latent periods during which numerous endogenous and exogenous factors can interact to obfuscate causality (Herrera and Ostrosky-Wegman, 2001). Fibiger's worm, *S. neoplastica* (later renamed *Gongylonema neoplastica*), provided a unique scientific tool for the experimental induction of cancer in mammalian hosts. Fibiger's work was later criticized for the gastric tumours not being true cancers, but merely worm-induced hyperplasia associated with vitamin A deficiency. Today, the International Agency for Research on Cancer (IARC) has in part vindicated Fibiger's belief in parasite-induced cancer by labelling *S. haematobium* and *Opisthorchis viverrini* as definitely carcinogenic (group 1) and *Clonorchis sinensis* as probably carcinogenic (group 2) to humans (IARC, 1994).

Helminth infections are of enormous importance worldwide, and the number of people exposed or infected is overwhelming. Pisani *et al.* (1997) used mathematical models to estimate the percentage of all cancers that were due to infections, and called this number the 'attributable fraction'. They showed that 15.6% of all cancer cases worldwide were attributable to infections including those due to schistosomes and liver flukes. The infectious origin of a cancer implies that the cancer is in fact preventable (Kuper *et al.*, 2000). Pisani *et al.* (1997) found that if the infections were prevented, by education and public health initiatives, the result would be 9% fewer cancer cases in developed countries and 21% fewer in developing countries. The 'attributable fraction' accounted for 1,375,000 preventable cancer cases per year.

Two biomedical scientists with diverse backgrounds have written this review. The first (D.A.M.) is a general and vascular surgeon who developed an interest in clinical parasitology as a surgical resident at Cornell University from 1973 to 1978. The second (B.F.) is an experimental parasitologist who became interested in medical parasitology during his tenure as a Louisiana State University medical fellow in the Central American tropics in 1967. Until now, there has been no comprehensive review on the causal relationship between helminth infection and carcinogenesis. Information on the topic was formerly scattered worldwide in diverse medical and scientific journals. We organize the available literature according to the carcinogenic effects of the following trematodes: *S. haematobium*, *Schistosoma mansoni* and *Schistosoma japonicum*, *O. viverrini*, *C. sinensis* and *Fasciola hepatica*. Research on the carcinogenicity of selected Cestoidea and Nematoda is not covered in the text but is summarized in tabular format. A section summarizing the current understanding of the mechanisms of helminth-induced carcinogenesis follows. We conclude by identifying areas for future research. It is our sincere hope that this review increases awareness of helminth infections as a preventable cause of human cancer.

2. TREMATODA

2.1. *S. haematobium*

Mayer and Fried (2002) reported that more than 200 million people in 77 countries have contracted schistosomiasis, also known as snail fever or bilharziasis. Globally, 1 in 30 people are infected. In 1851, Theodore Bilharz identified the blood fluke responsible for endemic haematuria during an autopsy of a patient in Cairo, Egypt. In the early twentieth century, Sir Patrick Manson described the three major disease-causing species. Infection occurs when cercariae of *S. mansoni*, *S. haematobium* or *S. japonicum* are shed into fresh water by intermediate snail hosts and penetrate the skin of a definitive human host. The schistosomula migrate through pulmonary capillaries into the left heart and systemic circulation. *Schistosoma* is the only bisexual genus of the class Trematoda in humans. Male and female worms pair off with the long thin female (14 mm $\times$ 0.2 mm) residing within the gynecophoric canal of the shorter male (10 mm $\times$ 1.1 mm). The male uses an oral sucker to attach to the endothelial wall of blood vessels; *S. mansoni* and *S. japonicum* prefer the mesenteric veins, while *S. haematobium* prefers the venules of the urinary bladder. Adult worms do little damage to the human host. Damage is created by the eggs deposited by the parasite as they lodge in host tissue to form a granulomatous response. Eggs are also shed in the stool or urine, which may hatch upon reaching fresh water to become miracidia capable of penetrating the intermediate snail host. Since adult worm pairs do not multiply within the definitive human host, intensity of infection depends upon the degree of contact with cercariae.

Ferguson (1911) was the first to link bilharziasis with bladder cancer in Egypt. Chen and Mott (1989) observed that the endemicity of the parasite supports the association of bladder cancer with schistosomiasis. The incidence of bladder cancer in the Middle East and Africa is greater in areas with higher *S. haematobium* prevalence and less in areas with lower prevalence. Mostafa *et al.* (1999) noted that 60% of the Egyptian population is still at risk for infection with *S. haematobium*, with rural children of school age at particular risk because of their proximity to infected water. The overall prevalence of *S. haematobium* infection in Egypt remains at 37–48%. Bladder cancer represents 30.8% of the total cancer incidence in Egypt and is the most common type of cancer in males and the second most common, after breast cancer, in females. This compares to bladder cancer's 5th to 7th ranking in males and 7th to 14th ranking in females in schistosome-free nations such as the United States, United Kingdom, Germany and Turkey. The National Cancer Institute in Cairo reported that 7,746 (30.8%) of 25,148 new cancer cases indexed from 1970 through 1981 were bladder cancers presumably induced by *S. haematobium* infection. In Iraq, Zambia, Malawi and Kuwait where the intensity of *S. haematobium* infection is high, bladder cancer is the most common malignancy. Talib (1970) noted that in Iraq, *S. haematobium* is the only species of *Schistosoma* that infects inhabitants. Since the South is more rural and agricultural than the North, the incidence of infection rises from 10% to 90% as one travels from Northern to Southern Iraq. The frequency of bladder cancer parallels the intensity of bilharzial infection, with inhabitants of agricultural marshlands being at greatest risk.

In schistosome-free nations, bladder cancer is a disease whose incidence peaks in the sixth or seventh decade of life. Mostafa *et al.* (1999) noted that in Egypt, Iraq, Zambia, Zimbabwe, Malawi and Sudan, the incidence of bilharzial bladder cancer peaks from age 40 to 49. The gender ratio (male to female) for cancer of the bladder is 5:1 in endemic and 3:1 in non-endemic areas. This is predictable since agricultural workers, mostly men, in endemic areas have daily exposure to water infected with the cercariae of *S. haematobium*. Makhyoun *et al.* (1971) reported that in the Nile Delta, 99% of bladder cancers found in male agricultural labourers (fellahin) were associated with *S. haematobium* infection, as opposed to only 52% in males with lower-risk occupations. The association between urinary schistosomiasis and bladder cancer is also supported by histopathologic findings. In endemic areas with high worm burdens, squamous cell carcinoma of the bladder is most frequent, while transitional cell carcinoma predominates in areas of lower endemicity such as North America and Europe. El-Bolkainy *et al.* (1981) noted a 54–81% incidence of squamous cell cancer among all bladder cancers in endemic areas, as opposed to 3–10% in Western countries. Tricker *et al.* (1989) postulated that the higher incidence of squamous cell cancer might be due to exposure to carcinogens such as *N*-nitroso compounds that are present in larger amounts in bilharzial urine.

The IARC (1994) analyzed seven case–control studies of the association between cancer of the bladder and *S. haematobium* infection. Intensity of infection was measured by urinary egg counts, pelvic X-rays, bladder and rectal biopsies and examination of bladder tissue following digestion and centrifugation. Possible confounding of results by smoking, a recognized cause of bladder cancer in non-endemic countries, was considered in only one study but thought not to affect the validity of their results. Six of the seven studies showed a strong association between bladder cancer and *S. haematobium* infection with odds ratios ranging from 2 to 14 (using all other cancer cases as controls, relative to no such history and adjusted for age, smoking history, education, occupation and geographic area of origin). *S. haematobium* infection did predispose to bladder cancer, especially the squamous cell variety. The more heavily infected individuals were, the more likely they were to develop bladder cancer, and at a younger age.

Parra *et al.* (1991) noted that the majority of the pathological findings of schistosomiasis are due to an inflammatory and immunological response to egg deposition. Granulomatous areas form around eggs and induce an exudative cellular response consisting of lymphocytes, polymorphonuclear leukocytes and eosinophils. Smith and Christie (1986) described the early stage of *S. haematobium* infection, which is characterized by egg deposition at the lower ureters and bladder. Resultant perioval granulomas, fibrosis and muscular hypertrophy are seen histologically. In the ureter, lesions can cause stenosis, leading to hydronephrosis (dilatation of the renal calyces). In the bladder, masses of large granulomatous inflammatory polyps containing eggs are found at the bladder apex, dome, trigone and posterior wall. Polyps may ulcerate and slough, producing haematuria. Smith and Christie (1986) found hyperplasia of the urothelium in 38% of autopsied *S. haematobium* cases as opposed to 21% in non-infected cases, metaplasia in 31.6% versus 11.5% and dysplasia in 27.2% versus 8.5%.

Smith and Christie (1986) characterized late-stage infection as schistosomal bladder ulcers and sandy patches, irregular thickened or atrophic mucosa in the posterior bladder or trigone area. Histologically, fibrosis with some round cell infiltration is seen, with old granulomas containing calcified or disintegrating eggs. Cheever *et al.* (1978) noted that the inflammatory and fibrotic response to egg deposition could lead to calcification of the bladder, infection and stone disease. These late changes are most frequently associated with bladder cancer. El-Bolkainy *et al.* (1981) found *S. haematobium* eggs in 902 (82.4%) of 1,095 bladder cancer cases in Egypt. Patients with egg deposition in the bladder wall predominately developed squamous cell subtypes, and at an earlier age. The authors found 798 cases of squamous cell cancer, 691 of which occurred in *S. haematobium* positive bladders. The spectrum of hyperplasia, metaplasia, dysplasia and squamous cell cancer was associated with late-stage schistosomal infection. Christie *et al.* (1986) found that the ova burden surrounding bladder tumours was nearly twice that of non-tumourous areas of the bladder. Hussein *et al.* (2005) studied bilharzial granulomas and transformation of the urothelium in response to *S. haematobium* infection. They found that CD3+ T cells and CD68+ histiocytes were the predominate cell populations in these lesions. Alterations in the deposition of fibronectin and laminin basement membrane proteins were characteristic of schistosomal bladder changes. Fibronectin deposition increased as lesions progressed from cellular loose fibrillary networks to fibrocellular dense fibrillary networks, and finally to fibrotic tight conglomerates. Normal and metaplastic urothelium had continuous basement membranes, while breaks in basement membrane continuity appeared in some samples of dysplastic urothelium.

Inflammation appears to be a common theme in understanding schistosome-induced carcinogenesis. Rosin *et al.* (1994a) found that chronic inflammation and bladder irritation associated with schistosomiasis causes cancer induction at the sites of the inflammatory lesions. Marletta (1988) noted that inflammatory cells including neutrophils and macrophages release oxygen-derived free radicals that aid in the formation of *N*-nitrosamines and other carcinogenic compounds. Shaeter *et al.* (1988) reported mutations, sister chromatid exchanges, DNA strand breaks and other genotoxic effects caused by inflammatory cells. Dizdaroglu *et al.* (1993) identified hydroxyl radicals released from inflammatory cells as the probable causative agent of these changes. O'Brien (1988) noted that inflammatory cells were involved in activation of polycyclic aromatic hydrocarbons, aromatic amines and similar procarcinogens to their active carcinogenic forms. Aromatic amines are especially carcinogenic to bladder tissue, an effect enhanced by the increased inflammatory cells in *S. haematobium* infection.

Abdel *et al.* (2000) suggested the following sequence of carcinogenesis. Chronic infection leads to schistosome eggs becoming trapped in the bladder wall. Proliferation of cells in bladder mucosa results from the constant irritation and inflammation. Clones of neoplastic cells develop stimulated by *N*-nitrosamines and other environmental carcinogens (cigarette smoke and pesticides). Makhyoun (1974) noted that the low level of smoking among patients with *S. haematobium*-associated bladder cancer argues against a strong causative relationship. Raziuddin *et al.* (1993) found that tumour necrosis factor-α levels are greatly elevated in patients with *S. haematobium*-induced squamous cell bladder cancer. Monocytes were found to be the source of tumour necrosis factor-α secretion from these tumours. The role of activated monocytes in cytokine production, and their effect on immunomodulation and progression of malignancy is not fully understood. Rosin *et al.* (1994b) proposed that inflammation, augmented cell proliferation and genetic instability of the urothelium are central concepts in current theories of carcinogenesis for *S. haematobium* infection. The micronucleus test was used to detect chromosomal breakage in bladder tissue of schistosome-infected patients. *In vitro* studies were also done by incubating bladder cells with activated neutrophils in conditions simulating inflammation. Results demonstrated that activated inflammatory cells induced micronuclei in bladder cells. This effect was related to breakage of chromosome 11, commonly damaged during initiation of bladder cancer.

Mostafa *et al.* (1999) noted that the staple diet of Egyptian farmers consisting of fava beans, raw salted fish, cheeses and spices contains small amounts of *N*-nitrosamines and their metabolites. The ingestion of *N*-nitrosamines in the diet coupled with their endogenous production by bacterial synthesis in the bladder has stimulated interest in exploring the role of these compounds in the induction of bladder cancer. Mostafa *et al.* (1994) reported high levels of volatile and non-volatile nitrosamines, nitrites and nitrates in the urine of schistosome-infected patients. These levels were highest with *S. haematobium*-infected cases. Stuehr and Marietta (1985) identified that monocyte accumulation from chronic inflammation was the probable cause of the elevated *N*-nitroso compounds. Badawi *et al.* (1992) found evidence of promutagenic methylation damage in bladder DNA from patients with bladder cancer associated with schistosomiasis. *N*-nitroso compounds, formed endogenously by activated macrophages, can generate reactive by-products that can alkylate tissue components, such as DNA.

Sheweita *et al.* (2003) studied the effects of *S. haematobium* infection on drug-metabolizing enzymes in human bladder cancer tissue. The mixed function oxidase system includes phase I and phase II drug oxidation enzymes. Phase I drug oxidation proteins include *N*-nitrosodimethylamine-*N*-demethylase (NDMA), cytochrome b5 and aryl hydrocarbon hydrolase (AHH). Phase II agents include glutathione and glutathione-*S*-transferase (GST). Phase I and phase II agents as well as free radicals, such as thiobarbituric acid-reactive substances, were studied in three groups: normal bladder tissue, bladder cancer and *S. haematobium*-induced bladder cancer. In schistosome-infected bladder tissue, AHH activity increased by 50%, while NDMA and GST activity decreased by 65% and 56%, respectively. Infected bladder tissue also showed a 29% increase in reduced glutathione, and a 57% increase in free radicals over non-infected bladder cancer samples. The majority of free radicals are formed from the breakdown of hydrogen peroxide, and cause lipid peroxidation and subsequent organ damage. Lipid peroxidation is central to the inflammatory effect of schistosome infection, and results in instability of cell membranes. Sheweita *et al.* (2003) concluded that bladder tissues infected with *S. haematobium* showed significant alterations in the activity of phase I and II drug-metabolizing enzymes. This change could impair the ability of the bladder to detoxify endogenous substances. It might also potentiate the action of bladder carcinogens such as *N*-nitrosamines, already present in elevated quantities in the urine of schistosome-infected patients. Schistosome-infected bladders could then activate *N*-nitrosoamines into active intermediary compounds capable of binding with DNA producing methylation damage.

Badawi *et al.* (1992) noted that analysis of samples of bladder tissue and bladder cancer from Egyptian patients infected with *S. haematobium* revealed high levels of promutagenic DNA lesions such as 06-methyldeoxyguanosine. Mostafa *et al.* (1999) suggested that the presence of 06-methyldeoxyguanosine was consistent with the continuous exposure of bladder tissues to alkylating agents found in abundance in the urine of schistosome-infected individuals. Badawi *et al.* (1994) found that bladder tissue in general has a diminished capacity for repair of DNA damage, along with decreased 06-alkylguanine-DNA-alkyltransferase activity. Whether endogenously derived through activated macrophages, or exogenously through nitrosation reactions in the bladder, promutagenic unrepaired DNA damage may be responsible for the initiation of bladder cancer in *S. hematobium* infection. Gutierrez *et al.* (2004) studied CpG for the initiation of bladder cancer in *S. haematobium* infection. Methylation-specific polymerase chain reaction was applied to 12 cancer-related genes in 41 samples of Egyptian bladder cancer tissue. All but two cases had at least one methylated gene, and 45% had three or more. *S. haematobium*-associated bladder cancers had higher methylation indices, confirming increased epigenetic changes of the urothelium. More recently, Saad *et al.* (2006) confirmed that N7-methylguanine, a marker of exposure to methylation from carcinogenic *N*-nitroso compounds, was present in 93% of DNA isolates from Egyptian bladder tumour patients. This suggests a role for gene methylation in the initiation and progression of schistosome-induced bladder cancer.

Badawi *et al.* (1995) reported high levels of carcinogenic tryptophan metabolites in the urine of *S. haematobium*-infected bladder cancer patients. Abdel-Tawab *et al.* (1986) studied excretion of tryptophan metabolites from the kynurenine pathway and noted that 64% of schistosome-associated bladder cancer patients metabolized tryptophan abnormally. Niagi and Bender (1990) found reduction of urinary excretion of kynurenine, kynurenic acid and methyl pyradone carboxamide in a mouse model for schistosome infection. They suggested that removal of the amino acid from the bloodstream, through uptake by the parasite or its eggs, was a possible mechanism for the observed impairment of tryptophan metabolism. Earlier, Teulings *et al.* (1973)

had demonstrated that the presence of the disordered tryptophan metabolite 3-hydroxyanthranilic acid appeared due to the bladder cancer itself, and was reversible upon successful eradication of the cancer. Whether abnormal tryptophan metabolism has an initiating or promoting role in the development of schistosome-associated bladder cancer is unknown.

El-Mouelhi and Mansour (1990) suggested a link between elevated levels of β-glucuronidase in the urine of *S. haematobium*-infected patients and the development of cancer of the bladder. The source of the enzyme in the urine is unknown. Theories include production from the schistosome worms themselves aided by secondary bacterial infection, from cellular debris originating from the worm-induced bladder lesions or from granulocytes present in the urine. Gentile *et al.* (1985) suggested that β-glucuronidase might release active carcinogens by hydrolyzing carcinogenic glucuronide conjugates. Syrian golden hamsters infected with *S. haematobium* had enhanced metabolic activation of the bladder carcinogens 3,3′-dichlorobenzidine and 2-acteylaminofluorene in the presence of elevated urinary β-glucuronidase levels. Fripp (1988) postulated that the parasite might secrete an inactive glucuronide that becomes an active carcinogenic metabolite when hydrolyzed by β-glucuronidase in the bladder.

Gentile (1985) documented the importance of urinary retention, whether from fibrosis and obstruction of the bladder neck or from voluntary causes such as pain on urination, in prolonging the exposure of the bladder mucosa to various exogenous and endogenous carcinogens. Schistosome-induced urinary stasis allows increased absorption of carcinogens and thus plays an integral role in carcinogenesis. Recurrent bacterial urinary tract infections are known to be associated with squamous cell carcinomas of the bladder, even in the absence of concomitant *S. haematobium* infection. El-Hawey *et al.* (1989) found that 39–66% of hospitalized patients with schistosomiasis had urinary tract bacterial infections. Penaud *et al.* (1983) hypothesized that the bacteria may become fixed on the tegument or in the intestinal ceca of the worms in a symbiotic type relationship. Mikhail *et al.* (1982) showed in an animal hamster model that bacterial growth from dual infections of both schistosomes and *Salmonella paratyphi* was greater than in control animals with single infections of *Salmonella* only. Mostafa *et al.* (1994) isolated nitrate-reducing bacteria such as *Staphlococcus aureus*, *Staphlococcus albus*, *Proteus mirabilis*, *Klebsiella* species and *Escherichia coli*, from urine samples of *S. haematobium*-infected patients. Hill (1988) found that schistosome-related bacterial cystitis facilitates nitrosation of secondary amines from both exogenously ingested and endogenously produced nitrite resulting in the generation of carcinogenic *N*-nitrosamines.

Badawi *et al.* (1995) described the process of carcinogenesis as involving an initiating and promoting effect on target cells. First damage occurs to the DNA template which, unless repaired, can lead to irreversible changes in the complementary strand of DNA produced during the S-phase of the cell cycle. Somatic mutation results when the altered strand is used as a template. The promotion phase is characterized by stimulation of cell proliferation. Oncogenes and tumour supressor genes have been associated with many cancers in humans, and recently efforts have been made to study the specific genes involved in the induction of schistosome-associated bladder cancer. Oncogenes release protein products that directly influence cell cycle regulation. Alterations in oncogenes or their by-products can lead to uncontrolled cell proliferation and carcinogenesis.

Strohmeyer and Slamon (1994) studied the ras oncogene and the potential association with bladder cancer. The ras oncogene encodes a 21-kDa protein that affects signal transmission between the nucleus and tyrosine kinase receptors. The relationship between ras oncogenes and cancer of the bladder is unclear. Estimates of H-ras activation in bladder cancer range between 7% and 17%, with its expression being similar with or without concurrent schistosomal infection. The p53 tumour suppressor gene, located on the short arm of chromosome 17, encodes a protein that regulates DNA damage repair and controls aspects of the cell cycle involving cellular apoptosis and senescence. Mutation of p53 results in a reduction of DNA damage surveillance leading to instability of the genome and malignant transformation. Habuchi *et al.* (1993) found that 86% of Egyptian patients with schistosome-associated bladder cancer had p53 mutations in exons 5, 6, 8 and 10. Badawi (1996) noted that as tumour grade increased, so did the p53 mutation rate. Inactivation of p53 occurred in 0–38% of early-stage bladder cancers, compared to 33–86% in late-stage disease. Ramchurren *et al.* (1995) identified multiple inactivation events at the locus of p53 in *S. haematobium*-induced bladder cancer that may be linked to abnormal tryptophan metabolites and *N*-nitrosamines. Lozano *et al.* (1994) reported that nitric oxide produced by the inflammation from schistosomal egg deposition could cause p53 mutations through stimulation of endogenous *N*-nitroso compound generation resulting in DNA alkylation. Fadl-Elmula *et al.* (2002) investigated chromosomal imbalances in bilharzial-associated bladder cancer. DNA was obtained from 20 archival samples of schistosome-induced bladder lesions and was studied using comparative genomic hybridization methods. Six carcinomas and one granuloma had evidence of chromosomal imbalances, with loss of 9p being most frequent in bilharzial-associated cases.

Alterations in cell cycle control are central to the current theory of carcinogenesis. Cyclin-dependant kinases are enzymes that control the cell cycle from G1 through the M phase. Tamimi *et al.* (1996) found that deletions of p16INK4, a cyclin-dependant kinase inhibitor, occurred in 53% of samples from patients with schistosome-associated bladder cancer. Elissa *et al.* (2004) described the roles of p16INKA and p15INKB as tumour suppressors whose inactivation or deletion can lead to malignancy. The p21 gene inhibits cyclin-dependant kinases resulting in growth arrest at the G1/S transition. Tissue was harvested from 132 Egyptian patients with bladder cancer and compared to 50 normal samples. P21 testing was done with Western blot, and p16 and p15 gene deletions were tested by polymerase chain reaction. Cyclin D1, thought to control the cell cycle from G1 to S, was also tested by Western blot. P16 and p15 were deleted in 38.7% and 30.2% of bladder cancers, respectively. The deletions were associated with schistosomiasis and poor tumour differentiation. Cyclin D1 was expressed in 57.5% of tumours, and was associated with schistosomiasis and early stage, well differentiated cancers. Levels of p21 were lower in tumour than in normal tissue. Cyclin D1 overexpression was thought to be an early trigger of bladder carcinogenesis in schistosome-infected patients. Gonzalez-Zulueta *et al.* (1995) also identified gene deletions (of CDKN2 on chromosome 9p) in 92% of squamous cell carcinoma of the bladder cases.

Chaudhary *et al.* (1997) investigated the overexpression of the Bcl-2 gene in patients with schistosome-associated bladder cancer. The Bcl-2 gene was first found to be associated with B-cell leukaemia and lymphoma. When expressed, it overrides programmed cell apoptosis increasing the risk of genomic instability. Bcl-2 also interacts with various proto-oncogenes facilitating tumourigenesis. Bcl-2 was found upregulated in squamous cell but not in transitional cell cancer of the bladder. The overexpression of Bcl-2 in *S. haematobium*-induced bladder cancer is consistent with the predominance of squamous variety tumours. Mutations of p53 were found in 73% of tumours, Bcl-2 expression in 32% and abnormalities of both p53 and Bcl-2 in 13%. Loss of the normal reciprocal control mechanism for apoptosis was suggested in the subset of patients with overexpression of both p53 and Bcl-2. More recently, Swellam *et al.* (2004) studied Bcl-2 expression in 118 cases of bladder cancer from Cairo, Egypt. Sixty had schistosomiasis. Methods included enzyme assay, Western blot and immunodot blot. Bcl-2 was significantly overexpressed in schistosome-associated squamous cell cancer, suggesting a role in carcinogenesis. Haitel *et al.* (2001) investigated the relationship of apoptosis-related proteins to the clinical outcome of schistosome-induced bladder cancer. For squamous cell cancer, p53, MIB-1, Bcl-x and Bax were independent predictors of prognosis. Mao *et al.* (1996) predicted that microsatellite analysis for genomic instability might allow early detection of bladder cancer in the future.

El-Sheikh *et al.* (2001) studied the expression of cyclooxygenase enzymes in schistosome-associated bladder cancer. An immunohistochemistry protocol employing mouse monoclonal antibodies was applied to 60-paired samples of tumour and adjacent normal urothelium. Results demonstrated that cyclooxygenase-2 is overexpressed in *S. haematobium*-induced bladder cancer. The quantitative relationship between cyclooxygenase-2 expression and tumour grade was statistically significant. The authors proposed the following role for cyclooxygenase-2 in the complex multi-stage process of schistosome-associated bladder carcinogenesis. Pro-inflammatory cytokines such as interleukin-1, tumour growth factor-B and tumour necrosis factor-α, are generated by activated macrophages in the inflammatory lesions. These cytokines and growth factors are potent inducers of cyclooxygenase-2 production. The up-regulation of cycloxygenase-2 activates environmental carcinogens such as *N*-nitrosamine and benzopyrene that initiate DNA damage and carcinogenesis. By-products of uncontrolled cyclooxygenase activity together with endogenous genotoxins produce oxidative and nitrosative stress creating lipid peroxidation by-products. Additional mutations are induced: p53, H-ras, deletion of p16 and p15, increased epidermal growth factor receptor, c-erB-2 and tumour necrosis factor-α. Increased prostaglandin production up-regulates cyclooxygenase-2, decreases killer T-cell activity, increases bcl-2 and glutathione-*S*-transferase. These changes increase tumourgenicity by decreasing cell apoptosis, creating immunosupression and resistance to chemotherapeutic drugs. Prostaglandin products of cyclooxygenase-2 cause tumour progression and eventual metastasis by down-regulating adhesion molecules, increasing degradation of extracellular matrix and increasing angiogenesis. Okajima *et al.* (1998) also noted the overexpression of cyclooxygenase-2 in bladder cancer and studied the chemopreventive effect of nimesulide, a cyclooxygenase-2 inhibitor, on the development of nitrosamine-induced bladder tumours in rats. Thought to be effective by inducing apoptosis, drugs as readily available as aspirin and the non-steroidal anti-inflammatory agents may provide both simple and affordable cycloxygenase-2 inhibition for endemic areas.

Badawi *et al.* (1995) noted that other malignancies have been reported in association with *S. haematobium* infection. These include squamous cell cancers of the female genitals, cervical cancers, ovarian cystadenocarcinomas, teratomas and Brenner tumours, uterine leiomyosarcomas, male breast cancers, hepatocellular carcinomas, lymphomas, bladder sarcomas, rectal carcinoid tumours and renal cell carcinomas. The numbers of patients with these additional malignancies are relatively small leaving their association with schistosomiasis unproven. Evidence for the causative role of *S. haematobium* infection in bladder carcinogenesis is sufficient for the IARC to classify the parasite as group 1: definitely carcinogenic to humans. The literature in support of this classification is voluminous, with some of the work appearing in relatively obscure journals. Therefore, this section of our review is, by necessity, selective rather than exhaustive.

Having concluded that *S. haematobium* is carcinogenic to humans, it follows that elimination or control of the parasite could provide a means of cancer prevention. Mayer and Fried (2002) noted that control of schistosomiasis is twofold: control of population morbidity and control of transmission. Application of mass chemotherapy to targeted populations can cure 75% of those infected with a single dose of praziquantel, and reduce egg release by 90–95%. This large-scale treatment of endemic areas is expensive, with a cost of US$ 0.35 per dose. Metrifonate is a less-costly alternative that is active against *S. haematobium*, but must be given in two or three doses at 2-week intervals. Snail control is effective in interrupting transmission. The molluscicide niclosamide is most frequently used, but because snail populations can reestablish themselves in 3 months, repeated applications are needed. Biologic snail control by introducing natural snail predators (ducks, fish, turtles) or snail parasites (fungi, bacteria viruses) are other effective methods. Modification of the environment to create conditions detrimental to the intermediate host may involve increasing the rate of water flow in irrigation canals, cementing over or enclosing canals and burying snails while digging irrigation ditches. These environmental modifications work best when combined with public health measures including education and improved hygiene, reduction of contact with contaminated water and improvement of the standard of living. Since schistosomiasis remains endemic in over 75 countries, anti-schistosomal vaccines are needed to complement existing control measures. To date, the immune responses elicited by experimentally defined antigens are inadequate for reliable human protection against infection. Dupre *et al.* (1999) reported a promising synergistic effect in the addition of a schistosomal GST DNA vaccination to standard praziquantel therapy. The praziquantel induced unmasking of native GST at the surface of the worms, allowing neutralization of the parasite by antibodies raised by the DNA immunization. Modification of the diet to include high doses of antioxidants to combat the DNA damage required for carcinogenesis to proceed has also been suggested. El-Desoky *et al.* (2005) identified a slow acetylator genotype, NAT2*5/*5, as a risk factor or marker for schistosome-induced bladder cancer. Further research in this area is necessary to develop biomarkers for population screening and identification of at-risk individuals.

2.2. *S. mansoni*

Mostafa *et al.* (1999) noted that there are no definitive reports linking the geographical occurrence of cancer with the prevalence of *S. mansoni* infection. Edington (1979) failed to find a strong association between

S. mansoni and the occurrence of liver cancer in Africa and South America. Parkin (1986) found uniform geographic distribution of colorectal cancer in Africa despite wide variations in the prevalence of *S. mansoni* infection. Murray (1967) studied colorectal tumour specimens throughout Africa and found that evidence of *S. mansoni* infection was no more common than could be predicted by known prevalence of infection.

The IARC (1994) reported cases of liver cancer associated with *S. mansoni* infection in Egypt, Mozambique, Nigeria, Saudi Arabia, Puerto Rico and Brazil. Cases of colorectal cancer associated with *S. mansoni* have been reported in Egypt and Lebanon. Andrade and Abreu (1971) and later Paes and Marigo (1981) described a total of 14 giant follicular lymphomas found in 1577 spleens removed in patients with portal hypertension due to *S. mansoni* infection. The IARC (1994) noted that *S. mansoni* infection has also been linked to prostate cancer, rectal carcinoid tumour, renal cell carcinoma, cervical cancer and leiomyosarcoma of the uterus.

Cheever *et al.* (1978) noted that acute *S. mansoni* infection is characterized by a severe inflammatory response to mature parasite eggs in tissues. Large granulomas develop with microscopic features of round cell and eosinophilic infiltration, areas of necrosis and fibrosis. Kamel *et al.* (1978) described the hepatic disease of late-stage *S. mansoni* infection. Portal radicles are obstructed by granulomas forming around schistosome eggs. Inflammation progresses to fibrosis, with the end-stage clay pipe stem fibrosis resulting in portal hypertension. Smith *et al.* (1977) described colonic inflammatory pseudopolyposis in 30 Egyptian males infected with *S. mansoni*, *S. haematobium* or both. The majority of the pseudopolyps were found in the rectosigmoid. Microscopic findings included round cell and eosinophilic infiltration, proliferative changes in the colonic glands and ulcers on the polyp surfaces.

Animal experimental models exist for *S. mansoni* that mimic infection and pathogenesis as it occurs in the human host. Warren (1973) studied the murine host and found that *S. mansoni* infection in mice produced granuloma formation and fibrosis. Sixty-three percent of eggs produced by the parasite were retained in the porto-mesenteric venous system where a delayed hypersensitivity granulomatous reaction ensued in the intestine and liver. Hepatomegaly and obstruction to portal blood flow produced portal hypertension similar to human infection. Olds *et al.* (1989) noted that the granulomas were eventually replaced by fibrosis when the lymphocytes, mononuclear cells and eosinophils were displaced by ingrowth of scar tissue. Lukacs and Boros (1993) found that interleukins 2 and 4 as well as interferon-gamma were cytokines involved in the induction of granulomas. Warren (1973) also noted that chimpanzees and baboons could be experimentally infected with *S. mansoni* producing disease paralleling human infection.

Shimkin *et al.* (1955) studied mice injected with lyophilized immature *S. mansoni* worms but found no tumours at the injection sites. There were three hepatomas and one pulmonary tumour among nine surviving mice necropsied at 24 months, similar to tumour incidence among non-infected controls. The IARC (1994) cited six studies between 1967 and 1985 in which there were no greater incidence of liver tumours in mice infected with *S. mansoni* as compared with uninfected control groups. For this reason, investigators have looked at the effects of *S. mansoni* infection in combination with known carcinogen exposure. El-Aaser *et al.* (1978) treated six groups of Swiss albino mice: the first group with *S. mansoni* infection by immersion in water containing 20–30 cercariae per milliliter for 1 h, the second group with 2-naphthylamine in the diet, the third group with *S. mansoni* and 2-naphthylamine exposure, the fourth group with 2-acetylaminofluorene in the diet, the fifth group with *S. mansoni* and 2-acteylaminofluorene exposure and a sixth group of untreated controls. All of the mice infected with *S. mansoni* had portal granulomas and eggs in the faeces. The investigators found that no liver or bladder tumours were noted in mice from any study group. Kakizoe (1985) compared 109 mice divided into 3 groups: those given intraperitoneal injections of 20 cercariae of *S. mansoni* and fed 4 weeks later with 2-acetylaminofluroene, those infected with *S. mansoni* fed normal diets and uninfected mice fed 4 weeks later with 2-acteylaminofluorene. After 40 weeks, the mice infected with *S. mansoni* alone exhibited no liver tumours. Uninfected mice exposed to the carcinogen had a 6.3% incidence of hyperplastic liver nodules. In the combined treatment group, 20% had hyperplastic nodules and 27% had evidence of hepatocellular carcinoma.

The IARC (1994) cited six studies between 1973 and 1984 evaluating the effects of *S. mansoni* infection in conjunction with administration of anti-schistosomal agents. Uninfected and infected groups of mice and hamsters were given hycanthone, niridazole and SQ 18506 and studied for possible tumour induction. Experimental endpoints included hyperplastic liver nodules, hepatomas and stomach tumours. Although results of these studies were inconclusive, some reports of lower tumour incidences might presumably be due to reduction or elimination of schistosomal infection in the experimental animals. Oettle *et al.* (1959) infected 200 rats (*Mastomys natalensis*) with intraperitoneal injection of *S. mansoni* cercariae and followed them up to 2.5 years until end of life. Presence of infection was confirmed by stool egg counts and post-mortem by examination of the liver and intestines for adult worms and egg granulomas. At study end, 106 animals had confirmed infection. Gastric carcinoid tumours were found in 23/106 (22%) similar to the 20% rate in non-infected control animals. Hepatomas were found in 22/106 (21%) while being notably absent among controls. Two animals (2%) showed evidence of reticulum cell sarcoma of the colon and ileum at the sites of schistosomal granulomas. Abe *et al.* (1993) reported a 12-year-old chimpanzee with hepatocellular carcinoma and coexisting *S. mansoni* infection as demonstrated by a granulomatous inflammatory reaction in the liver with schistosomal egg remnants.

Badawi *et al.* (1993) infected mice with *S. mansoni* cercariae and found promutagenic methylation damage in the form of 06-methyldeoxyguanosine in liver DNA. They also demonstrated that levels of hepatic 06-methyldeoxyguanosine rose with increasing intensity of schistosomal infection. The authors postulated that *S. mansoni* may have an activating effect on murine macrophages stimulating them to produce *N*-nitroso compounds capable of producing the DNA methylation damage. Mostafa *et al.* (1993) studied mice infected with cercariae from *S. mansoni* by necropsying the animals from 15 to 75 days post-infection and analyzing their livers for changes in the ability to metabolize carcinogens. Increases in the cytochromes p450 and b5 as well as in arylhydrocarbon hydrolase peaked 30 days following *S. mansoni* infection and continued at steady elevated levels throughout the study period. These changes during the early stages of schistosomal infection could result in the generation of active metabolites capable of acting as proneoplastic initiating agents. Habib *et al.* (1996) investigated the effect of *S. mansoni* infection on the ability of the human liver to metabolize carcinogens. Clinical data showed an increase in alanine aminotransferase, alkaline phosphatase and aspartate aminotransferase by 74%, 82% and 100%, respectively, in infected patients. Hepatic levels of cytochrome p450, cytochrome b-5 and (nicotinamide adenine dinucleotide phosphate) NADPH cytochrome C reductase were reduced by 52%, 61% and 72%, respectively. *In vitro* studies showed that these enzymatic alterations caused metabolites of aflatoxin to increase by 308%, supporting a potential role for *S. mansoni* in potentiating the deleterious effects of environmental carcinogens by impairing the detoxification capacity of the infected liver.

Zalata *et al.* (2005) investigated the expression pattern of p53, Bcl-2 and C-Myc in 75 Egyptian colorectal cancer specimens with 36 lymph node metastases by immunohistochemistry techniques with monoclonal antibodies. The patients were separated into two groups: the first with proven S. mansoni infection and the second without evidence of such infection. Although expression of p53 and CMyc was similar in both groups, 58.3% of *S. mansoni*-infected colorectal cancer patients were Bcl-2 positive as compared to only 33.3% of the non-infected group. Apoptotic activity was greater in the tumours of the non-infected group. Zalata *et al.* (2005) concluded by theorizing that genotoxic agents produced endogenously during *S. mansoni* infection might be involved in the pathogenesis of colorectal cancer with the overexpression of Bcl-2 leading to a reduction in programmed cell death in potential latent tumour foci. Thors *et al.* (2006)

tested the hypothesis that sera from patients with schistosomiasis cross-reacts with carcinomas. Eight-week-old mice were injected with 100 *S. mansoni* cercariae. Animals were killed 8–11 weeks post-infection and intravascular worms were recovered, livers were harvested for immunohistological staining and pooled serum was collected. The authors looked for carcinoma markers in schistosome adult worms and eggs using special monoclonal antibodies raised against various tumour antigens such as MUC1, Tn and TF (oncofetal Thomsen-Friedenreich antigen). The TF epitope has been associated with increased tumour grade, likelihood of relapse and tumour aggression in human cancers. Immunohistochemical staining revealed TF on adult worms and eggs. TF immunoreactivity was concentrated in the tegument and in the cytons of the worms, as well as in and outside eggs trapped in the liver. The localization of the immunofluorescence suggested that the TF epitope is expressed by excretory–secretory products of *S. mansoni*. Thors *et al.* (2006) also demonstrated that mice infected with *S. mansoni* produced antibodies against TF. Due to the complexities of the host–parasite relationship the significance of this finding is unclear. However, such an anti-schistosomal response on the part of the host may interfere with competitive binding of parasite and tumour TF glycans on host vascular endothelial and hepatocyte carbohydrate receptors, possibly inhibiting the progression of carcinogenesis.

Although reports of cancer of the liver, colon and lymphatic system have been confirmed in association with *S. mansoni* infection, the IARC (1994) concluded that there is inadequate evidence to support the carcinogenicity of *S. mansoni* infection in humans (group 3). Much of the more recent work remains inconclusive. Further study is required to identify a definitive role for *S. mansoni* infection in carcinogenesis.

2.3. *S. japonicum*

Dimmette *et al.* (1959) first suggested an association between *S. japonicum* infection and the development of colon cancer in humans. Since this early report, subsequent evidence has been inconsistent in confirming the carcinogenicity of *S. japonicum*. Although the geographical occurrence of cancer in endemic areas has been well studied, difficulties arise when investigators are forced to account for varying distributions of other known causes of the same malignancies, such as viral hepatitis, aflatoxins and other dietary carcinogens.

Liu *et al.* (1983) studied the correlation between the mortality from schistosomiasis and the mortality from cancers of the liver, colon, oesophagus and stomach in 24 Chinese provinces of varying *S. japonicum* prevalence and in 6 provinces with high infectivity. Correlation was found between mortality from schistosomiasis and from liver cancer in two provinces of high endemicity. Guo *et al.* (1984) found no such correlation in their analysis of 24 provinces in China, including the highly endemic Jiangsu Province. Inaba *et al.* (1977) noted that the mortality from liver cancer in the endemic Yamanashi Prefecture, Japan, was significantly higher as compared to non-endemic areas (95% confidence level). Liu *et al.* (1983) did find a significantly positive correlation between mortality from schistosomiasis and death from cancers of the oesophagus and stomach in the Jiangxi Province, but results from the other 29 provinces studied were inconclusive. Guo *et al.* (1984) were unable to find a positive correlation between the prevalence of *S. japonicum* infection and mortality rates from oesophageal and stomach cancers in the Jiangsu Province.

Liu *et al.* (1983) demonstrated a positive correlation between mortality from colorectal cancer and mortality from *S. japonicum* infection in 24 Chinese provinces ($r = 0.695$ in males and 0.625 in females). Guo *et al.* (1984) found a positive correlation between colorectal cancer mortality and endemicity in seven counties of Jiangsu Province, China ($r = 0.63$). Xu and Su (1984) confirmed the correlation between colorectal cancer mortality and prevalence of infection in Zhejiang Province, China. Li (1988) noted the correlation between colorectal cancer mortality and the incidence of *S. japonicum* infection in 12 Southern Chinese Provinces ($r = 0.71$), although possible confounding by low levels of dietary selenium was present. Chen *et al.* (1990) proved the correlation between colorectal cancer and *S. japonicum* mortality rates in 65 rural Chinese counties ($r = 0.89$, $p < 0.001$) with results adjusted for dietary micronutrient variables. Guo *et al.* (1993) confirmed the strong association between mortality from colorectal cancer and mortality from schistosomiasis, and attributed the continued high incidence of colonic malignancy in endemic areas to persistent large populations of chronically infected individuals, despite public health initiatives to eradicate the disease.

Kojiro *et al.* (1986) presented a series of 59 autopsy cases of hepatocellular carcinoma associated with *S. japonicum* infection proven by positive skin testing or direct histological evidence. Kitani and Iuchi (1990) documented cases of liver cancer in patients with schistosomal-induced cirrhosis, with the assumption that cirrhosis is a risk factor for hepatocellular malignancy. Case series of cancer of the stomach associated with evidence of *S. japonicum* infection have been reported in Japan, most recently by Zhou (1986). Case series of colorectal cancer associated with *S. japonicum* have been documented also in Japan, most recently by Sekiguchi *et al.* (1989). Similar to findings with *S. haematobium* infection, patients in the above series tended to be of younger age than did similar patients with liver, stomach and colorectal cancer without *S. japonicum* infection. Additional malignancies reported in association with *S. japonicum* infection include breast, lung, skin and parotid cancers, as well as malignant schwannoma (IARC, 1994).

Iuchi *et al.* (1971) reported that 85.2% of 52 patients with hepatocellular carcinoma had positive antigen skin tests for *S. japonicum*, as compared to 68.2% among 217 hospitalized control patients without the diagnosis of malignancy. Guo and Lu (1987) presented data from a cohort study of 3 groups of 166 patients each: group 1—patients who died from hepatocellular carcinoma, group 2—patients who died from other cancers and group 3—patients in good health. All three cohorts were similar with respect to age, sex and prevalence of *S. japonicum* exposure. Group 1 had an odds ratio for schistosomal infection of 2.2 ($p < 0.01$) as compared to groups 2 and 3. The relative risk of cancer increased as did the time interval since the initial diagnosis of *S. japonicum* infection. Amano (1980) compared two cohorts each having surgery in Yamanashi Prefecture, Japan. In the first group, 362 patients had operative treatment of their stomach cancer. In the second group, 897 patients had surgery for non-malignant disease of the stomach or duodenum. The author found histological evidence of *S. japonicum* eggs 1.8 times more frequently in the group with gastric carcinoma than in the control group. Amano (1980) also noted *S. japonicum* eggs 1.2 times more frequently in 102 patients with proven colorectal cancer than among 96 matched controls.

Xu and Su (1984) studied three cohorts of patients: the first with colorectal cancer, the second with other types of cancer and the third group of healthy controls. In the endemic Jiangsu Province, China, an odds ratio (for *S. japonicum* infection) of 8.3 was determined for rectal cancer patients when compared to patients with other cancer types. An odds ratio of 4.5 was found when comparing rectal cancer patients to healthy controls. The authors could not prove a significant relationship between colon (excluding rectal) cancer cases and past exposure to schistosomiasis. Guo and Lu (1987) compared cohorts of patients who died from colon cancer with matched lung cancer and normal controls from the same endemic Chinese province. When the colon cancer group was compared to normal controls, an odds ratio of 2.4 was determined for early-stage *S. japonicum* infection, and an odds ratio of 5.5 for late-stage infection. Odds ratios of 2.4 and 5.7 were found for early and late-stage *S. japonicum* infection when comparing the colon cancer cohort with lung cancer controls. Colon cancer risk was noted to increase from 1.2 with less than 10 years of *S. japonicum* infection, to 4.3 with greater than 30 years of schistosomal disease.

The IARC (1994) noted that studies on the association between liver cancer and *S. japonicum* infection have encountered difficulty with confounding by concomitant hepatitis viral infection. Qiu *et al.* (2005) examined liver and colon cancers and their relationship with schistosomiasis japonica in Sichuan, China. One hundred and twenty seven liver cancer patients and 142 colon cancer patients were each paired with matched controls. Only viral hepatitis negative pairs were used in the study.

Prior *S. japonicum* infection was associated with liver cancer (odds ratio 3.7, 95% CI) and with colon cancer (3.3, 95% CI). Their results indicated that 24% of colon cancer and 27% of liver cancer is attributable to schistosomiasis japonica.

Amano and Oshima (1988) infected 395 mice by exposing the skin on their shaved abdomens to five or six *S. japonicum* cercariae. Eight to 10 weeks later, 163 showed evidence of eggs in the faeces indicating infection. Seventy of the infected animals survived until the study endpoint of 50 weeks. Upon necropsy, 9 of the 70 surviving infected micehad no histological evidence of eggs in the liver or intestine and were excluded from the study. Forty-eight of the remaining 61 infected mice had developed hepatomas. No hepatomas were noted in 61 non-infected controls. Miyasato (1984) studied 77 mice infected by immersion of their tails in water containing 40 *S. japonicum* cercariae and later fed a diet containing the known carcinogen 2-acetylaminofluorene for 40 weeks. Results were compared to a control group of 86 non-infected mice exposed to the dietary carcinogen for 40 weeks. Upon necropsy, liver tumours were found in 24 of 77 (31%) carcinogen-treated infected mice as compared to 6 of 86 (7%) carcinogen-treated control animals.

Kurniawan *et al.* (1976) described the host granulomatous response to eggs of *S. japonicum* as differing from that seen in schistosomiasis mansoni. Granulomas developed around nests or groupings of eggs, rather than around individual eggs. In early-stage infection, lesions were filled with eosinophils, histiocytes and lymphocytes and resembled abscesses with areas of central necrosis. In late-stage disease, histiocytes predominated in the granulomas, and multi-nucleated giant cells were seen phagocytosing egg shells trapped in host tissues. The mature granuloma is fibrotic with features of hyaline degeneration.

Chen *et al.* (1980) studied 289 cases of *S. japonicum*-associated colorectal cancer and compared them to 165 colorectal cancer cases without documented schistosomiasis. Cases of *S. japonicum*-associated colorectal cancer were more often well-differentiated (91.6%) as compared to the control group (69.1%). The control group had a 29% incidence of benign papillary and adenomatous polyps as compared to 6.4% of the cases with schistosomiasis. Chen *et al.* (1981) later examined 60 colectomy specimens from patients with schistosomiasis japonica. Thirty-six specimens exhibited dysplasia, most often occurring in pseudopolyps or in the regenerative mucosa at the periphery of colonic ulcers, changes seen in association with chronic schistosomal colitis.

Irie and Iwamura (1993) used *in situ* hybridization techniques with P32-labelled probes in mice to demonstrate host DNA sequences in both the subtegumental layer and inner tissues of *S. japonicum* adult worms. Matsuoka *et al.* (1989) reported reduced levels of the liver enzyme cytochrome p450 in mice with schistosomiasis japonica, possibly impairing the ability of their livers to metabolize carcinogens. Arimoto *et al.* (1992) used the carcinogen Trp-P-2 as a substrate to demonstrate that hepatic homogenates from *S. japonicum*-infected mice had lower mutagen activating capacity than did liver fractions from non-infected control mice. Hasler *et al.* (1986) found that microsomes from infected mice were less effective in binding aflatoxin B1 than were microsomes from control animals. Aji *et al.* (1994) showed that mice infected with *S. japonicum* maintained higher levels of Trp-P-2 (following intravenous administration of the chemical) than did uninfected control mice. This suggested a lowered metabolism and increased mutagen retention in animals with schistosomiasis japonica. Haematin-containing pigments in the host liver produced during *S. japonicum* infection were efficient absorbents for Trp-P-2. These pigments may function as reservoirs for mutagens, prolonging exposure of the liver to the deleterious effects of these agents. Ishii *et al.* (1989) studied the mutagenicity and tumour-producing activity of *S. japonicum* parasite extracts. In bacterial mutagenicity testing using *Salmonella typhimurium* and *E. coli*, schistosomal soluble egg antigen and adult worm homogenates failed to show any positive responses. Schistosomal soluble egg antigen did have a weak but significant induction of Epstein–Barr virus expression in *in vitro* cultures of human lymphoblastoid cells. The authors concluded that the possibility that soluble egg antigens from *S. japonicum* possess tumour-promoting activity warrants further study.

Zhang *et al.* (1998) examined 44 Chinese patients with rectal cancer for p53 tumour suppressor gene mutations. The 44 patients were divided into 22 cases with advanced schistosomiasis japonica and 22 without *S. japonica* infection. Thirteen mutations were found in 10 patients with schistosomal-associated rectal cancer. Of 11 base substitutions, 7 were at CpG dinucleotides, a location associated with endogenous mutagenic processes. In the non-schistosomal rectal cancer patients, 13 mutations were found in 9 cases, with only 3 of the base pair substitutions being located at CpG dinucleotides. Arginine missense mutations also occurred with greater frequency in the infected group. The results suggested that the mutations found in schistosomal-associated rectal cancer arise from endogenous genotoxic agents produced during the natural course of schistosomiasis japonica. The authors concluded that the pattern of p53 mutation seen in schistosomal-associated rectal cancer suggests that *S. japonicum* is a special factor in carcinogenesis.

The IARC (1994) stated that there is limited evidence in humans for the carcinogenicity of infection with *S. japonicum*, classifying the parasite as possibly carcinogenic or group 2B. Although subsequent molecular biological work appears promising, further study is required to strengthen this association.

2.4. *O. viverrini*

Bunnag *et al.* (2000) described the life cycle of *O. viverrini* as being complex with two intermediate hosts, a first intermediate snail host and a second intermediate fish host. The adult liver fluke lives in the human biliary system, including the hepatic ducts, proximal and distal common bile duct, gallbladder and pancreatic duct. Adults of *O. viverrini* are transparent leaf-shaped worms, approximately 8–12 mm in length, that move about within the biliary tract producing mechanical irritation and finally pathologic changes in the ductal epithelium. The fluke produces eggs which are passed from the bile into the faeces. Upon reaching fresh water, miracidia hatch from eggs swallowed by various species of *Bithynia* snails. After 4–6 weeks, cercariae are released that penetrate cyprinoid fresh water fish. When man, the definitive host, consumes raw or improperly cooked fish containing metacercariae of *O. viverrini*, the worms excyst in the duodenum and migrate to the bile ducts where they mature within a month.

The association between infection with *O. viverrini* and carcinogenesis has been extensively studied. Viranuvatti and Mettiyawongse (1953) were the first to describe case reports of liver and bile duct cancer associated with *O. viverrini* infection, confirmed at autopsy in Thailand. Among patients from endemic areas with cancer of the liver or biliary tract, cases of cholangiocarcinoma predominate, as compared to a greater proportion of hepatocellular carcinoma in non-endemic regions.

Srivatanakul *et al.* (1988) found that the highest incidence data from the National Cancer Registry was in Khon Kaen Province in Northeastern Thailand, coinciding with the greatest prevalence of *O. viverrini* infection. The proportionate incidence ratio was 3.1 for cholangiocarcinoma and 1.2 for hepatocellular carcinoma (both at 95% confidence interval). Parkin *et al.* (1993) noted that the incidence of cholangiocarcinoma in Khon Kaen Province was 84.6 per 100,000 per year for men and 36.8 for women. This compares to control rates outside of Thailand of 0.2–2.8 for men and 0.1–4.8 for women. The incidence of cholangiocarcinoma, in the region of highest *O. viverrini* prevalence in Thailand, was more than 40 times greater than for any other geographic area. Vatanasapt *et al.* (1990) identified the highest incidence and mortality rates for liver cancer in three districts in Khon Kaen Province, two of which had high rates of *O. viverrini* prevalence and heavy infections (as determined by fluke eggs/g of stool).

Srivatanakul *et al.* (1991a) studied the correlation between the incidence of liver cancer and antibody titers to *O. viverrini* and faecal egg counts in five regions throughout Thailand. The correlation between cholangiocarcinoma incidence and the proportion of people with antibody titers >1:40 was 0.98 ($p = 0.0004$). There was a weaker association with faecal egg counts, possibly due to the introduction of population-based effective drug therapy. Antibody levels were believed to be more reliable indicators of past or present infection. Sriamporn *et al.* (2004) examined residents of 20 districts in the endemic Khon Kaen Province, Thailand, for the presence of *O. viverrini* infection using faecal egg counts. The incidence of cholangiocarcinoma was obtained for each district from the national cancer registry. Prevalence of infection varied from district to district (2.1–71.8%), for an overall prevalence of 24.5%. Incidence of cholangiocarcinoma also varied between districts, ranging from 93.8 to 317.6 per 100,000 patient years. A positive association was demonstrated between the prevalence of *O. viverrini* infection and the incidence of cholangiocarcinoma at the population level.

Kurathong *et al.* (1985) reported a case–control study of 551 patients from Northeast Thailand, all of whom were screened for the presence of *O. viverrini* eggs in the stool. Nineteen of 25 patients with cholangiocarcinoma and 9 of 12 with hepatocellular carcinoma had evidence of opisthorchiasis. Crude prevalence odds ratios were 1.3 for cholangiocarcinoma and 1.3 for hepatocellular carcinoma. Parkin *et al.* (1991) conducted a case–control study of 103 patients with cholangiocarcinoma diagnosed at 3 hospitals in Northeast Thailand. One control was matched to each case based on sex, age, residence and hospital, from a pool of patients with non-malignant disease unrelated to alcohol or tobacco abuse. The presence of *O. viverrini* infection was determined by ELISA testing for antibody titers. The matched odds ratio from their multi-variate model was 5.0 (95% CI, 2.3–11.0) for cholangiocarcinoma. No association with Hepatitis B infection or aflatoxin ingestion was found. *O. viverrini* infection was not found to be significantly associated with hepatocellular carcinoma.

Haswell-Elkins *et al.* (1994a) collected stool samples from 7727 subjects from Khon Kaen Province and from 4585 subjects from Maha Sarakham Province, both endemic areas in Northeast Thailand. Fifteen percent of 1807 uninfected or lightly infected (<3000 fluke eggs/g of stool) patients were selected randomly to undergo abdominal ultrasound exams. All subjects with higher intensity *O. viverrini* infection were also sent for ultrasound testing. In all, 78% of the subjects complied with ultrasound examination, and 44 cases of cholangiocarcinoma were identified. Among 410 uninfected patients, only 1 case of cholangiocarcinoma was found. The relationship of cholangiocarcinoma to *O. viverrini* infection was demonstrated by multi-variate prevalence odds ratios at the 95% CI as follows: 1.7 for subjects with up to 1500 fluke eggs/g of stool, 3.2 for 1501–6000 eggs/g and 14 for greater than 6000 eggs/g.

Honjo *et al.* (2005) noted that although *O. viverrini* infection is endemic in Northeast Thailand, less than 10% of the inhabitants develop cholangiocarcinoma. They suggested that although *O. viverrini* is associated with the development of bile duct cancer, other environmental and genetic factors may also play a role. They conducted a population-based case–control study in which age, sex and place of residence were matched individually. The greatest risk indicator for cholangiocarcinoma was elevated antibodies to *O. viverrini* >0.200 by ELISA (odds ratio 27.1 at 95% CI when compared to antibody levels <0.200). Polymorphism of the glutathione S-transferase, mu 1 (GSTM1) and glutathione S-transferase, theta (GSTT1) genes alone, were not risk factors for bile duct cancer. However, those subjects with elevated antibodies to *O. viverrini* had higher odds ratios for polymorphism of the GSTM1 gene. Additional independent risk factors for cholangiocarcinoma were alcohol ingestion (possibly due to an effect on pathways for endogenous and exogenous nitrosamines), smoking and consuming fermented fish.

Animal studies have provided limited evidence that infection with *O. viverrini* alone is carcinogenic. Infection with *O. viverrini* together with administration of known carcinogens was first studied by Thamavit *et al.* (1978). Male Syrian hamsters were divided into four groups: (1) untreated controls, (2) animals receiving *N*-nitrosodimethylamine in their drinking water, (3) animals treated with 100 metacercariae of *O. viverrini* administered by intragastric intubation and (4) animals infected with *O. viverrini* and 4 weeks later (when fluke eggs appeared in their stools) fed *N*-nitrosodimethylamine. Animals were killed at 23 weeks. All 21 infected hamsters that received the carcinogen (group 4) developed cholangiocarcinoma. Bile duct tumours were not found in the controls (group 1), nor in animals treated with either *O. viverrini* metacercariae (group 3) or *N*-nitrosodimethylamine alone (group 2). Results were significant at $p < 0.001$. Thamavit *et al.* (1988) noted that, under certain conditions, the ingestion of nitrite and aminopyrine can form *N*-nitrosodimethylamine in the stomach. In their study of 150 Syrian hamsters, the combined administration of nitrite and aminopyrine in the drinking water, together with 100 metacercariae of *O. viverrini* delivered by gastric intubation, was associated with significant increases in the occurrence of hepatocellular nodules (odds ratio = 8, $p < 0.05$), cholangiofibrosis (18, $p < 0.05$) and cholangiocarcinomas (14, $p < 0.01$). The carcinogenicity *N*-nitrosodimethylamine administered to hamsters infected with *O. viverrini* metacercariae was also confirmed by Flavell and Lucas (1983) and Thamavit *et al.* (1994a).

Moore *et al.* (1991) studied the effect of intraperitoneal injection of *N*-nitrosodihydroxydi-*n*-propylamine in hamsters infected with *O. viverrini* by intragastric exposure of 80 metacercariae/animal. A total of 100 hamsters were divided into 4 groups: 10 untreated controls, 20 infected with *O. viverrini*, 30 receiving 3 intraperitoneal injections of 500 mg/kg of the carcinogen at 16, 17 and 18 weeks, and 40 treated by both *O. viverrini* metacercariae and the carcinogen. All animals were killed at week 52 of the study, and were examined histologically.

Cholangiocarcinoma was found in 50% of the combined treatment group, and in 0% receiving the carcinogen alone ($p = 0.001$). Thamavit *et al.* (1993) in a multi-armed controlled study of 205 Syrian hamsters investigated the effect of infection with *O. viverrini* in combination with carcinogen exposure and administration of the antihelminthic drug, praziquantel. All animals were killed at 38 weeks post-treatment and studied pathologically. Of 16 Syrian hamsters given both *N*-nitrosodihydroxydi-*n*-propylamine and opisthorchiasis, 16 (100%) developed cholangiofibrosis, 8 (50%) developed cholangiocarcinomas and 16 (100%) developed hepatic nodules—with an occurrence of 13.6 nodules/cm^2. In a similar group of 44 animals with opisthorchiasis and carcinogen exposure, praziquantel reduced the incidence of cholangiocarcinoma to 4/22 (18%) if the drug was given 4 weeks following infection, and to 6/22 (27%) if given at 12 weeks. Praziquantel also reduced the multiplicity of liver nodules to 3.6 and 7.4 nodules/cm^2 in the 4 and 12 week dosage-groups, respectively. Cholangiofibrosis was a universal finding in all test animals treated with the carcinogen, *O. viverrini* metacercariae and praziquantel, with the exception of those treated with praziquantel at 4 weeks following infection which reduced the incidence to 8/22 (36%). Two of 18 (11%) hamsters treated with *O. viverrini* metacercariae alone had evidence of cholangiocarcinoma.

Riganti *et al.* (1989) described the pathological changes induced by *O. viverrini* infection. The flukes reside in the large and medium-sized bile ducts of the liver, and in cases of heavy infection, the gallbladder, common bile duct and pancreatic duct as well. Dilatation of the bile ducts with fibrosis of the periductal connective tissue was seen. Microscopically, hyperplasia, desquamation and adenomatous proliferation of the bile duct linings accompany infiltration with lymphocytes, monocytes, eosinophils and plasma cells. Granulomatous reaction to eggs and dying flukes lead to periductal and portal scarring and fibrosis. In patients with heavy worm burdens, cellular infiltration may also result from secondary bacterial infection. Suppurative cholangitis can lead to multiple liver abscesses. Should opisthorchiasis involve the gallbladder, cholecystitis without stones may occur. Sripa and Kaewkes (2002) investigated histological changes in the gallbladder and extrahepatic ducts in hamsters infected with metacercariae of *O. viverrini*. Changes were noted between 7 and 14 days post-infection (PI), with active inflammation reaching a plateau at 60 days PI. Chronic histological findings included fibrosis and mononuclear cell infiltration with lymphoid aggregation. Ductal dilatation and pathological changes were more severe in the extrahepatic ducts than in the gallbladder. Sripa and Kaewkes (2002) concluded that these changes can be extrapolated to human infection.

Gentile and Gentile (1994) noted that inflammatory responses in the infected host are believed to play a key role in carcinogenesis.

Polymorphonuclear leukocytes and other host defences produce nitrates, nitrites, nitrosating agents and reactive oxygen species to kill invading helminths. The resultant free radicles can induce genetic damage in host tissue adjacent to sites of inflammation, leading to DNA strand breaks and chromosome mutations. Haswell-Elkins *et al.* (1994b) demonstrated increased levels of urinary nitrates and salivary nitrites in *O. viverrini*-infected subjects in Northeast Thailand, and found decreased concentration of these substances following treatment with praziquantel. Srivatanakul *et al.* (1991b) showed a 10-fold greater potential for endogenous nitrosation among people living in endemic areas with positive antibody titers for *O. viverrini* as compared to uninfected controls. Urinary levels of *N*-nitrosoproline following proline ingestion were measured.

Tsuda *et al.* (1992) studied point mutations on the c-Ki-ras protooncogene in patients with cholangiocarcinoma with and without concomitant opisthorchiasis. Mutations were discovered at codon 12, in five of nine non-infected Japanese subjects with cholangiocarcinoma. Of six Thai patients with cholangiocarcinoma and *O. viverrini* infection, none had evidence of these mutations. Kiba *et al.* (1993) found, in a study of two similar sets of cholangiocarcinoma patients, that both groups shared a common mutation at the p53 tumour suppressor gene.

Wongratanacheewin *et al.* (1987) infected Syrian hamsters with *O. viverrini* and noted a depressed lymphoproliferative response to phytohaemaglutinin, indicative of immunosuppression. Makarananda *et al.* (1991) demonstrated that infection with *O. viverrini* enhances the host response to chemical carcinogens. Infected hamsters showed a significantly higher level of aflatoxin B1 metabolites than did uninfected control animals. Their results suggested increased expression of carcinogen-metabolizing enzymes. Kirby *et al.* (1994) found cytochrome p450 isozymes in the livers of hamsters infected with *O. viverrini*, concentrated in hepatocytes immediately adjacent to areas of inflammation. The cytochrome enzyme contributed to 50–60% of the metabolism of aflatoxin B1 and *N*-nitrosodiethylamine in infected male, and to 20–30% in infected female hamsters. Oshima *et al.* (1994) showed increased nitrosamine and nitrate production by nitric oxide synthase in hamsters infected with *O. viverrini*. Nitric oxide synthase was found in macrophages and eosinophils at sites of inflammation in liver tissue, at levels twice as great as in uninfected control animals.

More recently, Sithithaworn *et al.* (2002) found that tenascin, an extracellular matrix glycoprotein, is an integral factor in reciprocal interactions between the epithelium and mesenchyme during tumourigenesis. The expression of tenasin in the livers of Syrian golden hamsters was studied in a model of *O. viverrini*-associated cholangiocarcinoma. The glycoprotein was expressed in the walls of dilated bile duct and surrounding tissue, as well as in the stroma of cholangiocarcinoma nodules. The stroma of necrotic tumour nodules most strongly expressed tenascin. Stroma cells also exhibited an mRNA signal for tenascin production. Pinlaor *et al.* (2003) recalled that nucleic acid damage caused by reactive oxygen and nitrogen species associated with chronic infection and inflammation, may be carcinogenic. They studied 8-nitroguanine, 8-oxo-7,8-dihydro-2′-deoxyguanosine and nitric oxide production in hamsters with opisthorchiasis. Immunohistohemical testing showed evidence of damaged DNA in the cytoplasm and to a lesser extent in the nuclei of inflammatory cells and bile duct epithelium, reaching a peak on day 30 post-infection. The time profile of the appearance of the damaged nucleic acids paralleled that of plasma nitrate and nitrite. Pinlaor *et al.* (2004a) also demonstrated that in the livers of *O. viverrini*-infected hamsters, haem oxygenase-1 expression and subsequent iron accumulation may enhance oxidative DNA damage in the epithelium of small bile ducts. Pinlaor *et al.* (2004b) further postulated that the chronic inflammation induced by repeated infection with *O. viverrini* is a risk factor for cholangiocarcinoma development. Again using the hamster model, inducible nitric oxide synthase expression was increased in the bile duct epithelium at 90 days following repeated infection with *O. viverrini*. Proliferating cell nuclear antigen appeared in the bile duct epithelium 90 days following repeat infection, supporting the role of inflammation-mediated DNA damage in promoting cell proliferation. They concluded that repeated *O. viverrini* infections induce expression of nitric oxide synthase in inflammatory cells and bile duct epithelium that causes oxidative and nitrosative damage to nucleic acids. The resultant promotion of cell proliferation is an important step towards the development of cholangiocarcinoma. Pinlaor *et al.* (2005) investigated the mechanism by which *O. viverrini* induces inflammation in and adjacent to bile ducts leading to human cholangiocarcinoma. The expression of Toll-like receptors was examined in a RAW 264.7 macrophage cell line treated with antigen from *O. viverrini*, using immunohistochemistry and flow cytometry. Toll-like receptors are cell membrane receptors that actively participate in the immune response. The antigen exposure induced expression of Toll-like receptor-2, nuclear factor-kappaB, inducible nitric oxide synthase and cyclooxygenase-2 in a dose-dependent fashion. It appears that *O. viverrini* produces an inflammatory response through a Toll-like receptor pathway. This results in a nuclear factor-kappaB-mediated expression of inducible nitric oxide synthase and cyclooxygenase-2.

Thuwajit *et al.* (2004) postulated that in addition to the direct damage to the bile ducts by *O. viverrini*, excretory/secretory products released by the flukes may play an important role in carcinogenesis. To test this theory, a no-contact co-culture technique was used. Worms were kept in an upper chamber with no direct contact with a NIH-3T3 fibroblast cell line in the lower chamber. Results showed a marked increase in fibroblast cell proliferation compared to controls without parasites. Excretory/secretory products of *O. viverrini* stimulated expression of phosphorylated retinoblastoma and cyclin D1, both proteins that drive cells through the G1/S transition point in the S-phase of the cell cycle. Exposure to the excretory/secretory products also transformed the fibroblast cells into a more refractive and narrow shape, enabling more effective proliferation in the limited space of the culture area. The direct effect of excretory/secretory products on cell proliferation may clarify changes in the human bile ducts induced by *O. viverrini* during cholangiocarcinogenesis. Sripa *et al.* (2005) took fresh liver biopsy and bile specimens from a 65-year-old woman living in an endemic area in Northeast Thailand with cholangiocarcinoma of the porta hepatis. A cholangiocarcinoma cell line was established after cells were digested and cultured in Ham's F12 medium. The new cell line was tested by growth mechanics methodology, immunohistochemistry and cytogenetic techniques. Heterotransplantation into nude mice verified the tumourgenicity of the cell line. Malignant cells with *O. viverrini* eggs were identified in the bile specimens. Within 4 months, a cholangiocarcinoma cell line KKU-100 was established, characterized by expression of cytokeratin, epithelial membrane antigen (EMA), Carcinoembryonic antigen (CEA) and Cancer antigen 125 (CA125). Cells were aneuploid with a modal chromosome number of 78 with noticeable structural changes. A transplantable poorly differentiated carcinoma similar to the original tumour was produced by inoculation of KKU-100 cells into nude mice. The discovery of an egg-proven, opisthorchiasis-associated cholangiocarcinoma cell line provides an excellent experimental model for further study.

Pinlaor *et al.* (2006) studied the preventative effect of praziquantel on *O. viverrini*-induced cholangiocarcinogenesis. In a hamster model, a dramatic reduction in inflammatory cell infiltration was noted following a 1-week subtherapeutic course of praziquantel. Double immunofluorescence examination showed that drug treatment almost completely eliminated *O. viverrini*-induced, inducible nitric oxide synthase-dependent DNA damage, as measured by 8-nitroguanine and 8-oxo-7,8-dihydro-2′-deoxyguanosine levels in the livers of test animals. Drug treatment not only reduced chemical evidence of DNA damage, but also reduced levels of nuclear factor-kappaB and inducible nitric oxide synthase in the bile duct epithelium. Overall nitrate and nitrite in the plasma and liver were also significantly lowered following treatment. The authors concluded that praziquantel treatment can act in a preventative manner against *O. viverrini*-induced cholangiocarcinogenesis by inhibiting inducible nitric oxide synthase-dependent DNA damage. The effect may be due to both anti-parasitic and anti-inflammatory factors. Loilome *et al.* (2006) employed a hamster cholangiocarcinoma model, with cancer induction by *O. viverrini* infection and exposure to the carcinogen

N-nitrosodimethylamine, to investigate the molecular processes of carcinogenesis. Methods to analyze genes included differential display-polymerase chain reaction and reverse Northern macroarray blot. Specimens from cholangiocarcinoma cells were compared to controls from normal liver and bile duct epithelium. Up-regulated gene expression included signal transduction protein kinase A regulatory subunit alpha (Prkar1α), myristoylated alanine-rich protein kinase C substrate, transcriptional factor LIM-4, oxysterol-binding protein, splicing regulatory protein 9, ubiquitin conjugating enzyme, P tubulin, P actin and collagen type VI. Prkar1a expression increased as lesions in the bile ducts progressed from hyperplasia to precancerous to cancer.

The IARC (1994) concluded that infection with *O. viverrini* is carcinogenic to humans (group 1). Subsequent work has provided us with a better understanding of the molecular processes involved in *O. viverrini*-induced cholangiocarcinogenesis, which should stimulate the search for improved chemotherapeutics against this preventable malignancy.

2.5. *C. sinensis*

Bunnag *et al.* (2000) noted that the life cycle of *C. sinensis* is similar to that of *O. viverrini*. Adults of *C. sinensis* are 10–25 mm in length and 3–5 mm in width. The first intermediate hosts are various species of hydrobiid snails, and more than 100 species of freshwater fish serve as second intermediate hosts. A month following human ingestion of metacercariae, adult flukes mature in the biliary tract, move up and down the ducts by attaching and detaching their two suckers and contracting and extending their bodies, and produce eggs which are passed in the stool. Data relevant to the carcinogenicity of *C. sinensis* will be reviewed.

Watson-Wemyss (1919) reported the first case of carcinoma of the liver associated with infection by *C. sinensis* in a Chinese subject. Gibson (1971) found clonorchiasis on gross examination in 11 of 17 (65%) cases of cholangiocarcinoma and in 24 of 83 (29%) cases of hepatocellular carcinoma among 1484 autopsies performed in Hong Kong. The anticipated percentages for *C. sinensis* infection were 38% and 35% respectively, making the odds ratios 3.1 for cholangiocarcinoma and 0.73 for hepatocellular carcinoma (95% CI). Kim *et al.* (1974) examined autopsy and surgical specimen records from a hospital in Seoul, Korea with a low prevalence of clonorchiasis, and a second hospital in Pusan, Korea with a high incidence of *C. sinensis* infection. There were 386 cases of primary liver cancer among 1447 patients with liver disease in the non-endemic area, and 109 cases among 396 subjects in the endemic area. Infection with *C. sinensis* was determined by liver examination and stool sample analysis. Results demonstrated a weak positive association between liver cancer and *C. sinensis* infection (odds ratio 1.7, 95% CI). The association was stronger for cholangiocarcinoma (6.5 based on 54 cases) than for hepatocellular carcinoma (1.2 based on 423 cases). Chung and Lee (1976) found that 206 of 368 (56%) cases of primary liver cancer in the highly endemic area of Pusan, Korea had evidence of *C. sinensis* infection on stool sample analysis. A matching control group of 559 patients with diseases not affecting the liver was examined for clonorchiasis by faecal egg counts. A strong positive association between cholangiocarcinoma and *C. sinensis* infection was found (odds ratio 6.0, 95% CI), with a weaker association for hepatocellular carcinoma (1.1).

Jang *et al.* (1990) reported that the livers of F344 rats given intragastric delivery of 60 metacercariae of *C. sinensis* and exposed 4 weeks later to *N*-nitrosodimethylamine at 25 mg/L** in the drinking water for 8 weeks, had significantly increased GST P-positive hepatic foci as compared to untreated control animals ($p < 0.05$). The intrahepatic presence of the P-positive or placental form of GST has been implicated in hepatocarcinogenesis. Animals that were infected with *C. sinensis* during or after exposure to *N*-nitrosodimethylamine did not exhibit such an effect. Lee *et al.* (1993) divided 48 Syrian golden hamsters into 4 groups: (1) 12 animals exposed to *N*-nitrosodimethylamine at 15 mg/L in the drinking water for 8 weeks, and 7 days later to 10 metacercariae of *C. sinensis* by intragastric intubation, (2) 12 receiving *N*-nitrosodimethylamine alone, (3) 12 exposed to the metacercariae alone and (4) 12 untreated controls. At 11 weeks, six of eight (75%) surviving infected animals exposed to *N*-nitrosodimethylamine developed cholangiocarcinomas, and all eight had evidence of cholangiofibrosis. Five of 12 (42%) animals exposed only to the fluke developed cholangiofibrosis. No control animals developed liver lesions.

Lee *et al.* (1994) divided 90 Syrian golden hamsters into 6 groups of 15 animals each: (1) exposure to *N*-nitrosodimethylamine at 15 mg/L in the drinking water for 4 weeks, then delivery of 15 metacercariae of *C. sinensis* by intragastric intubation, followed by treatment with praziquantel at 200 mg/kg daily for 3 days, (2) infection with *C. sinensis* metacercariae, treatment with praziquantel for 3 days, then *N*-nitrosodimethylamine exposure, (3) simultaneous exposure to the metacercariae and the carcinogen, (4) *N*-nitrosodimethylamine exposure alone, (5) metacercariae exposure alone and (6) untreated controls. When animals were killed at 13 weeks, group 3, with concomitant *C. sinensis* infection and carcinogen treatment, had evidence of cholangiocarcinomas in 11/15 (73%), cholangiofibromas in 3/15 (20%) and cholangiofibrosis in 1/15 (7%). In group 1 treated with the carcinogen and later infected with *C. sinensis* and given praziquantel, 3/15 had cholangiocarcinomas (20%), 3/15 cholangiofibromas (20%) and 6/15 (40%) cholangiofibrosis. Group 2 given clonorchiasis, praziquantel and carcinogen exposure in that order, 11/15 (73%) had cholangiofibrosis. Group 4 given the carcinogen alone had 4/15 (27%) animals with cholangiofibromas and 5/15 (33%) with cholangiofibrosis. Group 5 infected with *C. sinensis* alone had 12/15 (80%) animals developing cholangiofibrosis. No animals in the untreated control group had evidence of precancerous or cancerous bile duct lesions.

Iida (1985) infected a group of 60 female Syrian golden hamsters with 40 metacercariae of *C. sinensis* delivered orally, and fed them diets containing 0.03% 2-acetylaminofluorene for 40 weeks. Another group of 50 hamsters was exposed to the carcinogen alone. After 40 weeks, surviving animals from both groups were fed normal diets. Small numbers of animals were killed every 3–4 weeks until the study was terminated at 54 weeks. For animals surviving beyond 25 weeks, the incidence of cholangiocarcinoma was significantly greater for the infected (11/14, 79%) than for the uninfected (6/17, 35%) groups ($p < 0.05$). Cholangiocarcinoma first appeared at 25 weeks in the infected group and at 35 weeks in the uninfected group. Only infected animals with cholangiocarcinoma exhibited metastatic spread to other organs. Park *et al.* (2000) studied 22 male Syrian golden hamsters for histopathologic changes in their biliary tracts induced by infection with *C. sinensis* and treatment with *N*-nitrosodimethylamine. Animals were divided into four groups: (1) untreated controls, (2) treatment with 10 metacercariae of *C. sinensis*, (3) treatment with 15 ppm of *N*-nitrosodimethylamine and (4) infection with *C. sinensis* and treatment with the carcinogen. Animals were killed after 12 weeks, and liver and biliary tissues were examined both by light microscopy and by immunohistochemistry stain for (proliferating cell nuclear antigen) PCNA and Bcl-2. Group 2 infected with *C. sinensis* alone had no cholangiocarcinoma, but evidence of cholangiofibrosis was present. Group 3 treated with the carcinogen alone had evidence of dilatation and hyperplasia of the bile ducts, but again no cholangiocarcinoma. Group 4 treated with both metacercariae of *C. sinensis* and *N*-nitrosodimethylamine had findings of bile duct hyperplasia, cholangiofibrosis, cholangiofibroma, goblet cell metaplasia and cholangiocarcinoma. The PCNA index was higher in cholangiocarcinoma and in premalignant lesions. There was no difference in expression of Bcl-2 protein in malignant and benign tissues. The study demonstrated that *N*-nitrosodimethylamine and *C. sinensis* had a synergistic effect on cholangiocarcinoma development in the hamster model. The higher PCNA index in bile duct hyperplasia, cholangiofibrosis, cholangiofibroma and goblet cell hyperplasia suggests that these are premalignant or precursor lesions for cholangiocarcinoma.

Lee *et al.* (1995) noted that the stem cell hypothesis for the development of liver cancer states that bipolar liver stem cells may differentiate into either hepatocytes or biliary cells. Under carcinogenic stimuli, the stem cell is initiated and gives rise to either hepatocellular carcinoma or cholangiocarcinoma. By choosing certain carcinogenic promoters, the incidence of either type of liver cancer can be experimentally modulated. In this study, 90 Syrian golden hamsters were divided into 6 groups of 15 animals each: (1) treatment with 15 ppm *N*-nitrosodimethylamine in drinking water for 4 weeks, then infection with 15 metacercariae of *C. sinensis*, followed by praziquantel at 200 mg/kg for 3 days, (2) infection with *C. sinensis*, praziquantel at week 5, followed 1 week later by carcinogen administration, (3) simultaneous infection and carcinogen exposure, (4) *N*-nitrosodimethylamine alone, (5) *C. sinensis* infection alone and (6) untreated controls. All hamsters were killed at 13 weeks. Cholangiocarcinomas were found in 3/15 (20%) of group 1 (carcinogen, *C. sinensis* infection, praziquantel) and in 11/15 (73%) of group 3 (simultaneous carcinogen and *C. sinensis* infection). No other groups had evidence of cholangiocarcinoma, although findings of cholangiofibroma, cholangiofibrosis and bile duct hyperplasia were present in varying degrees throughout the other treatment groups. Of note was the observation that cholangiocarcinoma did not develop when animals were infected with *C. sinensis* for 5 weeks, given praziquantel, then 1 week later administered *N*-nitrosodimethylamine (group 2). Infection with *C. sinensis* alone produced oval cell and bile duct hyperplasia, but not cholangiocarcinoma. These results suggested that cholangiocarcinomas were promoted by *C. sinensis* infection following induction by *N*-nitrosodimethylamine. The bile duct proliferation stimulated by *C. sinensis* infection may act as a promoter for the carcinogen-initiated bipolar liver stem cells. Although hamsters treated with *C. sinensis* alone do not develop bile duct cancers, in humans, clonorchiasis-induced bile duct hyperplasia may undergo malignant transformation through a stage of dysplasia. Lee *et al.* (1995) concluded that increased cell proliferation is required for the development of cancer from initiated cells by the production of genetic errors and altered growth control of malignant cell lines.

Hou (1956) reported an autopsy study of 28 patients with concomitant clonorchiasis and cholangiocarcinoma. Finding that carcinoma occurred in the presence of adenomatous changes in the wall of the bile duct, he observed all transitional stages in the transformation of the bile duct from hyperplasia to cholangiocarcinoma. Kim (1984) found that the epithelium lining the bile duct may undergo carcinomatous transformation through a stage of dysplasia when exposed to irritation from adult worms of *C. sinensis* or to biochemically altered bile. Choi *et al.* (2004) summarized the pathophysiology of human infection with *C. sinensis*. Findings include inflammatory response around the biliary tree, severe epithelial cell hyperplasia, mucin-producing goblet cell metaplasia of the bile duct epithelium and progressive periductal fibrosis. Besides biliary hyperplasia, the lining of the bile ducts can become edematous and desquamation is often seen in proximity to the flukes. Infiltrates of mononuclear cells are seen in the periductal tissues. Metaplasia of the biliary epithelial cells into mucin-producing goblet cells leads to small gland-like areas in the mucosa. Elevated mucin content characterizes the bile during *C. sinensis* infection, and combines with the worms and eggs to cause cholestasis, intrahepatic stone formation and secondary bacterial infection. The chronic infection triggers an increase in the amount of fibrous tissue, which may encase some of the proliferating glands to create cholangiofibrosis. Lee *et al.* (1978) described the histopathological response of albino rats to *C. sinensis* infection over a period of 12 weeks. The first phase consisted of desquamation and edema of the bile duct epithelium. This was followed by epithelial hyperplasia, pseudostratification of the ductal epithelium and mucin-producing goblet cell metaplasia. Glandular proliferation now appeared, suggestive of adenomatous hyperplasia. Early eosinophilic infiltration seen in the first 2 weeks post-infection, gave way to plasma cell, lymphocytic and mononuclear cell infiltration at week 12. The final phase was periductal fibrosis. Shim *et al.* (2004) reported a rare subtype of cholangiocarcinoma found in a 69-year-old Korean man with *C. sinensis* infection. Histologically, 70% of the tumour mass had extensive intra- and extracellular mucin production, and lymphovascular invasion of tumour cells that indicated a poor prognosis. The liver adjacent to the cancer contained eggs of *C. sinensis*, and the bile duct lining showed epithelial hyperplasia, mucinous metaplasia and adenomatous proliferation of intramural glands.

Lee *et al.* (1997) noted that cancers occur from interruption of cellular differentiation during tissue development, renewal or repair. In the liver, small intraportal 'oval' stem cells are not only believed to be the progenitors of hepatocytes and bile duct cells, but also of hepatocellular and cholangiocarcinomas. This study examined the immunohistologic phenotypes and ultrastructural appearance of small 'oval' cells in the livers of Syrian hamsters during cholangiocarcinogenesis induced by *N*-nitrosodimethylamine and promoted by *C. sinensis* infection. Forty female golden hamsters were divided into 2 groups: (1) 25 animals were infected with 15 metacercariae of *C. sinensis* orally, followed by administration of 15 ppm *N*-nitrosodimethylamine in the drinking water for 28 days and (2) 15 untreated control animals. Five hamsters in the experimental group and three in the control group were killed at 14, 35, 49, 70 and 105 days post-metacercarial infection. Histological, immunohistochemical and electron microscopy examination of liver tissues were then performed. Three distinct 'oval' cells were identified in the portal or periportal areas: (a) primitive oval cells which were small in size, possessed abundant heterochromatin and scant cytoplasm, and were negative for (alpha-fetoprotein) AFP, (cytokeratin 19) CK19, mouse anti-OV-6 monoclonal antibody (OV-6) and glutathione-S-transferase Pi (GST-p), (b) hepatocyte-like oval cells which were glycogen rich, positive for AFP but negative for CK19, OV-6 and GST-p and (c) ductular-like oval cells which were small in size, possessed desmosomes and basement membranes and were positive for CK19, OV-6 and GST-p, but negative for AFP. The type of liver cancer that develops appears to depend on the type of promotion chemical or infectious initiators effect on the putative liver stem cells. Lee *et al.* (1997) proposed that the ductular-like oval cells were precursors of dysplastic ductular cells that gave rise to cholangiocarcinoma following *N*-nitrosodimethylamine exposure. Primitive oval cells were bipolar progenitors to both hepatocytes and biliary cells. In this study, infection with *C. sinensis* stimulated the biliary lineage after initiation by the carcinogen, thereby promoting the development of cholangiocarcinoma. There appear to be species to species differences, as in the rat model the same stimuli favour bipolar liver stem cells to commit to hepatocyte-like oval cells and hepatocellular carcinomas.

Choi *et al.* (2004) proposed the following mechanism of carcinogenesis for *C. sinensis* infection. Bile duct irritation induced by the worms plays a key role in promotion of cholangiocarcinoma through the stage of dysplasia. The proliferating ducts may then be susceptible to tumour initiation by carcinogens, even at levels too low to produce cancers in non-infected individuals. How the parasite causes biliary hyperplasia and metaplasia is poorly understood. Hong *et al.* (1993) stained rat biliary epithelium with bromodeoxyuridine to observe changes during *C. sinensis* infection. Epithelial proliferation was confined to hepatic regions containing worms, suggesting that cell proliferation resulted from direct stimulation from the parasites. Adults of *C. sinensis* do provide mechanical stimulation with their suckers and tegument contacting the duct lining during movement. The ductal epithelium may also receive unknown immunologic or chemical stimuli in addition to mechanical promotion. Only a small percentage of individuals infected with *C. sinensis* actually develop cholangiocarcinoma, so it is likely that other factors are required for carcinogenesis to proceed. The two-step theory of carcinogenesis for *C. sinensis* infection starts with initiation, a rapid, permanent change in host tissue caused by exposure to a carcinogen. The next stage, promotion is a slow process with a progression of reversible effects. *C. sinensis* acts as a promoter in the two-stage model of carcinogenesis, which requires an initiator stimulus, most likely from an exogenous carcinogen.

Choi *et al.* (2004) also noted that biliary stone formation, a common finding in clonorchiasis, may enhance carcinogenesis. Kowalewski and Todd (1971) reported that hamsters with cholesterol pellets implanted in their gallbladders were more susceptible to induction of gallbladder carcinoma by chemical carcinogens. In humans, *C. sinensis* infection is linked to biliary obstruction, stones, recurrent pyogenic cholangitis and cholangiocarcinoma. Biliary calculi as a promoter of carcinogenesis remains an area yet to be fully explored. Kim *et al.* (2004) investigated the immunohistochemical expression of cyclooxygenase-2 in 102 patients with cholangiocarcinoma with respect to various clinopathological characteristics, including *C. sinensis* infection. Results showed that *C. sinensis* infection was related to aberrant cyclooxygenase-2 expression. As mentioned earlier in this review, up-regulation of cyclooxygenase-2 can activate environmental carcinogens leading to oxidative and nitrosative damage to the host genome.

The IARC (1994) determined that *C. sinensis* is probably carcinogenic to humans (group 2A). Recent work has only strengthened the evidence for *C. sinensis* as a promoting agent in carcinogenesis.

2.6. *F. hepatica*

Bunnag *et al.* (2000) noted that infection with *F. hepatica*, also known as sheep liver fluke disease, is common in sheep and other domestic livestock. Human infection can occur from ingesting watercress from sheep-raising areas. It is estimated that at least 2 million people are infected worldwide, mostly in temperate climates. Adult worms of *F. hepatica* live in the major bile ducts. The flat, leaf shaped flukes have scale-like spines on their anterior aspects, and range in size from 20 to 30 mm in length by 8–13 mm in length. Eggs are passed in the stool, miracidia hatch in fresh water and penetrate various species of lymnaeid snails, the first intermediate host. Mature cercariae emerge from the snail host within 4–7 weeks and encyst on various forms of aquatic vegetation to become metacercariae. Once ingested, the metacercariae excyst in the duodenum, migrate through the bowel wall into the peritoneal cavity. Next, the larval flukes penetrate the capsule of the liver and migrate through the liver parenchyma to reach the bile ducts. This prepatent period may take 3–4 months. The life span of adult *F. hepatica* is between 9 and 13 years. Bunnag *et al.* (2000) also described the pathologic changes accompanying fascioliasis. In addition to hepatic parenchymal necrosis and abscess formation from migrating larvae, adult flukes cause hyperplasia, thickening, desquamation and dilatation of the bile ducts. Biliary fibrosis may be due to the proline produced by adult worms. Human infection is often characterized by infection of extrahepatic sites, most commonly heart, brain, intestinal wall, lungs and skin.

The oncogenic potential of *F. hepatica* remains unproven at present, with a portion of the available evidence being contradictory. Experimental data supports two opposite effects, that is, tumour growth stimulating and tumour growth inhibiting. Fascioliasis is rarely complicated by neoplastic development as compared to other helminthoses. Galvez and Maglajlic (1956) first reported cases of liver cancer in *F. hepatica*-infected cattle. Vitovec (1974) found cases of hepatocellular carcinoma in cattle with biliary cirrhosis due to *F. hepatica* infection. Cornick (1988) reported a llama with metastatic squamous cell carcinoma of the stomach and fascioliasis. Sriurairatana *et al.* (1996) found isolates of the human cholangiocarcinoma cell line HuCCA-1 from a tumour of the bile duct in a patient infected by *F. hepatica*. Tsocheva-Gaytandzhieva (2005) noted the occurrence of cholangiocarcinoma in association with biliary cirrhosis, fibrosing cholangitis and pericholangitis of fasciolar origin.

The stimulation of tumour growth or hepatocyte proliferation has been observed during the acute phase of fascioliasis. The acute phase is defined as the period in which larval flukes migrate through the liver, with symptoms lasting for several weeks to months. Chronic infection is defined as the period in which the flukes reside within the biliary passages. Ginovker (1979) found that treatment of rats with *F. hepatica* extract or liver implant stimulated mitotic activity and proliferation of hepatocytes. Clark *et al.* (1981) infected mice with *F. hepatica*, allowed the parasite to undergo normal development in the host and then killed the animals to prepare liver homogenates. Results showed that in vitro, liver homogenates from *F. hepatica*-infected mice activated the promutagens aflatoxin B1 and methyl-nitrosourea-3,3′-dichlorobenzidene, producing 20× more colonies on the Ames salmonella/microsome test than did liver homogenates from uninfected control animals. Krustev *et al.* (1987) isolated a growth stimulating factor from the metabolic products of *F. hepatica* and demonstrated its proliferative effect on cell cultures. Tsocheva *et al.* (1992) noted that tumour growth in the rat liver following *N*-nitrosodimethylamine administration was stimulated if the carcinogen was given during the acute phase of *F. hepatica* infection. Gentile and Gentile (1994) proposed the following explanation for the observation of enhanced mutagen and carcinogen activity in the *F. hepatica*-acutely infected host. The fluke produces an intense inflammatory response due to the release of antigens from surface components and excretory products, with concomitant cellular proliferation in the area of infection. Inflammatory cells including macrophages and eosinophils, once activated, can produce reactive oxygen species like hydrogen peroxide that produce genetic instability by inducing DNA damage. Interactions between the inflammatory response and cell proliferation may create an increased frequency of spontaneous mutations. Proliferation of hepatocytes, a consequence of inflammation and cell death, also allows xenobiotics to interact with host DNA and to fix mutations that could play a role in carcinogenesis. Some of these mutations could lead to activation of oncogenes or the inactivation of tumour suppressor genes.

Gentile *et al.* (1998) used the lamda/LacI Big Blue transgenic mouse model to determine whether genetic damage in liver tissue could be produced by fascioliasis. Seven mice were each infected with two metacercariae of *F. hepatica* by oral intubation. Six uninfected animals served as controls. All animals were killed on day 23 of the experiment. Infection was confirmed by visual inspection of the liver and the presence of adult worms. DNA was isolated from the livers of infected and uninfected mice, and analyzed for mutations of the lacI gene by the transgenic Big Blue mouse assay. The average mutation frequency was 2.1×10^{-5} in the control group and 4.3×10^{-5} in the infected group. The results showed *F. hepatica* produced mutagenic events, most likely due to liver cell damage and associated inflammatory response. Montero *et al.* (1999) found increased activity of CYP2A5 isozyme in mouse liver specimens harvested 5 and 10 days following treatment with metacercariae of *F. hepatica*. Inflammation and proliferation in liver tissue was seen at the same time that CYP2A5 activity increased. CYP2A5 belongs to the CYP450 family of hepatic enzymes important in the metabolism of carcinogens such as aflatoxin B1, nitrosamines and butadiene. *F. hepatica* infection causes liver injury, inflammatory response and, during the acute phase of infection, an increase in metabolizing enzymes in the liver rendering the host more susceptible to the activation of exogenous carcinogens.

Inhibition of tumour growth has been observed during the chronic phase of fascioliasis. Retarding chemical carcinogenesis can be accomplished either by inhibiting the metabolism of the carcinogen or by activating the detoxification process. Hadjiolov (1984) found that *N*-nitrosodimethylamine-initiated malignant transformation can be prevented by inhibiting the metabolism of the carcinogen. Facino *et al.* (1989) later showed that fascioliasis retards drug metabolism in the rat liver. Gentile and Gentile (1994) noted that the ability of chronic *F. hepatica* infection to reduce drug metabolizing activity may be due to several mechanisms, including impairment of the mixed oxidase system, uncoupling of oxidative phosphorylation in mitochondria and respiratory inhibition. Tsocheva *et al.* (1992) studied the combined effect of *F. hepatica* infection and *N*-nitrosodimethylamine carcinogenesis on the rat monooxygenase system. Forty-eight male albino rats were divided into 5 groups: (1) 24 healthy untreated control animals, (2) 5 animals infected with *F. hepatica* twice, on the 1st day and the 10th week of the experiment, (3) 6 animals treated with *N*-nitrosodimethylamine 8 times at 7-day intervals beginning at the sixth week of the experiment, (4) 6 animals infected with *F. hepatica* on the first day of the experiment, then treated with the carcinogen as per group 3 protocol and (5) 6 animals infected as per group 2 protocol, then exposed to the carcinogen as per group 3 protocol. Infection and reinfection with *F. hepatica* was performed by introducing 15 metacercariae by mouth. *N*-nitrosodimethylamine was injected intraperitoneally at a dose of 100 mg/kg. The study end point was 27 weeks, after which animals were killed and liver microsomal fractions were obtained by centrifugation. Liver parameters studied included ethylmorphine-*N*-dimethylase (EMD), nitrosamine-*N*-dimethylase (NA), anilinehydroxylase (AH), microsomal haem, cytochrome b5, cytochrome p450.

Results were calculated as percentages of control values. Group 2 showed that double infection with *F. hepatica* inhibited EMD (46%), NA (60%), AH (27%), microsomal haem (35%), cytochrome p450 (27%) at $p < 0.05$. Cytochrome b5 was decreased 14% ($p > 0.10$). Group 3 showed that with *N*-nitrosodimethylamine treatment alone, a weak inhibition of EMD (6%), NA (6%), AH (11%), cytochrome b5 (14%) and microsomal haem (16%). Cytochrome p450 showed the greatest reduction (18%) at $p < 0.05$. Group 4 showed that treatment with fascioliasis and the carcinogen had a combined effect reducing activities of EMD (31%), NA (50%), cytochrome b5 (22%), microsomal haem (22%) and cytochrome p450 (24%) at $p < 0.05$. AH activity was near control values. Group 5, combining double infection with carcinogen exposure, had a similar but more pronounced inhibition than did group 4, with decreases in EMD (30%), NA (50%), cytochrome b5 (32%), microsomal haem (37%) and cytochrome p450 (54%). AH remained near control values. Tsocheva *et al.* (1992) believed the changes in the host liver monooxygenase system were largely due to *F. hepatica* infection, with *N*-nitrosodimethylamine having a modulating effect. The inhibition of liver drug metabolism may be due to direct effect of excretory/secretory products of *F. hepatica* on hepatocytes and cell organelles (specifically the endoplasmic reticulum where drug metabolizing enzymes are located), or to the release of substances from infection-activated Kupffer cells damaging the mixed-function oxidase system. The suppression of *N*-nitrosodimethylamine-induced carcinogenesis during the chronic phase of *F. hepatica* infection in the rat model appears to be the result of parasitic inhibition of biotransformation of the carcinogen in the liver. Finally, cell immunity-dependent zones were found in the spleens of animals chronically infected with fascioliasis. Cell immunity has a known antitumour effect. In acutely infected animals, no such zones were found.

Thamavit *et al.* (1994a) noted that proliferation of the linings of the bile ducts has been attributed to fluke-induced mechanical obstruction, chemical effects or immunological injury. In experimental animals, bile duct dilatation and proliferation can also be induced by intraperitoneal implant of *F. hepatica*, where the worms are not in direct contact with the biliary tree. Proline administered intraperitoneally has been shown to cause bile duct hyperplasia in rats and mice. Proline is an excretory product of *F. hepatica* and is produced in large quantities by the parasite. Thamavit *et al.* (1994b) attempted to show whether proline is a promoter in driving initiated bile duct cells to cancer. A total of 140 Syrian hamsters were divided into 8 groups: (1) *N*-nitrosodimethylamine (DMN) + proline intraperitoneally (ip), (2) DMN + proline subcutaneously (sc), (3) DMN + saline ip, (4) DMN + saline sc, (5) proline ip, (6) proline sc, (7) saline ip and (8) saline sc. DMN was administered at 20 mg/kg ip. Either 1 ml of 2M l-proline or 1 ml of saline was administered by ip or sc injection 3 × weekly for 20 weeks following DMN treatment. At week 40 of the experiment animals were killed and their livers examined. The incidences of hepatocellular nodules were similar for all groups (1–4) treated with DMN with or without ip or sc proline administration. The occurrence of cystic bile ductules was lower in the DMN + ip proline group (1) than in the DMN + ip saline group (3) $p < 0.05$, but severity of lesions was mild in all cases. Animals treated with proline alone failed to develop any bile duct changes. Proline administration at a dosage known to cause biliary hyperplasia in rats appears to have no such effect in the Syrian hamster. Also absent was a promoting effect on DMN-initiated liver carcinogenesis.

Gentile and DeRuiter (1981) identified *F. hepatica* as an ideal laboratory model since it is easily cultured in mice and infection progresses from metacercarial exposure to adult worms within 5 weeks. Future research is necessary in order to clarify the complex host–parasite interactions and to better define the oncogenic implications of *F. hepatica* infection.

3. CESTOIDEA AND NEMATODA

This section provides information on helminths not discussed in the body of the text. We have listed in Table 4.1, 12 references (8 on cestodes, 4 on nematodes) for which roles in carcinogenesis have been suggested.

4. MECHANISMS OF CARCINOGENESIS

Carcinogenesis is a complex process in which both xenobiotics and endogenous factors interact to modify normal cell growth. Shacter and Weitzman (2002) noted that the establishment of a mutation is the sine qua non of cancer development. The developmental stages of tumourigenesis have been referred to as initiation, promotion and progression. Initiation involves a primary DNA mutation, often the result of genotoxic carcinogens. Promotion involves the stimulated expansion of a clone of initiated preneoplastic cells by both genetic and epigenetic mechanisms. Progression occurs when an early neoplastic clone of cells is transformed into a fully malignant phenotype resulting in uncontrolled cell growth. Permanent changes in DNA are present in all cancer cells, and the transformed phenotype is passed on through multiple cell divisions. For a normal cell, mutation of a key gene results in loss of homeostasis and untimely cell death. Proneoplastic mutations, resulting in increased expression of oncogenes or decreased expression of tumour suppressor genes, confer a selective survival or growth advantage to the cell. Such phenotypic changes include a decreased need for metabolites and growth factors, aberrant signal transduction, abnormal expression of receptors for available growth factors (epidermal growth factor receptor and

TABLE 4.1 The role of selected cestodes and nematodes in carcinogenesis

Parasite	Association with carcinogenesis	References
Cestoidea		
Taenia solium	Immunodepressive effect of a *T. solium* cysticercus factor on cultured T and B lymphocytes stimulated with phytohaemagglutinin	Molinari *et al.* (1989)
T. solium	Immune response impairment, genotoxicity and transformation of cell morphology induced by *T. solium* metacestode	Herrera *et al.* (1994)
T. solium	Neurocysticercosis as a risk factor for the development of haematological malignancies	Herrera *et al.* (1999)
T. solium	Increased DNA damage in lymphocytes in neurocysticercosis patients	Herrera *et al.* (2000)
T. solium	Increased translocation frequency of chromosomes 7, 11 and 14 in lymphocytes from patients with neurocysticercosis	Herrera *et al.* (2001)
T. solium	Soluble factor secreted by *T. solium* metacestodes induces DNA damage in human lymphocytes	Herrera *et al.* (2003)
T. solium	Extracts of *T. solium* metacestodes produce genotoxicity in *Drosophila melanogaster*	Silva *et al.* (2006)

(*continued*)

TABLE 4.1 *(continued)*

Parasite	Association with carcinogenesis	References
T. taeniaeformis	Infection of rats with metacestodes increases hepatic CYP 450 and increases genotoxicity of procarcinogens	Montero *et al.* (2003)
Nematoda		
Trichostrongylus vitrinus; Trichostrongylus colubriformis; Cooperia curticei; Nematodirus battus; Teladorsagia circumcincta; Haemonchus contortus	Excretory/secretory products of six nematode species affect the proliferation of HT29-D4 and HGT-1 cell lines	Huby *et al.* (1995)
Spirocerca lupi	The role of *S. lupi* in oesophageal sarcomas and osteosarcomas in dogs may be due to production of nitrous oxide and other reactive oxygen intermediates	Melendez and Suarez-Pellin (2001)
S. lupi	Case report of a *S. lupi*-induced fibroma of the oesophagus in a dog	Banga *et al.* (2005)
Gongylonema pulchrum	Case report of *G. pulchrum* infection and oesophageal squamous cell carcinoma in a lemur	Blier *et al.* (2005)

Her2/neu), dysregulated cell–cell communication and resistance to apoptosis. Cells are normally able to turn on and off genes to help them survive toxic signals, such as induction of p450 enzymes to metabolize carcinogens. Kirby *et al.* (1994) found elevated levels of hepatic p450 isozymes in *O. viverrini* infection. Although this adaptation is usually transient, under conditions of prolonged stress, as found in chronic inflammation, a mutation may lock in the growth-advantaged phenotype selecting proneoplastic cells lines.

Herrera and Ostrosky-Wegman (2001) noted that chronic inflammation in host tissues is a common feature of helminth infections. The longer the inflammation persists, the higher the risk of associated carcinogenesis. Inflammatory cells such as macrophages and eosinophils generate prostaglandins, cytokines and free radicals in the form of reactive oxygen and nitrogen species, which can lead to genetic instability and malignant transformation. Herrera *et al.* (2005) summarized the events during the inflammatory response to parasitic infection that can lead to carcinogenesis. Although free radicals protect the host by killing invading pathogens, they also can induce cycles of cell destruction and proliferation, preferentially selecting clones of mutated cells. Reactive oxygen and nitrogen species can oxidise and damage DNA. Activated inflammatory cells use plasma membrane-associated nicotinamide adenine dinucleotide phosphate oxidase to reduce oxygen to the free radical superoxide anion which through the Haber–Weiss reaction generates the much more reactive hydroxyl radical. The superoxide radical can also form hydrogen peroxide which reacts with transition metals bound to DNA to form activated complexes with site-specific damage. Oxidative damage causes single and double strand DNA breaks, point and frame shift mutations and chromosome abnormalities. Activated phagocytosis produces nitric oxide that, when oxidized to nitrogen dioxide, induces DNA damage. Nitric oxide can also interact with the superoxide ion forming peroxynitrite. The cytotoxic peroxynitrite may decompose into the hydroxyl radical and nitrogen dioxide. It may also oxidise sulfhydral groups inducing lipid peroxidation. Products from lipid peroxidation such as malondialdehyde and 4-hydroxynonenal, interact with DNA in a mutagenic and genotoxic manner. El-Sheikh *et al.* (2001) found that inflammatory cytokines and growth factors up-regulated cyclooxygenase-2 that initiated activation of environmental carcinogens and stimulation of reparative and adaptive changes leading to DNA damage and carcinogenesis. T-lymphocytes and their cytokine products such as tumour necrosis factor-α are important in the malignant transformation of chronically inflamed tissue. Herrera *et al.* (2005) also noted that inflammatory cells participate in activating environmental carcinogens like aflatoxins and polycyclic aromatic hydrocarbons, and can form carcinogenic nitrosamines from nitric oxide. Inflammation also induces altered host xenobiotic metabolism that increases susceptibility to toxic agents. Watanapa and Watanapa (2002) postulated that the combination of inflammatory genotoxic products acting on proliferating cells results in malignant change. Once DNA mutations have occurred, active cell turnover favours formation of clones of transformed cells.

Modulation of the host immune system with impairment of immunologic surveillance has been proposed as another mechanism for helminth-induced carcinogenesis. The theory of immunologic surveillance holds that the immune system destroys neoplastic cells as they appear, limiting the development of cancer. Schatten *et al.* (1984) discovered tumour-specific antigens, and found that carcinogenesis was inhibited in mice immunized against those antigens. Herrera *et al.* (2005) observed that the high incidence of cancer in immunosuppressed transplant patients and in patients with congenital or acquired immune deficiency suggests a role for the immune system in controlling malignant cell transformation. Helminths have evolved mechanisms to avoid immune detection by the host. Toledo *et al.* (2006) noted that antibodies bound to the surface of the intestinal trematode *Echinostoma caproni* are rapidly lost *in vivo.* Additionally, newly excysted juvenile and adults of *E. caproni* shed surface antigens rapidly over 8–15 min. These adaptations make the fluke inaccessible to immune attack. The antigenic makeup of *E. caproni* also changes over time, enabling the parasite to avoid immune recognition. White *et al.* (1997) suggested mechanisms parasites employ to avoid immune detection: production of antigenic variants, inhibition of host histocompatibility antigens, disruption of antigen-processing pathways and inactivation of complement and antibodies. Pettit *et al.* (2000) stated that the result of these parasitic adaptations is immunopathological changes affecting host immune surveillance that contribute to clonal expansion of transformed cells. Further evidence of parasite ability to modulate host immune response was provided by Reddy and Fried (2007) who noted that long-lived helminths are able to down-regulate host immunity, possibly through induction of the T-helper 2 response. Exposure of patients with inflammatory bowel (Crohn's) disease to *Trichuris suis* (pig whipworm) therapy has reduced disease activity in clinical studies.

Malignancy associated with helminth-induced chronic inflammation usually occurs at or near the site of inflammation. Examples are bladder cancer with *S. haemotobium* infection and bile duct cancer with *O. viverrini* infection. Evidence that systemic cancers might also develop from chronic parasite infection would implicate mechanisms not related to the local inflammatory process. Herrera *et al.* (2000) studied the relationship between human larval *T. solium* infection and the occurrence of haematologic malignancies. Results indicated a higher frequency of genetic damage in bone marrow cells and peripheral lymphocytes of neurocysticercosis patients as compared to uninfected or treated controls. If these mutations are not repaired, the persistent antigenic stimulus from the parasitic infection could stimulate proliferation of clones of damaged cells promoting haematologic malignancy.

Trosko and Ruch (1998) noted that during evolution from single to multi-cellular organisms, new biological functions appeared: contact inhibition to control cell proliferation, the process of differentiation of committed stem cells of various tissues and the need for programmed apoptosis. Many new genes appeared to control these various functions, including a gene coding for a membrane-associated protein channel, also termed the gap junction. This gap junction allows passive transfer of ions and low molecular weight proteins. A group of connexin genes codes for connexin proteins. A hexameric unit of connexins (a connexon) in one cell couples with a corresponding connexon in an adjoining cell effectively joining cytoplasms. This process synchronizes metabolic functions within tissues. The majority of normal solid tissue cells have working gap junctions for intracellular communication. Distinct characteristics of cancer cells have been identified: origin from a stem-like cell, loss of contact inhibition, uncontrolled cell growth, inability to terminally differentiate and alteration of programmed cell death. A cancer cell is also characterized by dysfunctional gap junctions and abnormal cell–cell communication, that is, a process controlling the ability of a cell to proliferate, differentiate, respond adaptively and apoptose. Disruption of intracellular communication can lead to cell proliferation, aberrant differentiation, decreased apoptosis and malignant transformation.

Herrera *et al.* (2005) suggested that events leading to the inhibition of cell–cell communication and to the disruption of proliferation–antiproliferation pathways in parsite-infected tissues could play a pivotal role in the clonal expansion of initiated cells producing tumours. Parasites not only present physical barriers to intracellular communication, but also can interrupt communications chemically through secreted substances. McKerrow (1989) noted that helminths secrete soluble factors whose interactions with host cells aid the parasite to survive for long periods of time in a disadvantageous hostile environment. Some of these secreted molecules may modify host cell homeostasis, increasing the chance of malignancy. Herrera *et al.* (1994) reported an RNA molecule secreted by cysticerci of *T. solium* that transformed Syrian hamster embryo cells *in vitro* and induced chromosome damage in human lymphocyte cultures. The secreted RNA factor appeared capable of inducing genetic instability in infected individuals, possible by a direct interaction with host genomic DNA. Parasite-secreted factors may also favour tumour promotion and progression. Ishii *et al.* (1989) found that *S. japonicum* soluble egg antigen had a significant tumour-promoting effect. Konno *et al.* (1999) noted that excretory/secretory proteins of *Taenia taeniaeformis* induced hyperplasia in the stomachs of heavily infected rats. Phares (1996) reported accelerated host body growth in the presence of infection with the plerocercoid stage of *Spirometra mansonoides*. This effect was caused by a parasite-secreted protein that interacted with mammalian growth hormone receptors.

The induction of genomic instability is the sentinel event in helminth-induced carcinogenesis, allowing initiated cells to transform into fully malignant ones. Chromosome alterations can be most easily detected by the micronucleus method. Rosin *et al.* (1994a) found that urothelial cells from the urinary bladder of patients with *S. haematobium* infection had a

6.1-fold higher frequency of micronucleated cells than did uninfected controls. The incidence of micronucleated cells decreased following anti-parasitic treatment. Mostafa *et al.* (1999) noted that although researchers have identified some of the molecular events causing parasite-induced malignant transformation (inactivation of p53, deletion of p16INK4 and overexpression of Bcl-2 for *S. haematobium*), the majority of such observations have been made in advanced tumour stages. This limits our understanding of the sequence involved in the multi-stage process of human carcinogenesis. Studies demonstrating the presence of viral particles in parasites have suggested the possibility of genetic exchange with host tisues and the promotion of carcinogenesis. Laclette (1990) found crystals of virus-like particles in metacestodes of *T. solium*. Khramtsov (2000) discovered the presence of double stranded RNA in *Cryptosporidium parvum* isolates. Parasites may also induce proliferation of viral-infected cells, shortening the period for cancer to develop. Gabet (2000) noted that asymptomatic human T-cell leukaemia carriers developed clinical T-cell leukaemia sooner if they had concomitant infection with *Strongyloides stercoralis*.

Trosko and Ruch (1998) recalled a major controversy in theoretical carcinogenesis: whether all cells or only a few special cells can give rise to cancer. The dedifferentiation theory states that for cancer to develop a 'mortal' cell must be 'immortalized' by a differentiated cell reverting back to an early progenitor or pluripotent cell type. The stem cell theory holds that a pleuripotent stem cell, already 'immortalized', gives rise to cancer, remaining 'immortalized' in the process. Totipotent stem cells can give rise to all cell types within an organism. Pleuripotent stem cells are derived from the totipotent ones, and are committed to a finite number of cell divisions and a specific lineage of cell types. The daughter cells of the pleuripotent stem cells are called progenitor cells and can produce one cell type resulting in terminally differentiated cells. The stem cell theory is supported by three lines of evidence. First, a stem cell divides asymetrically into a daughter cell committed to terminally differentiate and a daughter cell destined to remain immortal or stem cell-like. This is consistent with the first step in carcinogenesis, namely to prevent the mortalization of an immortal stem cell. Second, Nakano *et al.* (1985) found that only few cells from a population of Syrian baby hamster cells were susceptible to neoplastic transformation. Third, Kao *et al.* (1995) noted that presumptive pleuripotent stem cells isolated from human breast tissue had no connexin genes or functional gap junction intracellular communication, a phenotype similar to that of cancer cells. Block (1984) stated that the fact that some cancer cells can be induced to terminally differentiate demonstrates that they behave like pleuripotent stem cells. The work of Lee *et al.* (1995) showed that bile duct proliferation stimulated by *C. sinensis* promoted carcinogen-initiated bipolar liver stem cells to differentiate along biliary rather than hepatocyte lineage, favouring the development of cholangiocarcinoma. Our understanding of the role of the stem cell hypothesis in helminth-induced malignant transformation remains incomplete at present, and is an area for future research.

5. CONCLUDING REMARKS

The purpose of this review is to explore the role of helminth infections in carcinogenesis. The idea is not new, as in the mid-nineteenth century Rudolph Virchow observed a higher frequency of cancer of the human urinary bladder in areas of North Africa endemic for *S. haematobium* (Parsonnet, 1999). Difficulty arises in proving causality between parasitic infection and cancer due to the complex natural histories and long latent periods of both conditions during which numerous exogenous and endogenous factors interact. Although numerous helminths have been implicated, only two, *S. haematobium* and *O. viverrini*, have been shown to be definitely carcinogenic to humans. Approximately 16% of all cancer cases worldwide are attributable to infectious agents, including viruses, bacteria, schistosomes and liver flukes. This equates to 1,375,000 preventable cancers per year. The known mechanisms of parasite-induced cancer include chronic inflammation, modulation of the host immune system, impairment of immunologic surveillance, inhibition of cell–cell communication, disruption of proliferation–antiproliferation pathways, induction of genomic instability and stimulation of malignant stem cell progeny.

Public health measures for the prevention of helminth-induced cancer begin with education and improvement of socioeconomic status to reduce prevalence of infection. Control of population morbidity is achievable through mass chemotherapy for those at risk. For example, a single dose of praziquantel can cure 75% of people infected with *S. haematobium*, and can reduce egg release by 90–95%. El-Desoky *et al.* (2005) identified a slow acetylator phenotype at risk for *S. haemotobium*-induced bladder cancer. Future research to identify other biomarkers for population screening of at-risk individuals is needed. Control of transmission begins with treatment of endemic areas to eradicate intermediate hosts. The molluscicide niclosamide is effective, as is biologic control using natural predators, fungi, bacteria and viruses to reduce the snail population. Environmental modifications detrimental to the intermediate host are most effective when combined with improved hygiene and reduction of contact with contaminated water. Due to the enormity of helminthic infection worldwide, the development of effective anti-parasitic vaccines is of paramount importance to supplement the above measures. Dupre *et al.* (1999)

reported promising results from a schistosomal GST DNA vaccine supplementing praziquantel therapy.

Earlier diagnosis of helminth-induced malignancy allows more effective therapy with a greater chance for cure. Prempracha *et al.* (1994) identified a soluble 200-kDa glycoprotein antigen which was markedly elevated in the serum and bile of animals with *O. viverrini*-associated cholangiocarcinoma. Future work is needed to discover other clinically useful tumour markers for parasite-realated cancer. There has also been recent interest in modulating the parasitic inflammatory response as a means of cancer prevention. Feig *et al.* (1994) noted the role of reactive oxygen species in mutagenesis. Dietary antioxidants may play a role in slowing tumourigenesis sufficiently to prevent clinical cancer during a patient's lifespan. Shacter and Weitzman (2002) suggested that non-steroidal anti-inflammatory drugs and selective cyclooxygenase-2 inhibitors may reduce carcinogenesis by inhibiting prostaglandins, cytokines and angiogenic factors.

It is our hope that this review increases awareness of helminth infections as a preventable cause of human cancer. In 1926 Johanes Fibiger won the only Nobel Prize awarded for helminthology. Although doubt was initially cast on his conclusions, his belief in parasite-induced cancer now appears to have been validated.

REFERENCES

Abdel, M. M., Hassan, A., and El-Sewedy, S. (2000). Human bladder cancer, schistosomiasis, N-nitroso compounds and their precursors. *Int. J. Cancer* **88,** 682–683.

Abdel-Tawab, G. A., Aboul-Azm, T., and Ebid, S. A. (1986). The correlation between certain tryptophan metabolites and the N-nitrosamine content in the urine of bilharzial bladder cancer patients. *J. Urol.* **135,** 826–830.

Abe, K., Kagei, N., Teramura, Y., and Ejima, H. (1993). Evidence for the segregation of a major gene in human susceptibility/resistance to infection by *Schistosoma mansoni*. *J. Med. Primatol.* **22,** 237–239.

Aji, T., Matsuoka, H., Ishii, A., Arimoto, S., and Hayatsu, H. (1994). Retention of a mutagen 3-amino-1-methyl-5H-pyrido[4, 3-b]indole (Trp-P-2) in the liver of mice infected with *Schistosoma japonicum*. *Mutat. Res.* **305,** 265–272.

Amano, T. (1980). Clinicopathological studies on the gastrointestinal schistosomiasis in the endemic area of Yamanashi Prefecture with special reference to the carcinogenicity of schistosome infection. *Jpn. J. Parasitol.* **29,** 305–312.

Amano, T., and Oshima, T. (1988). Hepatoma formation in ddY mice with chronic *schistosomiasis japonica*. *Jpn. J. Cancer Res.* **79,** 173–180.

Andrade, Z. A., and Abreu, W. N. (1971). Follicular lymphoma of the spleen in patients with hepatosplenic *schistosomiasis mansoni*. *Am. J. Trop. Med. Hyg.* **20,** 237–243.

Arimoto, S., Matsuoka, H., Aji, T., Ishii, A., Wataya, Y., and Hayatsu, H. (1992). Modified metabolism of a carcinogen 3-amino-1-methyl-5H-pyrido[4, 3-b]indole (Trp-P-2), by liver S9 from *Schistosoma japonica* infected mice. *Mutat. Res.* **282,** 177–182.

Badawi, A. F. (1996). Molecular and genetic events in schistosomiasis-associated human bladder cancer: Role of oncogenes and tumor suppressor genes. *Cancer Lett.* **105,** 123–138.

Badawi, A. F., Mostafa, M. H., Aboul-Asm, T., Haboubi, N. Y., O'Connor, P. J., and Cooper, D. P. (1992). Promutagenic methylation damage in bladder DNA from patients with bladder cancer associated with schistosomiasis and from normal individuals. *Carcinogenesis* **13,** 877–881.

Badawi, A. F., Cooper, D. P., Mostafa, M. H., Doenhoff, M. J., Probert, A., Fallon, P., Cooper, R., and O'Connor, P. J. (1993). Promutagenic methylation damage in liver DNA of mice infected with *Schistosoma mansoni*. *Carcinogenesis* **14,** 653–657.

Badawi, A. F., Cooper, D. P., Mostafa, M. H., Aboul-Asm, T., Barnard, R., Margison, G. P., and O'Connor, P. J. (1994). 06-Alkylguanini-DNA-alkyl-transferase activity in schistosomiasis-associated human bladder cancer. *Eur. J. Cancer* **30,** 1314–1319.

Badawi, A. F., Mostafa, M. H., Probert, A., and O'Connor, P. J. (1995). Role of schistosomiasis in human bladder cancer: Evidence of association, aetiological factors, and basic mechanisms of carcinogenesis. *Eur. J. Cancer Prev.* **4,** 45–49.

Banga, H. S., Singh, G., Brar, A. P., and Brar, R. S. (2005). *Spirocerca lupi*-induced fibroma of esophagus in a dog. *Indian Vet. J.* **82,** 780–781.

Blier, T., Hetzel, U., Nat, R., Bauer, C., Behlert, O., and Burkhardt, E. (2005). *Gongylonema pulchrum* infection and esophageal squamous cell carcinoma in a vari (*Lemur macaco variegata*). *J. Zoo Wildl. Med.* **36,** 342–345.

Block, A. (1984). Induced cell differentiation in cancer therapy. *Cancer Treat. Rep.* **68,** 199–206.

Bunnag, D., Cross, J. H., and Bunnag, T. (2000). Liver fluke infections. *In* "Hunter's Tropical Medicine and Emerging Infectious Diseases" (G. T. Strickland, ed.), pp. 840–846. W.B. Saunders, Philadelphia.

Campbell, W. (1997). The worm and the tumor: Reflections on Fibiger's Nobel Prize. *Perspect. Biol. Med.* **40,** 498–504.

Chaudhary, K. S., Lu, K. S., Abel, P. D., Khandan, N., Shoma, A. M., El-baz, M., Stamp, G. W., and Lalani, E. N. (1997). Expression of bcl-2 and p53 oncoproteins in schistosomiasis-associated transitional and squamous cell carcinoma of the urinary bladder. *Br. J. Urol.* **79,** 78–84.

Cheever, A. W., Kamel, I. A., Elwi, A. M., Mosimann, J. E., Danner, R., and Sippel, J. E. (1978). *S. mansoni* and *S. haematobium* infection in Egypt. III. Extrahepatic pathology. *Am. J. Trop. Med.* **27,** 55–75.

Chen, M. G., and Mott, K. E. (1989). Progress in the assessment of morbidity due to *Schistosoma haematobium*: A review of the recent literature. *Trop. Dis. Bull.* **48,** 2643–2648.

Chen, M. C., Chuang, C. Y., Chang, P. Y., and Hu, J. C. (1980). Evolution of colorectal cancer in schistosomiasis. Transitional mucosal changes adjacent to large intestinal carcinoma in colectomy specimens. *Cancer* **46,** 1661–1675.

Chen, M. C., Chang, P. Y., Chuang, C. Y., Chen, Y. J., Wang, F. P., Tang, Y. C., and Chou, S. C. (1981). Colorectal cancer and schistosomiasis. *Lancet* **1,** 971–973.

Chen, J., Campbell, T. C., Li, J., and Peto, R. (1990). "Diet, Life-Style and Mortality in China. A Study of the Characteristics of 65 Chinese Counties." Oxford, UK: Oxford University Press, Oxford, UK.

Choi, B. I., Han, J. K., Hong, S. T., and Lee, K. H. (2004). Clonorchiasis and cholangiocarcinoma: Etiologic relationship and imaging diagnosis. *Clin. Microbiol. Rev.* **17,** 540–552.

Christie, J., Crouse, D., Kelada, A. S., Anis-Ishak, E., Smith, J. H., and Kamel, I. A. (1986). Patterns of *Schistosoma haematobium* egg distribution in the human lower urinary tract. III. Cancerous lower urinary tracts. *Am. J. Trop. Med.* **35,** 759–764.

Chung, C. S., and Lee, S. K. (1976). An epidemiological study of primary liver carcinomas in Busan area with clonorchiasis. *Korean J. Pathol.* **10,** 33–46.

Clark, D., DeRuiter, B., Johnson, M., and Blankespoor, H. (1981). The activation of promutagens in parasite-infected organisms. *Environ. Mutag.* **3,** 391.

Cornick, J. L. (1988). Gastric squamous cell carcinoma and fascioliasis in llama. *Cornell Vet.* **78,** 235–241.

Dimmette, R. M., Elwi, A. M., and Spratt, H. F. (1959). Relationship of schistosomiasis to polyposis and adenocarcinoma of the large intestine. *Am. J. Clin. Pathol.* **26,** 266–276.
Dizdaroglu, M. R., Olinski, J. H., Doroshow, J. H., and Akman, S. A. (1993). Modification of DNA bases in chromatin of intact target human cells by activated human polynuclear leukocytes. *Cancer Res.* **53,** 1269–1272.
Dupre, H. M., Schacht, A. M., Capron, A., and Riveau, G. (1999). Control of schistosomiasis by combination of Sm28GST DNA immunization and praziquantel treatment. *J. Infect. Dis.* **180,** 454–463.
Edington, G. M. (1979). Schistosomiasis and primary liver cell carcinoma. *Trans. R. Soc. Trop. Med. Hyg.* **73,** 351.
El-Aaser, A. A., Hassanein, S. M., El-Bolkainy, M. N., Omar, S., El-Sebai, I., and El-Merzabani, M. M. (1978). Bladder carcinogenesis using bilharzias-infested Swiss albino mice. *Eur. J. Cancer* **14,** 645–648.
El-Bolkainy, M. N., Mokhtar, M., Ghonim, M. A., and Hussein, M. H. (1981). The impact of schistosomiasis on the pathology of bladder carcinoma. *Cancer* **48,** 2643–2648.
El-Desoky, E. S., AbdelSalam, Y. M., Akkad, M. A., Atanasova, S., von Ahsen, N., Armstrong, V. W., and Oellerich, M. (2005). NAT2*5/*5 genotype is a potential risk factor for schistosomiasis-associated bladder cancer in Egypt. *Ther. Drug Monit.* **27,** 297–304.
El-Hawey, A., Massoud, A., Badr, D., Waheeb, A., and Abdel-Hamid, S. (1989). Bacterial flora in hepatic encephalopathy in bilharzial and non-bilharzial patients. *J. Egypt. Soc. Parasitol.* **19,** 797–804.
El-Mouelhi, M., and Mansour, M. M. (1990). Hepatic drug conjugation/deconjugation systems in hepatosplenic schistosomiasis. *Biochem. Pharmacol.* **40,** 1923–1925.
El-Sheikh, S. S., Madaan, S., Alhasso, A., Abel, P., Stamp, G., and Lalani, E. N. (2001). Cyclooxygenase-2: A possible target in *Schistosoma*-associated bladder cancer. *Br. J. Urol.* **88,** 921–927.
Elissa, S., Ahmed, M. I., Said, H., Zaghlool, A., and El-Ahmady, O. (2004). Cell cycle regulators in bladder cancer: Relationship to schistosomiasis. *IUBMB Life* **56,** 557–564.
Facino, R. M., Carini, M., Genchi, C., Tofanetti, O., and Casciarri, J. (1989). Participation of lipid peroxidation in the loss of hepatic drug-metabolizing activities in experimental fascioliasis in the rat. *Pharmacol. Res.* **21,** 549–560.
Fadl-Elmula, I., Kytola, S., Leithy, M. E., Abdel-Hameed, M., Mandahl, N., Elagib, A., Ibrahim, M., Larsson, C., and Heim, S. (2002). Chromosomal aberrations in benign and malignant bilharziasis-associated bladder lesions. *BMC Cancer* **2,** 5.
Feig, D. I., Reid, T. M., and Loeb, L. A. (1994). Reactive oxygen species in tumorigenesis. *Cancer Res.* **54,** 1890–1894.
Ferguson, A. R. (1911). Associated bilharziasis and primary malignant disease of the urinary bladder with observations on a series of forty cases. *J. Pathol. Bacteriol.* **16,** 76–94.
Flavell, D. J., and Lucas, S. B. (1983). Promotion of N-nitrosodimethylamine-initiated bile duct carcinogenesis in the hamster by the human liver fluke, *Opisthorchis viverrini*. *Carcinogenesis* **7,** 927–930.
Fripp, P. J. (1988). *Schistosoma mansoni* and B-glucuronidase. *Trans. R. Soc. Trop. Med. Hyg.* **82,** 351.
Gabet, A. S. (2000). High circulating proviral load with oligoclonal expansion of HTLV-1 bearing T cells in HTLV-1 carriers with strongyloidiasis. *Oncogene* **19,** 4954–4960.
Galvez, E., and Maglajlic, E. (1956). Fascioliasis associated cattle carcinoma polymorphocellularae primarium, hepatitis and cirrhosis. *Veterinaria* **2,** 235–239.
Gentile, J. M. (1985). Schistosome related cancers: A possible role for genotoxins. *Environ. Mutagen.* **7,** 775–785.
Gentile, J. M., and DeRuiter, E. (1981). Promutagen activation in parasite-infected organisms, preliminary observations with *Fasciola hepatica*-infected mice and aflatoxins B1. *Toxicol. Lett.* **8,** 273–282.

Gentile, J. M., and Gentile, G. J. (1994). Implications for the involvement of the immune system in parasite-associated cancers. *Mutat. Res.* **305,** 315–320.

Gentile, J. M., Brown, S., Aardema, M., Clark, D., and Blankespoor, H. (1985). Modified mutagen metabolism in *Schistosoma haematobium* infested organisms. *Arch. Environ. Health* **40,** 5–12.

Gentile, J. M., Gentile, G. J., Nannenga, B., Johnson, M., Blankespoor, H., and Montero, R. (1998). Enhanced liver cell mutations in trematode-infected Big Blue transgenic mice. *Mutat. Res.* **400,** 355–360.

Gibson, J. B. (1971). Parasites, liver disease and liver cancer. *In* "Liver Cancer" (IARC Scientific Publications No. 1), pp. 42–50. IARC, Lyon.

Ginovker, A. G. (1979). Influence of trematode helminths' extracts and liver implants on the hepatocytes mitotic activity. *Dokl. AN-USSR* **248,** 739–740.

Gonzalez-Zulueta, M., Shibata, A., Ohneseit, P. F., Spruck, C. H., Bush, C., Shamaa, M., Elbaz, M., Nichols, P. W., Gonzalgo, M. L., Malstrom, P. U., and Jones, P. A. (1995). High frequency of chromosome 9p allelic loss of CDKN2 tumor suppressor gene alterations in squamous cell carcinoma of the bladder. *J. Natl. Cancer Inst.* **7,** 1383–1392.

Guo, Z. R., and Lu, Q. X. (1987). Parasitic diseses. A case-control study on the relationship between schistosomiasis and liver cancer. *Chin. J. Parasitol. Parasitic Dis.* **5,** 220–223.

Guo, Z. R., Ni, Y. C., and Wu, J. L. (1984). Epidemiological study on the relationship between schistosomiasis and colorectal cancer. *Jiangsu Med. J.* **4,** 35.

Guo, W., Zheng, W., Li, J., Chen, J., and Blot, W. J. (1993). Correlations of colon cancer mortality with dietary factors, serum markers and schistosomiasis in China. *Nutr. Cancer* **20,** 13–20.

Gutierrez, M. I., Siraj, A. K., Khaled, H., Koon, N., El-Rifai, W., and Bhatia, K. (2004). CpG island methylation in *Schistosoma* and non-*Schistosoma*-associated bladder cancer. *Mod. Pathol.* **17,** 1268–1274.

Habib, S. L., Sheweta, S. A., Awad, A. T., Mashaal, N. M., Soliman, A. A., and Mostafa, M. H. (1996). Influence of *Schistosoma mansoni* infection on carcinogen-metabolizing capacities and *in vitro* aflatoxin B-1 metabolism in human liver. *Oncol. Rep.* **2,** 769–773.

Habuchi, T., Takahashi, R., Yamada, H., Kakehi, O., Ourga, K., Hamazaki, S., Toguchida, J., Ishizaki, K., and Yoshida, O. (1993). Influence of cigarette smoking and schistosomiasis on p53 gene mutation in urothelial cancer. *Cancer Res.* **53,** 3795–3799.

Hadjiolov, D. (1984). "DNA-Alteration and Cancer." p. 189. Medicina I Fizkultura, Sofia.

Haitel, A., Posch, B., El-Baz, M., Mokhtar, A. A., Susani, M., Ghoneim, M. A., and Marberger, M. (2001). Bilharzial related, organ conferred, muscle invasive bladder cancer: Prognostic value of apoptosis markers, proliferation markers, p53, E-cadherin, epidermal growth factor receptors and c-erB-2. *J. Urol.* **165,** 1481–1487.

Hasler, J. A., Siwela, A. H., Nyathi, C. B., and Chetsanga, C. J. (1986). The effect of schistosomiasis on the activation of aflatoxins B1. *Res. Commun. Chem. Pathol. Pharmacol.* **51,** 421–424.

Haswell-Elkins, M. R., Mairiang, E., Mairiang, P., Chaiyakum, J., Chamadol, N., Loapaiboon, V., Sithithaworn, P., and Elkins, D. B. (1994a). Cross-sectional study of *Opisthorchis viverrini* infection and cholangiocarcinoma in communities within a high risk area of Northeast Thailand. *Int. J. Cancer* **59,** 505–509.

Haswell-Elkins, M. R., Satarug, S., Tsuda, M., Mairiang, E., Esumi, H., Sithithaworn, P., Mairiang, P., Saitoh, S., Yongvanit, P., and Elkins, D. B. (1994b). Liver fluke infection and cholangiocarcinoma: Model of endogenous nitric oxide and extragastric nitrosation in human carcinogenesis. *Mutat. Res.* **305,** 241–252.

Herrera, L. A., and Ostrosky-Wegman, P. (2001). Do helminths play a role in carcinogenesis? *Trends Parasitol.* **17,** 172–175.

Herrera, L. A., Santiago, P., Rojas, G., Salazar, P. M., Tato, P., Molinari, J. L., Schifman, D., and Ostrosky-Wegman, P. (1994). Immune response impairment, genotoxicity and morphological transformation induced by *Taenia solium* metacestode. *Mutat. Res.* **305,** 223–228.

Herrera, L. A., Benita-Bordes, A., Sotelo, J., Chavez, L., Olvera, J., Rascon, A., Lopez, M., and Ostrosky-Wegman, P. (1999). Possible relationship between neurocysticercosis and hematological malignancies. *Arch. Med. Res.* **30,** 154–158.

Herrera, L. A., Ramirez, T., Rodriguez, U., Corona, T., Lorenzo, M., Ramos, F., Verdorfer, I., Gebhart, E., and Ostrosky-Wegman, P. (2000). Possible association between *Taenia solium* cysticercosis and cancer: Increased frequency of DNA damage in peripheral lymphocytes from neurocysticercosis patients. *Trans. R. Soc. Trop. Med. Hyg.* **94,** 61–65.

Herrera, L. A., Rodriguez, U., Gebhart, E., and Ostrosky-Wegman, P. (2001). Increased translocation frequency of chromosomes 7, 11 and 14 in lymphocytes from patients with neuroysticercosis. *Mutagenesis* **16,** 495–497.

Herrera, L. A., Tato, P., Molinari, J. L., Perez, E., Dominguez, H., and Ostrosky-Wegman, P. (2003). Induction of DNA damage in human lmphocytes treated with a soluble factor secreted by *Taenia solium* metacestodes. *Teratog. Carcinog. Mutagen.* **1,** 79–83.

Herrera, L. A., Benitez-Bribiesca, L., Mohar, A., and Ostrosky-Wegman, P. (2005). Role of infectious diseases in human carcinogenesis. *Environ. Mol. Mutagen.* **45,** 284–303.

Hill, M. J. (1988). N-nitroso compounds and human cancer. *In* ''Nitrosamine Toxicology and Microbiology'' pp. 90–102. Ellis Horwood, Chichester, UK.

Hong, S. T., Kho, W. G., Kim, W. H., Chai, J. Y., and Lee, S. H. (1993). Turnover of biliary epithelial cells in *Clonorchis sinensis* infected rats. *Korean J. Parasitol.* **31,** 83–89.

Honjo, S., Srivatanakul, P., Sriplung, H., Kikukawa, H., Hanai, S., Uchida, K., Todoroki, T., Jedpiyawongse, A., Kittiwatanachot, P., Sripa, B., Deerasamee, S., and Miwa, M. (2005). Genetic and environmental determinants of risk for cholangiocarcinoma via *Opisthorchis viverrini* in a densely infested area in Nakhon Phanom, northeast Thailand. *Int. J. Cancer* **117,** 854–860.

Hou, P. C. (1956). The relationship between primary carcinoma of the liver and infestation with *Clonorchis sinensis. J. Pathol. Bacteriol.* **72,** 239–246.

Huby, F., Hoste, H., Mallet, S., Fournel, S., and Nano, J. L. (1995). The effects of excretory/secretory products of six nematode species, parasites of the digestive tract, on the proliferation of HT29-D4 and HGT-1 cell lines. *Epithelial Cell Biol.* **4,** 156–162.

Hussein, M. R., Nassar, M. I., Kamel, N. A., Osman, M. E., and Georguis, M. N. (2005). Analysis of fibronectin expression in the bilharzial granulomas and of laminin in the transformed urothelium in *Schistosoma haematobium* infested patients. *Cancer Biol. Ther.* **4,** 676–678.

IARC (1994). Monograph on the evaluation of carcinogenic risks to humans: Schistosomes, liver flukes and *Helicobacter pylori. WHO: International Agency for Research on Cancer* **61,** 9–175.

Iida, H. (1985). Experimental study of the effects of *Clonorchis sinensis* infection on induction of cholangiocarcinoma in Syrian golden hamsters administered 0.03% N-2-fluorenylacetamide (FAA). *Jpn. J. Parasitol.* **34,** 7–16.

Inaba, Y., Takahashi, E. Y., and Maruchi, N. (1977). A statistical analysis on the mortality of liver cancer and liver cirrhosis in Yamanashi Prefecture, with special emphasis on the relation to the prevalence of schistosomiasis. *Jpn. J. Public Health* **24,** 811–815.

Irie, Y., and Iwamura, Y. (1993). Host-related DNA sequences are localized in the body of schistosome adults. *Parasitology* **107,** 519–528.

Ishii, A., Matsuoka, H., Aji, T., Hayatsu, H., Wataya, Y., Arimoto, S., and Tokuda, H. (1989). Evaluation of the mutagenicity and the tumor-promoting activity of parasite extracts: *Schistosoma japonicum* and *Clonorchis sinensis. Mutat. Res.* **224,** 229–233.

Iuchi, M., Nakayama, Y., Ishiwa, M., Yamada, H., and Chiba, K. (1971). Primary cancer of the liver associated with chronic *schistosomiasis japonica. Naika* **27,** 761–766.

Jang, J. J., Cho, K. J., Myong, N. H., and Chai, J. Y. (1990). Enhancement of dimethylnitrosamine-induced glutathione-S-transferase P-positive hepatic foci by *Clonorchis sinensis* infestation in F344 rats. *Cancer Lett.* **52,** 133–138.

Kakizoe, Y. (1985). The influence of *Schistosoma mansoni* infection on carcinogenesis of mouse livers initiated by N-2-fluorenylacetamide. *Kureme Med. J.* **32,** 169–178.

Kamel, I. A., Elwi, A. M., Cheever, A. W., Mosimann, J. E., and Danner, R. (1978). *Schistosoma mansoni and S. haematobium* infections in Egypt. IV. Hepatic lesions. *Am. J. Trop. Med. Hyg.* **27,** 931–938.

Kao, C.-Y., Nomata, K., Oakley, C. S., Welsch, C. W., and Chang, C. C. (1995). Two types of normal human breast epithelial cells derived from reduction mammoplasty: Phenotypic characterization and response to SV40 transfection. *Carcinogenesis* **16,** 531–538.

Khramtsov, N. (2000). Presence of double-stranded RNAs in human and calf isolates of *Cryptosporidium parvum. J. Parasitol.* **86,** 275–282.

Kiba, T., Tsuda, H., Pairojkul, C., Inoue, S., Sugimura, T., and Hirohashi, S. (1993). Mutations of the p53 tumor suppressor gene and the ras gene family in intrahepatic cholangiocellular carcinomas in Japan and Thailand. *Mol. Carcinog.* **8,** 312–318.

Kim, Y. I. (1984). Liver carcinoma and liver fluke infection. *Arzneimittelforschung* **34,** 1121–1126.

Kim, Y. I., Yang, D. H., and Chang, K. R. (1974). Relationship between *Clonorchis sinensis* infestation and *cholangiocarcinoma* of the liver in Korea. *Seoul J. Med.* **15,** 247–253.

Kim, H. J., Lee, K. T., Kim, E. K., Sohn, T. S., Heo, J. S., Choi, S. H., Choi, D. I., Lee, J. K., Paik, S. W., and Rhee, J. C. (2004). Expression of cyclooxygenase-2 in *cholangiocarcinoma:* Correlation with clinicopathological features and prognosis. *J. Gastroenterol. Hepatol.* **19,** 582–588.

Kirby, G. M., Prlkonen, P., Vatanasapt, V., Camus, A. M., Wild, C. P., and Lang, M. A. (1994). Liver fluke (*Opisthorchis viverrini*) infestation is associated with increased expression of CYP2A and carcinogen metabolism in male hamster liver. *Mol. Carcinog.* **11,** 81–89.

Kitani, K., and Iuchi, M. (1990). *Schistosomiasis japonica*: A vanishing epidemic in Japan. *J. Gastroenterol. Hepatol.* **5,** 160–172.

Kojiro, M., Kakizoe, S., Yano, H., Tsumagari, J., Kenmochi, K., and Nakashima, T. (1986). *Hepatocellular carcinoma* and *schistosomiasis japonica.* A clinicopathological study of 59 autopsy cases of *hepatocellular carcinoma* associated with chronic *schistosomiasis japonica. Acta Pathol. Jpn.* **36,** 525–532.

Konno, K., Oku, Y., Nonaka, N., and Kamiya, M. (1999). Hyperplasia of gastric mucosa in donor rats orally infected with *Taenia taeniformis* eggs and in recipient rats surgically implanted with the larva in the abdominal cavity. *Parasitol. Res.* **85,** 431–436.

Kowalewski, K., and Todd, E. F. (1971). Carcinoma of the gallbladder induced in hamsters by insertion of cholesterol pellets and feeding dimethylnitrosamine. *Proc. Soc. Exp. Biol. Med.* **136,** 482–486.

Krustev, L., Poljakova-Krusteva, O., Brodvarova, J., Aleksiev, B., Minchev, S., and Blagova, N. (1987). "A Device for Cell Stimulation and a Method for its Preparing." *Invention* No. 79202, Sofia.

Kuper, H., Adami, H. O., and Trichopoulos, D. (2000). Infections as a major preventable cause of human cancer. *J. Intern. Med.* **248,** 171–183.

Kurathong, S., Lerdverasirkul, P., Wongpaitoon, V., Pramoolsinsap, C., Kanjanapitak, A., Varavithya, W., Phuapradit, P., Bunyaratvej, S., Upatham, E. S., and Brockelman, W. Y. (1985). *Opisthorchis viverrini* infection and *cholangiocarcinoma.* A prospective case-control study. *Gastroenterology* **89,** 151–156.

Kurniawan, A. N., Hardjawidjaja, L., and Clark, R. T. (1976). A clinco-pathologic study of cases with *Schistosoma japonicum* infection in Indonesia. *Southeast Asian J. Trop. Med. Public Health* **7,** 263–269.

Laclette, J. (1990). Crystals of virus-like particles in the metacestodes of *Taenia solium* and *T. crassiceps*. *J. Invertebr. Pathol.* **56,** 215–221.
Lee, S. H., Shim, T. S., Lee, S. M., and Chi, J. G. (1978). Studies on pathological changes of the liver in albino rats infected with *Clonorchis sinensis*. *Korean J. Parasitol.* **16,** 148–155.
Lee, J. H., Rim, H. J., and Bak, U. B. (1993). Effect of *Clonorchis sinensis* infection and dimethylnitrosamine administration on the induction of *cholangiocarcinoma* in Syrian golden hamsters. *Korean J. Parasitol.* **31,** 21–29.
Lee, J. H., Yang, H. M., Bak, U. B., and Rim, H. J. (1994). Promoting effect of *Clonorchis sinensis* infection on induction of *cholangiocarcinoma* during two-step carcinogenesis. *Korean J. Parasitol.* **31,** 13–18.
Lee, J. H., Rim, H. J., and Sell, S. (1995). Promotion of cholangiocarcinomas by *Clonorchis sinensis* infection in hamsters after induction by dimethylnitrosamine. *J. Tumor Marker Oncol.* **10,** 7–15.
Lee, J. H., Rim, H. J., and Sell, S. (1997). Hererogeneity of the "oval-cell" response in the hamster during *cholangiocarcinogenesis* following *Clonorchis sinensis* infection and dimethylnitrosamine treatment. *J. Hepatol.* **26,** 1313–1323.
Li, Y. (1988). Geographical correlation analysis between schistosomiasis and large intestine cancer. *Chung Hua Liu Hsing Ping Hsueh Tsa Chih* **9,** 265–268.
Liu, B. C., Rong, Z. P., Sun, X. T., Wu, Y. P., and Gao, R. Q. (1983). Study of geographic correlation between colorectal cancers and schistosomiasis in China. *Acta Acad. Med. Sin.* **5,** 173–177.
Loilome, W., Yongvanit, P., Wongkham, C., Tepsiri, N., Sripa, B., Sithithaworn, P., Hanai, S., and Miwa, M. (2006). Altered gene expression in *Opisthorchis viverrini*-associated *cholangiocarcinoma* in the hamster model. *Mol. Carcinog.* **45,** 279–287.
Lozano, J. C., Nakazawa, H., Cros, M. P., Cabral, R., and Yamasaki, H. (1994). G to T mutations in p53 and H-ras genes in esophageal papillomas induced by N-nitromethoxybenzamine in two strains of rats. *Mol. Carcinog.* **271,** 33–39.
Lukacs, N. W., and Boros, D. L. (1993). Lymphokine regulation of granuloma formation in murine *schistosomiasis mansoni*. *Clin. Immunol. Immunopathol.* **68,** 57–63.
Makhyoun, N. A. (1974). Smoking and bladder cancer. *Br. J. Cancer* **30,** 577–581.
Makhyoun, N. A., El-Kashlan, K. M., Al-Ghorab, M. M., and Mokhles, A. S. (1971). Aetiological factors in bilharzial bladder cancer. *J. Trop. Med. Hyg.* **74,** 73–78.
Makarananda, K., Wild, C. P., Jiang, Y. Z., and Neal, G. E. (1991). Possible effect of infection with liver fluke (*Opisthorchis viverrini*) on the monitoring of urine by enzyme-linked immunosorbent assay for human exposure to aflatoxins. *In* "Relevance to Human Cancer of N-Nitroso Compounds, Tobacco Smoke and Mycotoxins" (I. K. O'Neill, J. Chen, and H. Bartsch, eds.), pp. 96–101. (IARC Scientific Publications No. 105). IARC, Lyon.
Matsuoka, H., Aji, T., Ishii, A., Arimoto, S., Wataya, Y., and Hayatsu, H. (1989). Reduced levels of mutagen processing potential in the *Schistosoma japonica*-infected mouse liver. *Mutat. Res.* **227,** 153–157.
Mao, L., Schoenberg, M. P., Sciccitano, M., Erozan, Y. S., Merlo, A., Schwab, D., and Sidransky, D. (1996). Molecular detection of primary bladder cancer by microsatellite analysis. *Science* **271,** 659–662.
Marletta, M. A. (1988). Mammalian synthesis of nitrite, nitrate, nitric oxide and N-nitrosating agents. *Chem. Res. Toxicol.* **1,** 249–257.
Mayer, D. A., and Fried, B. (2002). Aspects of human parasites in which surgical intervention may be important. *Adv. Parasitol.* **51,** 1–94.
McKerrow, J. H. (1989). Parasite proteases. *Exp. Parasitol.* **68,** 111–115.
Melendez, R. D., and Suarez-Pellin, C. (2001). *Spirocerca lupi* and dogs: The role of nematodes in carcinogenesis. *Trends Parasitol.* **17,** 516.

Mikhail, I. A., Higashi, G. I., and Edman, S. H. (1982). Interactions of *Salmonella paratyphi* and *S. mansoni* in the hamster. *Am. J. Trop. Med. Hyg.* **31,** 828–834.
Miyasato, M. (1984). Experimental study of the influence of *Schistosoma japonicum* infection on carcinogenesis of mouse liver treated with N-2-fluoroenylacetamide (2-FAA). *Jpn. J. Parasitol.* **33,** 41–48.
Molinari, J. L., Tato, P., Reynoso, O. A., and Cazares, J. M. (1989). Depressive effect of a *Taenia solium* cysticercus factor on cultured human lymphocytes stimulated with phytohaemagglutinin. *Ann. Trop. Med. Parasitol.* **84,** 205–208.
Montero, R., Gentile, G. J., Frederick, L., McMannis, J., Murphy, T., Silva, G., Blankspoor, H., and Gentile, J. M. (1999). Induced expression of CYP2A5 in inflamed trematode-infested mouse liver. *Mutagenesis* **14,** 217–220.
Montero, R., Serrano, L., Davila, V. M., Ito, A., and Plancarte, A. (2003). Infection of rats with *Taenia taeniformis* metacestodes increases hepatic CYP450, induces activity of CYP1A1, CYP2B1 and COH isoforms and increases the genotoxicity of the procarcinogens benzopyrene, cyclophosphamide and aflatoxins B-1. *Mutagenesis* **18,** 211–216.
Moore, M. A., Thamavit, W., Tiwawech, D., and Ito, N. (1991). Cell death and proliferation in *Opisthorchis viverrini*-DHPN induced carcinogenesis in the Syrian hamster hepatopancreatic axis. *In* "Chemical Carcinogenesis" (A. Columbiano, ed.), Vol. 2. pp. 503–510. Plenum Press, New York.
Mostafa, M. H., Sheweita, S. A., Elkoweidy, A. H., and Badawi, A. F. (1993). Alterations in the carcinogen metabolizing capacities of mouse liver during *Schistosoma mansoni* infection. *Int. J. Oncol.* **2,** 695–699.
Mostafa, M. H., Helmi, S., Badawi, A. F., Tricker, A. R., Spiegelhalder, B., and Preussman, R. (1994). Nitrate, nitrite and volatile N-nitroso compounds in the urine of *S. mansoni* infected patients. *Carcinogenesis* **15,** 619–625.
Mostafa, M. H., Shewwita, A., and O'Connor, P. J. (1999). Relationship between schistosomiasis and bladder cancer. *Clin. Microbiol. Rev.* **12,** 97–111.
Murray, J. F. (1967). Tumors of the alimentary tract in Africans. *Natl. Cancer Inst. Monogr.* **25,** 49–55.
Nakano, S., Ueo, H., Bruce, S. A., and Ts'o, P. (1985). A contact-insensitive subpopulation in Syrian hamster cell cultures with a greater susceptibility to chemically-induced neoplastic transformation. *Proc. Natl. Acad. Sci. USA* **82,** 5005–5009.
Niagi, E. N., and Bender, D. A. (1990). *Schistosoma mansoni*: Effects on tryptophan metabolism in mice. *Exp. Parasitol.* **70,** 43–54.
O'Brien, P. J. (1988). Radical formation during the peroxidase-catalised metabolism of carcinogens and xenobiotics. The reactivity of these radicals with GSH, DNA and unsaturated fatty lipid. *Free Radic. Biol. Med.* **4,** 216–226.
Oettle, A. G., de Maillon, B., and Lazer, B. (1959). Carcinomas of the glandular stomach and hepatomas in *Rattus (Mastomys) natalensis* infected with *bilharzia mansoni*. *Acta Unio Int. Contra Cancrum* **15,** 200–202.
Okajima, E., Denda, A., and Ozono, S. (1998). Chemopreventive effects of nimesulide, a selective cyclooxygenase-2 inhibitor, on the development of rat urinary bladder carcinomas initiated by N-butyl-N-(4-hydroxybutyl) nitrosamine. *Cancer Res.* **58,** 3028–3031.
Olds, G. R., El Meneza, S., Mahmoud, A. A., and Kresina, T. F. (1989). Differential immunoregulation of granulomatous inflammation, portal hypertension and hepatic fibrosis in murine *schistosomiasis mansoni*. *J. Immunol.* **142,** 3605–3611.
Oshima, H., Bandaletova, T. Y., Brouet, I., Bartsch, H., Kirby, G., Ogunbiyi, F., Vatansapt, V., and Pipitgool, V. (1994). Increased nitrosamine and nitrate biosynthesis mediated by nitric oxide synthase induced in hamster infected with liver fluke (*Opisthorchis viverrini*). *Carcinognesis* **15,** 271–275.

Paes, R. A., and Marigo, C. (1981). Giant follicular lymphoma and *schistosomiasis mansoni*. *Rev. Inst. Trop. Med. Sao Paulo* **23,** 287–292.

Park, H., Kang, D. W., Song, G. N., Cho, M., and Yang, U. (2000). Histopathological study on the carcinogenesis of *cholangiocarcinoma* experimentally induced in hamsters with *Clonorchis sinensis* and dimethylnitrosamine. *Gastroenterology* **118,** 2497.

Parkin, D. M. (1986). Cancer occurrence in developing countries. *Int. Agency Res. Cancer* **75,** 68–73.

Parkin, D. M., Srivatanakul, P., Khlat, M., Chenvidhya, D., Chotiwan, P., Insiripong, S., L'Abbe, K. A., and Wild, C. P. (1991). Liver cancer in Thailand: A case-control study of *cholangiocarcinoma*. *Int. J. Cancer* **48,** 323–328.

Parkin, D. M., Ohshima, H., Srivatanakul, P., and Vatanasapt, V. (1993). *Cholangiocarcinoma*: Epidemiology, mechanisms of carcinogenesis and prevention. *Cancer Epidemiol. Biomark. Prev.* **2,** 537–544.

Parsonnet, J. (1999). Introduction. *In* "Microbes and Malignancy" (J. Parsonnet, ed.), pp. 3–15. Oxford University Press, New York.

Parra, J. C., Gazzinelli, G., Goes, A. M., Moyes, R. B., Rocha, R., Colley, D. G., and Doughty, B. L. (1991). Granulomatous hypersensitivity to *S. mansoni* egg antigens in human schistosomiasis. *J. Immunol.* **147,** 3949–3954.

Penaud, A., Nourrit, J., Chapoy, P., Alessandrini, P., Louchet, E., and Nicoli, R. M. (1983). Bacterio-parasite interactions. Enterobacteria and schistosomes. *Trop. Med.* **43,** 331–340.

Pettit, S. J., Seymour, K., O'Flaherty, E., and Kirby, J. A. (2000). Immune selection in neoplasia: Towards a microevolutionary model of cancer development. *Br. J. Cancer* **82,** 1900–1906.

Phares, K. (1996). An unusual host-parasite relationship: The growth hormone-like factor from plerocercoids of spirometrid tapeworms. *Int. J. Parasitol.* **26,** 575–588.

Pinlaor, S., Yongvanit, P., Hiraku, Y., Ma, N., Semba, R., Oikawa, S., Murata, M., Sripa, B., Sithithaworn, P., and Kawaniahi, S. (2003). 8-nitroguanine formation in the liver of hamsters infected with *Opisthorchis viverrini*. *Biochem. Biophys. Res. Commun.* **309,** 567–571.

Pinlaor, S., Hiraku, Y., Yongvanit, P., Semba, R., Oikawa, S., Murata, M., Sripa, B., Sithithawron, P., and Kawaniahi, S. (2004a). Mechanism of NO-mediated oxidative and nitrative DNA damage in hamsters infected with *Opisthorchis viverrini*: A model of inflammation-mediated carcinogenesis. *Nitric Oxide Biol. Chem.* **11,** 175–183.

Pinlaor, S., Ma, N., Hiraku, Y., Yongvanit, P., Semba, R., Oikawa, S., Murata, M., Sripa, B., Sithithaworn, P., and Kawaniahi, S. (2004b). Repeated infection with *Opisthorchis viverrini* induces accumulation of 8-nitroguanine and 8-oxo-7, 8-dihydro-2''-deoxyguanine in the bile duct of hamsters via inducible nitric oxide synthase. *Carcinogenesis.* **25,** 1535–1542.

Pinlaor, S., Tada-Oikawa, S., Hiraku, Y., Pinlaor, P., Ma, N., Sithithaworn, P., and Kawanishi, S. (2005). *Opisthorchis viverrini* antigen induces the expression of Toll-like receptor 2 in macrophage RAW cell line. *Int. J. Parasitol.* **36,** 591–596.

Pinlaor, S., Hiraku, Y., Yongvanit, P., Tada-Oikawa, S., Ma, N., Pinlaor, P., Sithithaworn, P., Sripa, B., Murata, M., Oikawa, S., and Kawaniahi, S. (2006). iNOS-dependent DNA damage via NF-kappa B expression in hamsters infected with *Opisthorchis viverrini* and its suppression by the antihelminthic drug praziquantel. *Int. J. Cancer* **119,** 1067–1072.

Pisani, P., Parkin, D. M., Munoz, N., and Ferlay, J. (1997). Cancer and infection: Estimates of the attributable fraction in 1990. *Cancer Epidemiol. Biomark. Prev.* **6,** 387–400.

Prempracha, N., Tengchaisri, T., Chawengkirtikul, R., Boonpucknavig, S., Thamavit, W., and Duongchawee, G. (1994). Identification and potential use of a soluble tumor antigen for the detection of liver-fluke associated *cholangiocarcinoma* induced in the hamster model. *Int. J. Cancer* **57,** 691–695.

Qiu, D. C., Hubbard, A. E., Zhong, B., Zhang, Y., and Spear, R. C. (2005). A matched case-control study of the association between *Schistosoma japonicum* and liver and colon cancers in rural China. *Ann. Trop. Med. Parasitol.* **99,** 47–52.

Ramchurren, N., Cooper, K., and Summerhays, I. C. (1995). Molecular events underlying schistosomiasis-related bladder cancer. *Int. J. Cancer* **62,** 237–244.

Raziuddin, S., Masihuzzaman, M., Shetty, S., and Ibrahim, A. (1993). Tumor necrosis factor alpha production in schistosomiasis with carcinoma of the urinary bladder. *J. Clin. Immunol.* **13,** 23–29.

Reddy, A., and Fried, B. (2007). The use of *Trichuris suis* and other helminth therapies to treat Crohn's disease. *Parasitol. Res.* **100,** 921–927.

Riganti, M., Pungpak, S., Punpoowong, B., Bunnag, D., and Harinasuta, T. (1989). Human pathology of *Opisthorchis viverrini* infection: A comparison of adults and children. *Southeast Asian J. Trop. Med. Public Health* **20,** 95–100.

Rosin, M. P., Anwar, W. A., and Ward, A. J. (1994a). Inflammation, chromosomal instability and cancer: The schistosomiasis model. *Cancer Res.* **54,** 1929–1933.

Rosin, M. P., Zaki, S. S., Ward, A. J., and Anwar, W. A. (1994b). Involvement of inflammatory reactions and elevated cell proliferation in the development of bladder cancer in schistosomiasis patients. *Mutat. Res.* **305,** 283–292.

Saad, A. A., O'Connor, P. J., Mostafa, M. H., Metwalli, N. E., Cooper, D. P., Margison, G. P., and Povey, A. C. (2006). Bladder tumor contains higher N7-methylguanine levels in DNA than adjacent normal bladder urothelium. *Cancer Epidemiol. Biomarkers Prev.* **15,** 740–743.

Schatten, S., Granstein, R. D., Drebin, J. A., and Grene, M. I. (1984). Suppressor T cells and the immune response to tumors. *Crit. Rev. Immunol.* **4,** 335–379.

Sekiguchi, A., Shindo, G., Okabe, H., Aoyanagi, N., Furuge, A., and Oka, T. (1989). A case of metastatic lung tumor of colon cancer with ova of *Schistosoma japonicum* in the resected lung specimen. *Jpn. J. Thorac. Surg.* **42,** 1025–1028.

Shacter, E., and Weitzman, S. A. (2002). Chronic inflammation and cancer. *Oncology* **16,** 217–232.

Shaeter, E., Beecham, J. M., Covey, K., Kohn, K., and Potter, M. (1988). Activated neutrophils induce prolonged DNA damage in neighboring cells. *Carcinogenesis* **9,** 2297–2304.

Sheweita, S. A., El-Shahat, F. G., Bazeed, M. A., Abu El-Maati, M. R., and O'Connor, P. J. (2003). Effects of *Schistosoma haematobium* infection on drug-metabolizing enzymes in human bladder cancer tissues. *Cancer Lett.* **205,** 15–21.

Shim, H. S., Lim, B. J., Lee, W. J., Park, C., and Park, Y. N. (2004). Mucinous cholangiocarcinoma associated with *Clonorchis sinensis* infestation: A case report. *Korean J. Hepatol.* **10,** 223–227.

Shimkin, M. B., Mustacchi, P. O., Cram, E. B., and Wright, W. H. (1955). Lack of carcinogenicity of lyophilized *Schistosoma* in mice. *J. Natl. Cancer Inst.* **16,** 471–474.

Silva, L. P., Costa-Cruz, J. M., Spano, M. A., and Graf, U. (2006). Genotoxicity of vesicular fluid and saline extract of *Taenia solium* metacestodes in somatic cells of *Drosophila melanogaster*. *Environ. Mol. Mutagen.* **47,** 247–253.

Sithithaworn, P., Ando, K., Limviroj, W., Tesana, S., Pairojkul, C., Yutanawiboonchai, W., Chinzei, Y., Yoshida, T., and Sakakura, T. (2002). Expression of tenascin in bile duct cancer of hamster liver by combined treatment of dimethylnitrosamine with *Opisthorchis viverrini* infections. *J. Helminthol.* **76,** 261–268.

Smith, J. H., and Christie, J. D. (1986). The pathobiology of *Schistosoma haematobium* infection in humans. *Hum. Pathol.* **17,** 333–345.

Smith, J. H., Said, M. N., and Kelada, A. S. (1977). Studies on schistosomal rectal and colonic polyposis. *Am. J. Trop. Med. Hyg.* **26,** 80–84.

Sriamporn, S., Pisani, P., Pipitgool, V., Suwanrungruang, K., Kamsaard, S., and Parkin, D. M. (2004). Prevalence of *Opisthorchis viverrini* infection and incidence of *cholangiocarcinoma*. *Trop. Med. Int. Health* **9,** 588–594.

Sripa, B., and Kaewkes, S. (2002). Gallbladder and extrahepatic bile duct changes in *Opisthorchis viverrini*-infected hamsters. *Acta Trop.* **83,** 29–36.

Sripa, B., Leungwattanawanit, S., Nitta, T., Wongkham, C., Bhudhisawasdi, V., Puapairoj, A., Sripa, C., and Miwa, M. (2005). Establishment and characterization of an opisthorchiasis-associated *cholangiocarcinoma* cell line (KKU-100). *World J. Gastroenterol.* **11,** 3392–3397.

Sriurairatana, S., Tengchaisri, T., and Sirisinha, S. (1996). Ultrastructural characteristics of liver fluke-associated human *cholangiocarcinoma* cell lines. *Southeast Asian J. Trop. Med. Public Health* **27,** 57–62.

Srivatanakul, P., Sontipong, S., Chotiwan, P., and Parkin, D. M. (1988). Liver cancer in Thailand: Temporal and geographic variations. *J. Gastroenterol. Hepatol.* **3,** 413–420.

Srivatanakul, P., Parkin, D. M., Jiang, Y. Z., Khalat, M., Kao-Ian, U. T., Sontipong, S., and Wild, C. P. (1991a). The role of infection by *Opisthorchis viverrini,* Hepatitis B virus and aflatoxin exposure in the etiology of liver cancer in Thailand. A correlation study. *Cancer* **68,** 2411–2417.

Srivatanakul, P., Ohshima, H., Khlat, M., Parkin, M., Sukarayodhin, S., Brouet, I., and Bartsch, H. (1991b). Endogenous nitrosamines and liver flukes as risk factors for *cholangiocarcinoma* in Thailand. *In* "Relevance to Human Cancer of N-nitroso Compounds, Tobacco Smoke and Mycotoxins" (I. K. O'Neill, J. Chen, and H. Bartsch, eds.), pp. 88–95. (IARC Scientific Publications No. 105), IARC, Lyon.

Strohmeyer, T. G., and Slamon, D. J. (1994). Proto-oncogenes and tumor suppressor genes in human urological malignancies. *J. Urol.* **151,** 1479–1497.

Stuehr, D. J., and Marietta, M. A. (1985). Mammalian nitrate biosynthesis: Mouse macrophages produce nitrite and nitrate in response to *Escherichia coli* lipopolysaccharides. *Proc. Natl. Acad. Sci. USA* **82,** 7738–7742.

Swellam, M., Abd-Elmaksoud, N., Halim, M. H., and Khiry, H. (2004). Incidence of Bcl-2 expression in bladder cancer: Relation to schistosomiasis. *Clin. Biochem.* **37,** 798–802.

Tamimi, Y., Bringuier, P. P., Smit, F., Bokhoven, A., Abbas, A., Debruyne, F. M., and Schalken, J. A. (1996). Homozygous deletions of p16INK4 occur frequently in bilharzial-asociated bladder cancer. *Int. J. Cancer* **68,** 183–187.

Talib, H. (1970). The problem of carcinoma of bilharzial bladder in Iraq. *Br. J. Urol.* **42,** 571–579.

Teulings, F. G., Fokkens, W., Kaalen, J. A., and van der Werf-Messing, B. (1973). The concentration of free and conjugated 3-hydroxyanthranilic acid in the urine of bladder tumor patients before and after therapy, measured with an enzymatic method. *Br. J. Cancer* **27,** 316–322.

Thamavit, W., Bhamarapravati, N., Sahaphong, S., Vajrasthira, S., and Angsubhalkorn, S. (1978). Effects of dimethylnitrosamine on induction of *cholangiocarcinoma* in *Opisthorchis viverrini*-infected Syrian golden hamsters. *Cancer Res.* **38,** 4634–4639.

Thamavit, W., Moore, M. A., Hiasa, Y., and Ito, N. (1988). Generation of high yields of Syrian hamsters cholangiocellular carcinomas and hepatocellular nodules by combined nitrite and aminopyrine administration and *Opisthorchis viverrini* infection. *Jpn. J. Cancer Res.* **79,** 909–916.

Thamavit, W., Moore, M. A., Sirisinha, S., Shirai, T., and Ito, N. (1993). Time-dependent modulation of liver lesion development in *Opisthorchis*-infected Syrian hamster by an antihelminthic drug, praziquantel. *Jpn. J. Cancer Res.* **84,** 135–138.

Thamavit, W., Pairojkul, C., Tiwawech, D., Shirai, T., and Ito, N. (1994a). Strong promoting effect of *Opisthorchis viverrini* infection on dimethylnitrosamine-initiated hamster liver. *Cancer Lett.* **78,** 121–125.

Thamavit, W., Pairojkul, C., Tiwawech, D., Shirai, T., and Nobuyuki, I. (1994b). Lack of promoting effect of proline on bile duct cancer development in Dimethylnitrosamine-initiated hamster livers. *Teratog. Carcinog. Mutagen.* **14,** 169–174.

Thors, C., Jansson, B., Helin, H., and Linder, E. (2006). Thomsen-Friedenreich oncofetal antigen in *Schistosoma mansoni*: Localization and immunogenicity in experimental mouse infection. *Parasitology* **132,** 73–81.

Thuwajit, C., Thuwajit, P., Kaewkes, S., Sripa, B., Uchida, K., Miwa, M., and Wongkham, S. (2004). Increased cell proliferation of mouse fibroblast NIH-3T3 *in vitro* induced by excretory/secretory products from. *Opisthorchis viverrini. Parasitology* **129,** 455–464.

Toledo, R., Esteban, J.-G., and Fried, B. (2006). Immunology and pathology of intestinal trematodes in their definitive hosts. *Adv. Parasitol.* **63,** 285–365.

Tricker, A. R., Mostafa, M. H., Spiegelhalder, B., and Preussmann, R. (1989). Urinary excretion of nitrate, nitrite and N-nitroso compounds in schistosomiasis and bilharzial bladder cancer patients. *Carcinogenesis* **10,** 547–552.

Trosko, J. E., and Ruch, R. J. (1998). Cell-cell communication in carcinogenesis. *Front. Biosci.* **3,** 208–236.

Tsocheva-Gaytandzhieva, N. T. (2005). Fascioliasis and tumor growth. *Helminthologia* **42,** 107–113.

Tsocheva, N. T., Kaduska, M. B., Poljakova-Krusteva, O. T., Krustev, L. P., Yanev, S. S., and Stoytchev, T. S. (1992). Combined effect of fascioliasis and diethylnitrosamine carcinogenesis on the activity of the rat monoxygenase system. *Comp. Biochem. Physiol.* **101,** 475–479.

Tsuda, H., Satarug, S., Bhudhisawasdi, V., Kihana, T., Sugimura, T., and Hirohashi, S. (1992). *Cholangiocarcinomas* in Japanese and Thai patients: Difference in etiology and incidence of point mutation of the c-Ki-ras proto-oncogene. *Mol. Carcinog.* **6,** 266–269.

Vatanasapt, V., Tangvoraphonkchai, V., Titapant, V., Pipitgool, V., Viriyapap, D., and Sriamporm, S. (1990). A high incidence of liver cancer in Khon Kaen Province, Thailand. *Southeastern Asian J. Trop. Med. Public Health* **21,** 489–494.

Viranuvatti, V., and Mettiyawongse, S. (1953). Observations on two cases of *opisthorchiasis* in Thailand. *Ann. Trop. Med. Parasitol.* **43,** 291–293.

Vitovec, J. (1974). Hepatocellular carcinoma in cattle and its relationship to biliary cirrhosis of fasciolar origin. *Vet. Pathol.* **11,** 548–557.

Warren, K. S. (1973). The pathology of shistosome infections. *Helminthol. Abstr.* **42,** 592–633.

Watanapa, P., and Watanapa, W. B. (2002). Liver fluke-associated *cholangiocarcinoma. Br. J. Surg.* **89,** 962–970.

Watson-Wemyss, H. L. (1919). Carcinoma of the liver associated with infection by *Clonorchis sinensis. Edinburgh Med. J.* **22,** 103–104.

White, A. C., Robinson, P., and Kuhn, R. (1997). *Taenia solium* cysticercosis: Host parasite interactions and the immune response. *Chem. Immunol.* **66,** 209–230.

Wongratanacheewin, S., Rattanasiriwilai, W., Priwan, R., and Sirisinha, S. (1987). Immunodepression in hamsters experimentally infected with. *Opisthorchis viverrini. J. Helminthol.* **61,** 151–156.

Xu, Z., and Su, D. L. (1984). *Schistosoma japonicum* and colorectal cancer: An epidemiological study in the Peoples Republic of China. *Int. J. Cancer* **34,** 315–318.

Zalata, K. R., Nasif, W. A., Ming, S. C., Lotfy, M., Nada, N. A., El-Hak, N. G., and Leech, S. H. (2005). P53, Bcl-2 and C-Myc expressions in colorectal carcinoma associated with schistosomiasis in Egypt. *Cell. Oncol.* **27,** 245–253.

Zhang, R., Takahashi, S., Orita, S. I., Yoshida, A., Maruyama, H., Shirai, T., and Ohta, N. (1998). P53 gene mutations in rectal cancer associated with *schistosomiasis japonica* in Chinese patients. *Cancer Lett.* **131,** 215–221.

Zhou, X. X. (1986). Relationship between gastric schistosomiasis and gastric cancer, chronic gastric ulcer and chronic gastritis: Pathological analysis of 79 cases. *Chin. J. Pathol.* **15,** 62–64.

CHAPTER 5

A Review of the Biology of the Parasitic Copepod *Lernaeocera branchialis* (L., 1767) (Copepoda: Pennellidae)

Adam J. Brooker, Andrew P. Shinn, and **James E. Bron**

Contents

Institute of Aquaculture, University of Stirling, Stirling, Scotland FK9 4LA, United Kingdom

Advances in Parasitology, Volume 65
ISSN 0065-308X DOI: 10.1016/S0065-308X(07)65005-2

Abstract

This review concerns the parasitic marine copepod *Lernaeocera branchialis* (L., 1767) and provides an overview of current knowledge concerning its biology and host–parasite interactions. The large size and distinctive appearance of the metamorphosed adult female stage, coupled with the wide exploitation and commercial importance of its final gadoid hosts, means that this species has long been recognised in the scientific literature. The fact that the Atlantic cod, *Gadus morhua* L., is one of its key host species, and has itself had a major impact on the social and economic development of many countries bordering the North Atlantic for more than 10 centuries is also a factor in its widespread recognition. *L. branchialis* is recognised as a pathogen that could have major effects on the aquaculture industry and with gadoid (especially cod) farming expanding in several North Atlantic countries, there is considerable potential for this parasite to become a serious problem for commercial mariculture. The main subject areas covered are the parasite's taxonomy; the life history of the parasite including its life cycle, reproduction and host associations; parasite physiology; parasite seasonality and distribution; and the pathogenic effects of the parasite on its host.

1. INTRODUCTION

This review concerns the marine parasitic copepod *Lernaeocera branchialis* (L., 1767) and provides an overview of current knowledge concerning its biology and host–parasite interactions. The large size and distinctive appearance of the metamorphosed adult female stage, coupled with the wide exploitation and commercial importance of its final gadoid hosts, means that this species has long been recognised in the scientific literature. The fact that the Atlantic cod, *Gadus morhua* L., is one of its key host species, and has itself had a major impact on the social and economic development of many countries bordering the North Atlantic for more than 10 centuries (Kurlansky, 1998) is also a factor in its widespread recognition. *L. branchialis* is recognised as a pathogen that could have major effects on the aquaculture industry (Anon, 2005; Khan *et al.*, 1990) and with gadoid (especially cod) farming expanding in several North Atlantic countries, there is considerable potential for this parasite to become a serious problem for commercial mariculture.

The literature regarding *L. branchialis* is extensive, as it spans over a century; however, many papers are outdated or have been superseded by more recent work. Therefore, the canon of work concerning this potentially important and damaging pathogen is here re-examined to provide an up to date overview, which includes both aquaculture and wild fisheries perspectives.

2. TAXONOMY

L. branchialis is a pennellid copepod, whose final stage parasitises a range of gadoids. The Pennellidae are unique amongst parasitic copepods of fishes in having a two-host life cycle. Most members of the family are mesoparasitic as the thorax and abdomen become deeply embedded within the host's tissues, whereas the genital segment protrudes externally and bears egg sacs (Kabata, 1970). The most obvious feature of the Pennellidae compared with most other parasitic copepods (with the exception of the Sphyriidae) is their frequent large size. Once attached to the final host, females exhibit gigantism as a result of massive expansion in the length and girth of the genital complex and in some species the production of a substantial holdfast (Kabata, 1979). Most segmental boundaries are lost during this transformation (Sproston, 1942) and there is considerable morphological plasticity in the Pennellidae due to resistance encountered as the parasite grows through the host tissues. As a consequence, there are many different 'biological forms' which have resulted in the misidentification of some species and have generated much debate amongst taxonomists.

Within the genus *Lernaeocera* (Blainville, 1822), two valid species have been identified: *L. branchialis* (L., 1767) and *L. lusci* (Bassett Smith, 1896). Wilson (1917) described *Lernaeocera* from the American side of the Atlantic and Schuurmans-Stekhoven (1936a) concluded that this morphotype was distinct from *L. branchialis* and should be given the name *L. wilsoni*. However, careful examination of the differences between American and European forms led Kabata (1961) to conclude that the two forms are conspecific. The species *L. minuta* (Scott, 1900) was dismissed as a separate species by Van Damme and Ollevier (1995), who concluded that it is a miniature form of *L. lusci* and that size should not be used as a morphological characteristic due to its high dependence on host size, although this requires confirmation. *L. obtusa* (Kabata, 1957) was identified as a separate species parasitising older haddock [*Melanogrammus aeglefinus* (L.)] (Kabata, 1958), but was later found to be a larger form of *L. branchialis* (Kabata, 1979). Other *Lernaeocera* species proposed by various authors have been found to be invalid; these proposals were based on unusually shaped trunks or abnormally developed holdfasts (Kabata, 1961, 1979; Tirard *et al.*, 1993; Van Damme and Ollevier, 1995).

3. ADULT MORPHOLOGY OF THE FEMALE

The final and most characteristic stage of *L. branchialis* is the metamorphosed adult female. The cephalothorax comprises a holdfast composed of one dorsal and two lateral branches/antlers (Fig. 5.1i). The neck is thin and short (Kabata, 1979) and four pairs of vestigial swimming legs are located on the neck. The expanded genital complex has three points of flexure: between the anterior cylindrical region (or neck) and trunk, around the middle of the trunk and between the trunk and abdomen. *L. branchialis* f. *obtusa* has a longer and thicker neck than the previous form and has two points of flexure in its genital complex: in the middle of the trunk and between the trunk and abdomen. *L. branchialis* can be discriminated from *L. lusci* by the absence of antennary processes (Van Damme and Ollevier, 1995). As the adult female parasite grows, it develops a dark red colouration. The attachment site of the parasite on the final host is most commonly at the base of the third gill arch (Van den Broek, 1978, 1979b), where the cephalothorax penetrates the afferent branchial artery.

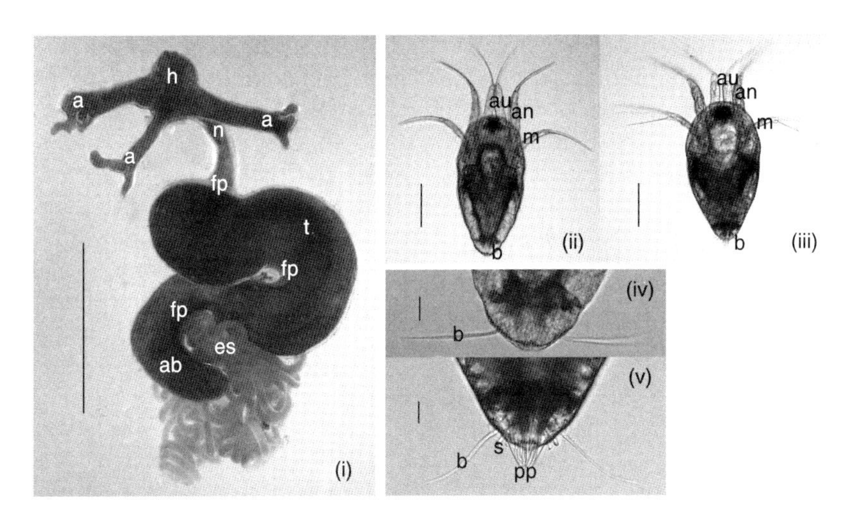

FIGURE 5.1 (i) The main anatomical features of an adult female *Lernaeocera branchialis* (scale bar: 5 mm), (ii) nauplius I—dorsal aspect (scale bar: 0.1 mm), (iii) nauplius II—dorsal aspect (scale bar: 0.1 mm), (iv) nauplius I—posterior extremity (scale bar: 10 μm) and (v) nauplius II (scale bar: 10 μm). Abbreviations: h, head; a, antlers; n, neck; fp, flexion points; t, trunk; ab, abdomen; es, egg strings; au, antennule; an, antenna; m, mandible; b, balancers; s, blunt spines; pp, posterior process.

From here it grows along the artery and the ventral aorta until it reaches the *bulbus arteriosus* and the holdfast develops (Smith *et al.*, 2007). Once embedded in the cardiac region, it begins feeding on host blood.

The morphology and size of adult females varies as a result of the interaction of several environmental and host-dependent factors. Most specimens are around 2 cm in length but occasionally they have been found to measure up to 5 cm (Kabata, 1979). Adult females living on cod and whiting have only a short distance to cover between the base of the gill arch and the *bulbus arteriosus* where feeding occurs, hence have a relatively short cylindrical neck. In contrast, the form *L. branchialis* f. *obtusa*, which parasitises on older haddock, tends to develop a longer, thicker neck, with a less acute angle where the neck merges into the trunk (Kabata, 1958). However, this form rarely penetrates as far as the cardiac region, but derives nourishment from one of the blood vessels, which does not appear to affect the parasite's survival rate. This form can also be distinguished by the angle of its genital flexure which is less acute than in *L. branchialis* f. *branchialis*.

4. LIFE CYCLE

L. branchialis was first described by Claus (1868a) as having a complex two-host life cycle. The life cycle comprises two free-swimming nauplius stages, one infective copepodid stage, four chalimus stages and the adult copepod, each separated by a moult stage (Fig. 5.2). Further descriptions were given by Scott (1901), Wilson (1917) and Sproston (1942). The following accounts rely principally upon the latter descriptions.

4.1. Nauplius I–II

The nauplius I has an average length of around 0.37 mm, although there is quite a wide range of sizes (0.35–0.41 mm) (Sproston, 1942) (Fig. 5.1ii). Whitfield *et al.* (1988) recorded a mean length of 0.41 mm (range 0.39–0.44 mm) and width of 0.33 mm (range 0.33–0.36 mm). Dark red chromatophores occur laterally around the middle two-third of the body and a darker, almost black, pigmentation surrounds the eye. Three pairs of appendages are apparent in nauplius I, with a pair of balancers at the posterior extremity. The antennules arise from the sides of a flattened frontal region and there is a short anterior spine and two long setae at the tip. Between the antennules is a small median papilla under the cuticle (Sproston, 1942). The antennae arise ventrally from shallow depressions on the anterior lateral margins. A short spine on the distal end of the basal segment gives rise to a four-jointed exopodite and a three-jointed endopodite. Each segment of the exopodite bears a long seta, four in total and

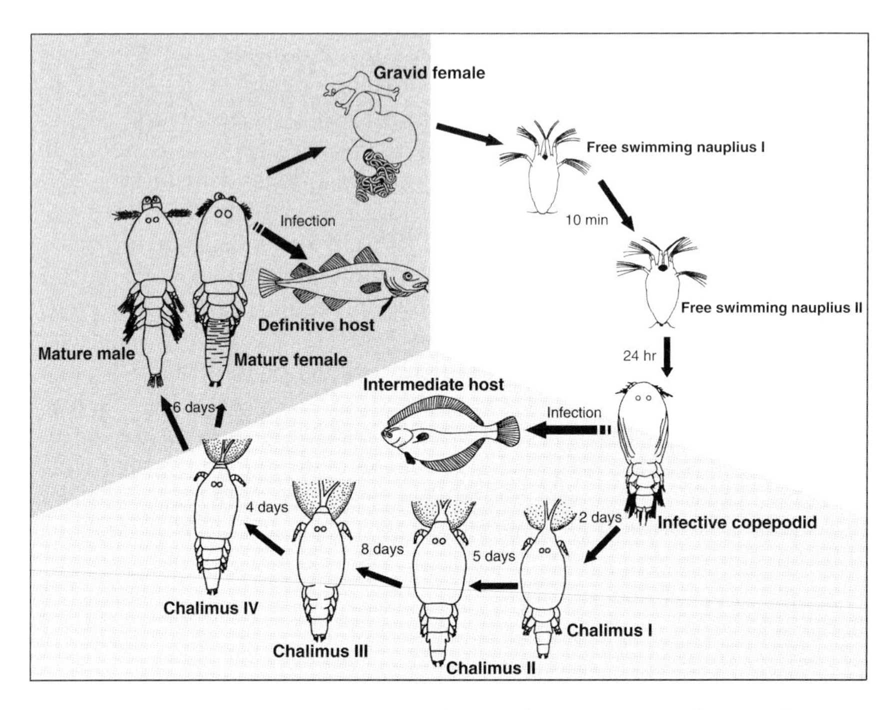

FIGURE 5.2 The life cycle of *Lernaeocera branchialis* (stage timings for 10 °C from Whitfield *et al.*, 1988).

the terminal segment of the endopodite bears two long setae. The mandibles are very similar to the antennae at this stage, although the basal segments are longer and more slender.

The second nauplius stage (Fig. 5.1iii) was not found by Sproston (1942) and a detailed description is lacking. From the author's personal observations, nauplius II appears to be similar to nauplius I except for a pair of conical processes at the posterior extremity, which appear to be developing caudal rami, and a pair of blunt, rounded spines at either side at its base (Fig. 5.1iv and v). Whitfield *et al.* (1988) describes the nauplius II as a meta-nauplius, indicating that it has more than three pairs of functional appendages, which would be the case if the paired processes at the posterior extremity are indeed caudal rami. Mean length of nauplius II was 0.54 mm (range 0.51–0.55 mm) and width 0.33 mm (range 0.24–0.27 mm) and whilst the body shapes of nauplius I and II have been described as 'ovoid' and 'bullet-shaped', respectively (Whitfield *et al.*, 1988), personal observations have shown that the body shapes of nauplius I and II are very similar and that differences may be due to individual variation (Fig. 5.1ii and iii).

The nauplius I moults to nauplius II about 10 min after hatching at 10 °C. This may account for nauplius II not being found by Sproston (1942); nauplius II being mis-identified as nauplius I. Nauplius I are positively phototaxic, but once moulted to nauplius II, they sink to the sea bed and exhibit negative phototaxis (Whitfield *et al.*, 1988).

4.2. Copepodid

The copepodid shows a large variation in size between individuals. Sproston (1942) found specimens varying from 0.39 to 0.63 mm with a mean of 0.48 mm (Fig. 5.3i). The cephalothorax is about five-eighth of the body length and in all the post-nauplius developmental forms show a strong ventral infolding along the lateral margins, particularly in the anterior half. The antennules consist of five segments, each equipped with setae and the distal segment bears one long, thick aesthetasc. The chelate antennae arise slightly ventral to the anterior margin of the cephalothorax and have a considerable degree of freedom—more than 180 °C (Sproston, 1942). Once the chelae grip a gill tip, they become the anchorage of the animal until the frontal filament is produced at the next moult. The copepodid has two pairs of biramous swimming legs: the first on the second thoracic somite, which is fused to the cephalothorax and the second on the third thoracic somite. Both rami of the first leg consist of two segments and both distal segments of the first and second legs bear five long setae, although only the exopodite bears a fine spine. Both rami of the second leg are one segmented. The fourth thoracic somite bears two stout spines, which represent the third pair of swimming legs; the fifth somite represents the pre-genital and genital segments; the sixth is the anal somite and bears two caudal rami, each with five long setae.

The moult to copepodid stage from nauplius II occurs after 48 h at 10°C (Whitfield *et al.*, 1988). The copepodid then seeks out a host and attaches to a gill tip using the chelate antennae (Sproston, 1942). They preferentially attach to the tips of the posterior gill filaments, but show no preference for the right or left gill chamber (Edwards, 1984). However, a preference is always shown for the posterior gill arches, rather than the anterior ones. This intermediate host is usually a pleuronectiform or other demersal fish. In the United Kingdom, this is usually the flounder, *Platichthys flesus* L.; but in more northern latitudes, the common sole, *Microstomus kitt*, Walbaum is the most common host (Kabata, 1979). Table 5.1 lists the recorded intermediate hosts. According to strict definition, the 'intermediate' host is effectively the definitive host, since *L. branchialis* matures and mates on this host. However, throughout this review, it will be referred to as the intermediate host and the final host will remain the definitive host in order to adhere to the terminology used by previous workers.

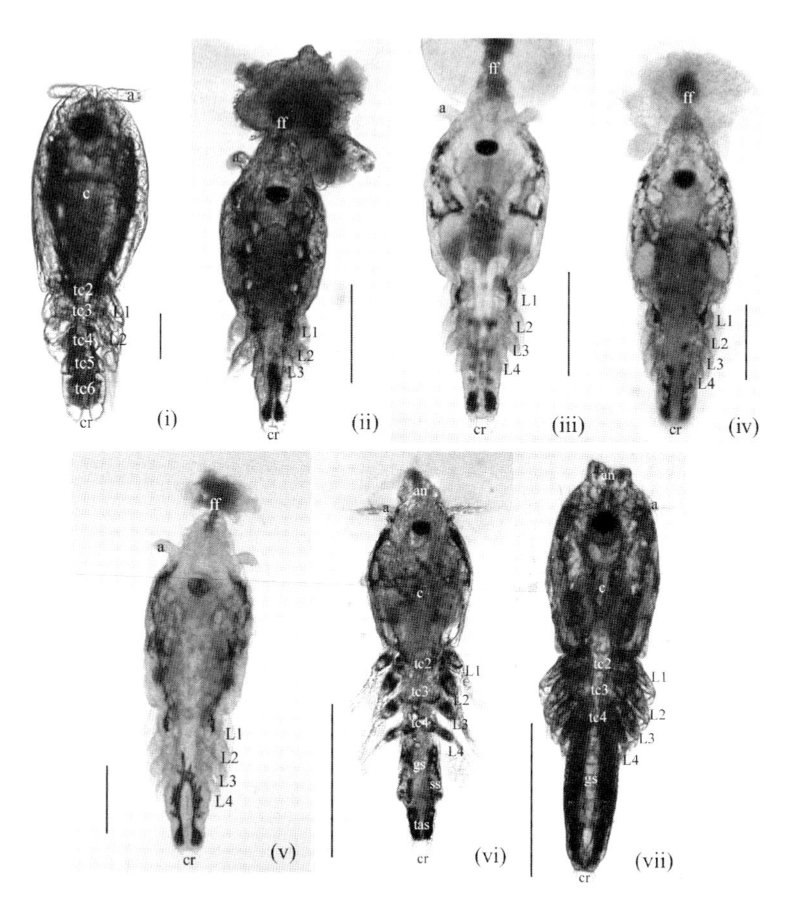

FIGURE 5.3 (i) Copepodid (scale bar: 0.1 mm), (ii) chalimus I (scale bar: 0.25 mm), (iii) chalimus II (scale bar: 0.25 mm), (iv) chalimus III (scale bar: 0.25 mm), (v) chalimus IV (scale bar: 0.25 mm), (vi) adult male (scale bar: 0.5 mm), (vii) adult female (scale bar: 0.5 mm) (all dorsal aspect). Abbreviations: a, antennule; ff, frontal filament; c, cephalothorax; L1, first swimming leg; L2, second swimming leg; L3, third swimming leg; L4, fourth swimming leg; tc2, second thoracic somite; tc3, third thoracic somite; tc4, fourth thoracic somite; tc5, fifth thoracic somite; tc6, sixth thoracic somite; gs, genital segment; tas, terminal abdominal somite; ss, spermatophore sac; cr, caudal rami.

4.3. Chalimus I–IV

During the moult to chalimus I, a secretion from the mid-frontal region is extruded as a thread between the antennae into the surrounding gill tissue, where it diverges into two filaments (Sproston, 1942). The rest of the secretion forms a hood dorsally and laterally, which is attached to the cephalothorax, and encloses in it the recently moulted antennae and the distal segment of the previous moult which remains attached to the gill.

TABLE 5.1 Intermediate hosts of *Lernaeocera branchialis* (adapted from Kabata, 1979)

Host species	Family	References
Platichthys flesus (L.)	Pleuronectidae	Boxshall, 1974; Claus, 1868b; Køie, 1999; Pedashenko, 1898; Polyanski, 1955; Scott, 1901; Van Den Broek, 1979a
P. flesus bogdanovi (Sandberg)	Pleuronectidae	Shulman and Shulman-Albova, 1953
Liopsetta glacialis (Pallas)	Pleuronectidae	Shulman and Shulman-Albova, 1953
Microstomus kitt (Walbaum)	Pleuronectidae	Kabata, 1957, 1958
Limanda limanda (L.)	Pleuronectidae	Shulman and Shulman-Albova, 1953
Pleuronectes platessa L.	Pleuronectidae	Boxshall, 1974; Gouillart, 1937; Van Oared-de Lint and Schuurmans Stekhoven, 1936
Solea solea (L.)	Soleidae	Sproston and Hartley, 1941a
Scophthalmus maximus (L.)	Bothidae	Gouillart, 1937
Cyclopterus lumpus L.	Cyclopteridae	Fleming and Templeman, 1951; Metzger, 1868; Polyanski, 1955; Reed and Dymond, 1951; Shulman and Shulman-Albova, 1953; Sproston and Hartley, 1941a; Templeman *et al.*, 1976
Agonus cataphractus (L.)	Agonidae	Kabata, 1958, 1979
Zoarces viviparous (L.)	Zoarcidae	Polyanski, 1955
Myoxocephalus scorpius (L.)	Cottidae	Boxshall, 1974; Kabata, 1958; Polyanski, 1955
Callionymus lyra L.	Callionymidae	Kabata, 1958
Zeugopterus punctatus (Bloch)	Scophthalmidae	Kabata, 1979

Soon after the moult is complete, the movement of the chalimus causes the hood to crack where it joins the head and occasionally it breaks off (Sproston, 1942). Each chalimus stage can be identified by counting the number of hoods, as a new one is secreted over the old ones at each moult (Claus, 1868b). This frontal filament structure is different from that of other siphonostomatoid parasites that have been studied. Unlike other siphonostomatoids that have a single filament, the filament of *L. branchialis* diverges into two as it enters the gill tissue. In the genus *Lepeophtheirus* (von Nordmann, 1832), a new filament can be produced at each moult (Bron *et al.*, 1991; Pike *et al.*, 1993), whereas in *Caligus elongatus* (von Nordman, 1832), the filament, once lost, cannot be replaced, although the filament base is extended at each moult (Piasecki, 1996). In *L. branchialis*, a new hood is added to the filament structure at each moult, suggesting that the mechanism may be similar to that of *C. elongatus* in that the filament is separate from the chalimus body and cannot be replaced once lost.

In the chalimus I (Fig. 5.3ii), the long swimming setae of the biramous legs have been lost and the rami have lost their broad flattened shape, becoming more rounded in transverse section, although the rami of the second leg now have two segments. There is a change in shape of the thoracic somites and a rudiment in place of the spines representing the third leg (Sproston, 1942). The caudal rami are smaller than in the copepodid and their setae are short and poorly developed. The moult to chalimus I occurs 2 days post-infection (p.i.) at 10 °C (Whitfield *et al.*, 1988).

There is very little change in the chalimus II, although segmental boundaries are less clear (Fig. 5.3iii). The antennae are swollen with segmentation shown only by folds and the leg segments are becoming more indistinct. There is an indication of the fourth thoracic somite dividing into the pre-genital and genital segments and the former shows a rudiment representing the fourth pair of legs. The labrum and labium of the mouth tube have not yet joined up and the mandibles are seen at the sides. The moult to chalimus II occurs around 7 days p.i. at 10 °C (Whitfield *et al.*, 1988).

The general morphology of chalimus III stage is similar to the second although some appendages are beginning to re-differentiate (Fig. 5.3iv). The free thoracic segments are less clearly marked but there is an indication of five. A fourth pair of legs has developed, which have two joints at this stage, although there are no setae on the distal segment. The moult to chalimus III stage occurs around 15 days p.i. at 10 °C (Whitfield *et al.*, 1988).

There is again little change in morphology after the moult to chalimus IV (Fig. 5.3v). However, the segmentation of all parts is better defined than in the previous stage. Although in this stage, the swimming legs are still not employed in swimming they are beginning to regain their characteristic form and the setae are slightly broader as are those of the antennules and caudal rami. The mouth tube is nearly closed at this stage and the maxillules can be seen at its sides. The maxillae of both sexes and maxillipeds of the male (the female lacks maxillipeds) are both swollen compared with those of the adult and bear no spines or claws. The moult to the chalimus IV stage occurs 19 days p.i. at 10 °C and after 25 days p.i. at 10°C, the parasite moults to the adult stage (Whitfield *et al.*, 1988).

4.4. Adult

4.4.1. Morphology

Once the chalimus IV has moulted to the adult, there is a considerable change in shape and the body form of the copepodid has returned (Fig. 5.3vi and vii), although the setae on the swimming legs are not quite so long (Sproston, 1942).

The antennules are not distinctly segmented but have a constriction separating the terminal segment, which has a rounded tip. They bear many fine setae, the terminal quarter bearing five stronger setae (Kabata, 1979) and one aesthetasc. The antennae are very similar to those of the copepodid although they are larger and more robust. They are three-jointed with the final joint bearing a short spine.

The first two pairs of swimming legs are biramous and each ramus has two joints. The third and fourth pairs of legs are uniramous and carry three joints. In the male, each ramus bears five long setae and one shorter one (Sproston, 1942). In the female, the distribution of the setae is as follows: leg 1–7 + spine (exopod), 6 + spine (endopod); leg 2–6 + spine (exopod), 6 (endopod); leg 3–6 + spine; leg 4–5 (Kabata, 1979). In the female, the coxa and the basis of the swimming legs are longer and more slender, and the setae are slightly longer than in the male (Sproston, 1942).

In the male, the genital segment is barrel-shaped and contains the spermatophore sacs, which open into the tube-like gonopore near the median line (Sproston, 1942). The genital segment represents two somites and bears the vestiges of the fifth and sixth legs. Immediately below the genital segment is the free abdominal somite which is approximately square and about half the length of the genital segment (Kabata, 1979). The caudal rami each bear four setae, the inner two being longer than the finer outer two. In the male, the caudal rami are much larger and the setae longer and thicker than in the female. In the female, the thoracic segments have lost the characteristic shape seen in the male and copepodid (Sproston, 1942). They are rounded in transverse section, except the pre-genital segment which is slightly quadrangular. The genital segment does not have the vestiges of the fifth legs seen in the male (Sproston, 1942) and is enormously elongated, equal to or longer than the rest of the body, with a transversely wrinkled cuticle (Kabata, 1979; Smith and Whitfield, 1988) which indicates the region of imminent expansion (see Section 4.4.4). Once the final chalimus has moulted to the adult, the parasite is capable of swimming and active movement within the gill chamber of the host.

4.4.2. Mouth and mouthparts

The mouth of *L. branchialis* consists of a buccal tube situated at the end of a buccal cone, which comprises three rings (Fig. 5.4i) (Kabata, 1962). The rings are bands of thick cuticle alternated with thinner bands, an arrangement which permits them to move in relation to one another and allows the buccal tube to shorten when pressed against the host, bringing the feeding appendages into contact with the host tissues (Boxshall, 1990b). The outer ring, the broadest and thinnest of the three, is divided by a thick plaque at the mid-anterior line (Fig. 5.4i). The outer part of this bilaterally symmetrical plaque is divided into two longitudinal parts, each attached

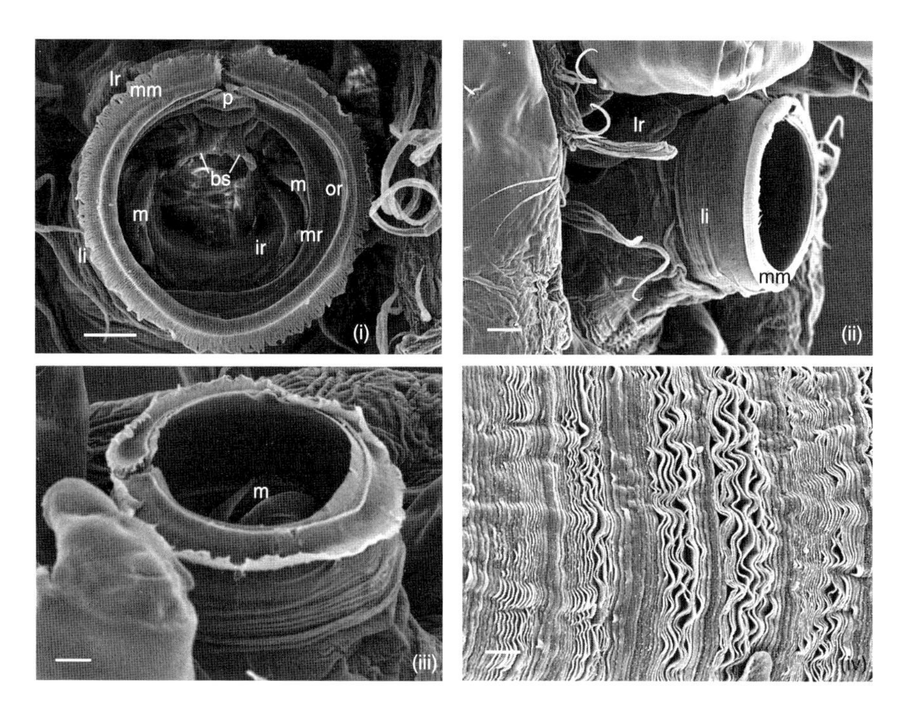

FIGURE 5.4 SEMs of the buccal tube of *Lernaeocera branchialis.* (i) Ventral aspect, (ii) lateral aspect, (iii) close-up of lateral aspect, (iv) SEM of the abdomen of pre-metamorposed adult female *L. branchialis* showing transverse cuticular folds. Scale bars: 10 μm. Abbreviations: mm, marginal membrane; m, mandible; or, outer ring; p, mid-anterior plaque of the outer ring; mr, middle ring; ir, inner ring; bs, buccal stylets; li, labium; lr, labrum.

to the thickened outer rim of the ring. The plaque is suggested to act as a spring imparting a degree of elasticity to the outer ring, and thereby maintaining pressure on the host tissue as it feeds (Kabata, 1962). The middle ring is thicker and narrower than the outer ring and the inner ring is the thickest and narrowest of the three.

At the tip of the outer ring is a flared marginal membrane (Fig. 5.4i), which encircles the opening of the mouth cone and may prevent the escape of macerated particles of tissue (Boxshall, 1990b; Capart, 1948; Kabata, 1962). A thin chitinous band is found around the line of attachment and is thickened at intervals by 'studs' of chitin (Kabata, 1962). In males, the chitinous band is broader than in the female and the studs are about twice the size but fewer in number. While the labrum forms the anterior wall of the buccal tube, it does not reach the opening and only the labium is connected to the marginal membrane (Boxshall, 1990b) (Fig. 5.4ii).

The mandible is a long, slender structure derived from the coxa (Huys and Boxshall, 1991) with a stylet-like gnathobase and its base dorsomedial to and in close proximity to the maxillae. It is broadest at the base and narrows towards the distal end, but broadens again at the tip, where its ventral margin bears eight pointed and slightly curving teeth (Fig. 5.4iii). The mandible curves into the buccal cone and the retraction of the cone allows the serrated edges of the mandible to come into contact with the host tissue, macerating it and allowing the parasite to feed (Kabata, 1962).

At the mid-anterior line of the inner ring is attached the paired base of the buccal stylets. These bilobed structures project obliquely down towards the buccal opening (Fig. 5.4i). The inner face of the base has a pair of laterally diverging processes, which are likely to be the insertion point of muscles involved with movement of the appendages (Kabata, 1962). When the buccal tube is retracted, the plaque on the outer ring can close against the buccal stylets which fit into grooves in the plaque and help maintain the position of the buccal tube.

The maxillule lies laterally outside the buccal tube (Boxshall, 1990b). It is a bilobed structure consisting of an inner and outer lobe, representing the praecoxal gnathobase and palp, respectively (Huys and Boxshall, 1991). The inner lobe bears two long setae at the terminal tip, whereas the outer lobe bears a single seta (Kabata, 1979).

The maxilla is posterior to the buccal tube and consists of two segments. The proximal segment, representing the praecoxa and coxa (Huys and Boxshall, 1991), is broad and robust and in the female bears two large unciform processes (Kabata, 1979). The distal segment is armed with rows of setules and its tip bears a short claw with parallel rows of setules.

4.4.3. Infection of the definitive host

Mating occurs on the intermediate host (see Section 5.2) and the fertilised female then leaves to search for the definitive host. This is usually a gadoid fish, with cod, whiting (*Merlangius merlangus* L.), pollack (*Pollachius pollachius* L.) and haddock being the most common hosts (Kabata, 1979) (Table 5.2). The occurrence of metamorphosed *L. branchialis* on plaice (*Pleuronectes platessa* L.), sole (*Solea solea* L.) and dab (*Limanda limanda* L.)

TABLE 5.2 Definitive hosts of *Lernaeocera branchialis* (adapted from Kabata, 1979)

Host species	Family	References
Gadus morhua L.	Gadidae	Boxshall, 1974; Hemmingsen *et al.*, 1995, 2000; Jones and Taggart, 1998; Kabata, 1957, 1958; Lee and Khan, 2000; Linderby and Thulin, 1983; Polyanski, 1955; Sanchez-Lizaso and Vasquez, 1987; Schuurmans Stekhoven, 1936b; Sherman and Wise, 1961; Shulman and Shulman-Albova, 1953; Sundnes *et al.*, 1997; Templeman and Fleming, 1963; Templeman *et al.*, 1976; Van Damme and Hamerlynck, 1992; Vazquez *et al.*, 1988; Wilson, 1917
Merlangius merlangus (L.)	Gadidae	Boxshall, 1974; Desbrosses, 1945; Kabata, 1957; Pilcher *et al.*, 1989; Potter *et al.*, 1988; Schuurmans Stekhoven, 1936b; Shotter, 1973a,b, 1976; Smith, 1969; Sproston and Hartley, 1941a; Van Den Broek, 1979a
Pollachius pollachius (L.)	Gadidae	Gouillart, 1937; Sproston and Hartley, 1941a
Pollachius virens (L.)	Gadidae	Boxshall, 1974
Gadus ogac Richardson	Gadidae	Wilson, 1917
Trisopterus minutus (L.)	Gadidae	Olsson, 1869

(*continued*)

TABLE 5.2 (*continued*)

Host species	Family	References
Boreogadus saida (Lepechin)	Gadidae	Shulman and Shulman-Albova, 1953
Eleginus navaga (Koelreuter)	Gadidae	Shulman and Shulman-Albova, 1953
Melanogrammus aeglefinus (L.)	Gadidae	Kabata, 1957, 1958; Polyanski, 1955
Phycis blennoides (Brünnich)	Gadidae	Fox, 1945
Molva molva (L.)	Gadidae	Kabata, 1965; Polyanski, 1955
Merluccius merluccius (L.)	Merluccidae	Machado Cruz, 1959; Osorio, 1892; Van Oared-de Lint and Schuurmans Stekhoven, 1936
Ammodytes tobianus L.	Ammodytidae	Schuurmans Stekhoven, 1936b
Dicentrarchus labrax (L.)	Serranidae	Van Oared-de Lint and Schuurmans Stekhoven, 1936
Serranus cabrilla (L.)	Serranidae	Radulescu *et al.*, 1972
Pholis gunnellus (L.)	Blennidae	Van Oared-de Lint and Schuurmans Stekhoven, 1936
Labrus mixtus L.	Labridae	Olsson, 1869
Callionymus lyra L.	Callionymidae	Hansen, 1923; Schuurmans Stekhoven, 1936b
Conger conger (L.)	Congridae	Van Oared-de Lint and Schuurmans Stekhoven, 1936
Pleuronectes platessa L.	Pleuronectidae	Markevich, 1956
Solea solea (L.)	Soleidae	Schuurmans Stekhoven, 1936b
Limanda limanda (L.)	Pleuronectidae	Begg and Bruno, 1999

may indicate that they can be opportunistic parasites, as these species are usually found to be intermediate hosts. It is possible that in the absence of any definitive host options, the fertilised adult female may infect and metamorphose on an intermediate host. The mechanism of host location and identification is unknown in *Lernaeocera*, although it is likely that chemosensory detection of host-derived cues is involved. In caligids, where chemotaxis has been clearly demonstrated, it is uncertain at present whether the major role of host chemo-reception is to bring the parasite into contact with an appropriate host or to assist host recognition once contact has been achieved. This mechanism was originally demonstrated in caligids by Fraile *et al.* (1993), where mucus from sea bass (*Dicentrarchus labrax* L.) was shown to alter the behaviour of *Caligus minimus* (Otto, 1821). It has been demonstrated more recently in the caligid *Lepeophtheirus salmonis* (Krøyer, 1837) where a specific chemical from Atlantic salmon (*Salmo salar* L.) elicits a chemotaxic response (Devine *et al.*, 2000; Ingvarsdottir *et al.*, 2002). It is likely that *L. branchialis* employs a number of mechanisms to bring it into contact with the host. *L. salmonis*, for instance, employs mechano-reception to detect host movement (Bron *et al.*, 1993; Heuch and Karlsen, 1997), chemo-reception to recognise host-associated chemical cues (Devine *et al.*, 2000; Ingvarsdottir *et al.*, 2002) and, in adult stages, a pronounced shadow response to passing fish (J. E. B., personal observation). The cues employed by *L. branchialis*, however, have yet to be established.

4.4.4. Metamorphosis

Following fertilisation, the female undergoes a dramatic transformation in size and shape and the segmental boundaries between the somites become obscured. This process begins on the intermediate host and is completed once attached to the definitive host. The metamorphosis has been divided into seven sub-stages by Sproston and Hartley (1941a) (Table 5.3). Examples of the sub-stages can be found in Khan (1988). Once a host has been located, the female preferentially attaches to the filaments of the third gill arch (Van den Broek, 1978, 1979b) and from there migrates to the ventral portion of the gill arch (P-stage) (Kabata, 1958). The long genital complex twists and dilates, the cuticle becomes thicker and the neck extends along the branchial artery (U–V-stage). Antlers grow from the cephalothorax, penetrating the host and anchoring the adult (W-stage). Metamorphosis is rapid and antlers have been found in the *bulbus arteriosus* 5 days after infection (Smith *et al.*, 2007).

The mouth is situated at the end of the buccal cone, at the centre of the antlers and is embedded in the blood vessel of the host (Sproston and Hartley, 1941a). Following the commencement of feeding, the swollen genital segment becomes darker red and two long, irregularly coiled uniseriate egg strings are extruded just above the abdomen (X-stage). The eggs have an extended hatching period with eggs hatching for around 12 days (Y-stage) (Whitfield *et al.*, 1988). Because of the persistence of parasite attachment structures and host tissue responses following the death of the parasite, signs of previous parasite infection have usually been characterised as stages. Due to the permanent mode of attachment, the antlers remain in the host tissues after the death of the parasite and have been termed as Z-stage (Khan, 1988). Similarly, if the parasite is successfully rejected by the host, a proliferation of gill arch tissue is found without the presence of the parasite and has been termed an R-stage.

TABLE 5.3 Classification of *Lernaeocera branchialis* adult female metamorphosis sub-stages found on whiting (based on Sproston and Hartley, 1941a; Van Damme and Hamerlynck, 1992)

Sub-stage	Definition
Pennella (P1)	Abdomen elongate but straight, genital region not swollen, recently attached to the host, fully pigmented
Pennella (P2)	One point of flexure, genital region not swollen, penetration of host tissue, rudiments of three holdfast processes present
Immature (U)	Two or three points of flexure, genital region not swollen, holdfast processes begin to elongate
Immature (V)	Torsion of abdomen complete—full sigmoid curvature, genital region partly swollen, holdfast processes with some secondary branching
Mature—pre-gravid (W)	Fully metamorphosed, genital region fully swollen, fully developed branched holdfast, no external egg strings
Mature—gravid (X)	External egg strings present
X1	Immature eggs
X2	Mature pigmented eggs
Mature—post-gravid (Y)	External egg strings partly or completely spent
Dead parasite (Z)	Remains of holdfast embedded in host tissue
Rejected parasite (R)	Absence of parasite specimen, proliferation of gill arch tissue

Although included by Sproston and Hartley (1941a) and Van Damme and Hamerlynck (1992), the dead parasite (Z-stage) and rejected parasite (R-stage) cannot strictly be described as parasite sub-stages. Estimates of the life span of the adult female vary from 8 weeks (Sproston, 1942) through 9–10 months (Khan, 1988) to 1 year or more (Capart, 1948; Schuurmans Stekhoven and Punt, 1937). No methods allow a direct estimation of the life span of the adult female in wild populations (Sundnes, 1970) and aquarium studies do not reflect the situation in the wild. Consequently, the life span of the adult female is an issue which remains unresolved.

The metamorphosis of *L. branchialis* from a free-swimming copepod to a worm-like protrusion from the gill cavity of its host without any further moults was first reported by Metzger (1868). The mechanism that allows the cuticle to increase in size with at least a 20-fold increase in the length and girth of the abdomen was formally elucidated using SEM and TEM ultrastructural analysis by Smith and Whitfield (1988), although Kabata (1979) described a pattern of 'transverse wrinkling' on the abdomen which is in fact a series of posteriorly directed transverse cuticular folds that allow expansion of the cuticle (Fig. 5.4iv). The mean density of the folds on unfertilised adult females is around 1.04 μm^{-1} of abdominal length; but once the female has been fertilised, the folds are pulled apart resulting in a linear decrease in the density of the folds with increasing abdominal length (Smith and Whitfield, 1988).

The development of the folded cuticle occurs before the chalimus IV to adult moult, beneath the cuticle of the chalimus IV. Using TEM, it has been shown that the cuticle consists of an outer epicuticle of about 0.08-μm thickness and an inner procuticle, which can be divided into two layers, an inner and an outer layer (Smith and Whitfield, 1988). Once secretion of the adult cuticle has begun, it is thrown into a series of first order folds around 4–6 μm deep and then later into large-scale second order folds about 8–10 μm deep, resulting in a complex super-folded cuticle. The adult females do not possess these second order folds suggesting that they straighten out during the moult from the chalimus IV, which would account for the 43% increase in abdominal length at the moult.

The pulling apart of the first order folds results in a sixfold increase in abdominal length (Smith and Whitfield, 1988). As the final elongation factor is about 20-fold, there must be other mechanisms involved. It is suggested that these may include continued and large-scale secretion after the final moult and a change in the properties of the inextensible cuticle to allow for further increases in girth and length.

5. REPRODUCTION

5.1. Mating strategies

In common with other parasitic copepods, for example, caligids, males reach maturity before females from the same brood and once they are mature, they begin to search for a mate (Anstensrud, 1989). Mate guarding by males is common in *L. branchialis* and they will establish pre-copula with all stages of developing females on the host, although they prefer females closer to maturation (Anstensrud, 1992; Heuch and Schram, 1996). This behaviour is common in other parasitic copepods such as *L. salmonis* (Hull *et al.*, 1998) and *C. elongatus*, which have been seen to move vigorously over the surface of a fish in search of females and to exhibit mate guarding if females are not mature (Piasecki and MacKinnon, 1995). However, males tend to show a preference for a chalimus female over an adult virgin female in pre-copula with another male, suggesting that a male has little chance of taking over a female from another male (Anstensrud, 1992).

Sex-specific signals are developed early in *L. branchialis* as significantly more mature males will choose chalimus I females than chalimus I males. However, mature males sometimes assume pre-copula with immature males, indicating that sex-specific signals may be weak and confusing at this stage (Heuch and Schram, 1996). The establishment of these male–male associations suggests that mate recognition may be based on tactile stimuli and not pheromones. Mature males always leave the immature parasite in this situation once the correct signals are not obtained. Males can also recognise an already inseminated female, which has attached spermatophores, and will refuse to mate with a non-virgin female if presented with other female options.

It is likely that males will guard the first female they encounter as many males are found guarding immature females, while some mature females are found without male company (Kabata, 1958). In most cases, the first male to take up the pre-copula position with a chalimus female also takes up the copula position once the female matures (Anstensrud, 1989, 1992). At high male:female ratios, males tend to surround a female or form clusters of males only, attached by their chelate antennae to a gill filament close together or attached to each other (Anstensrud, 1989, 1992).

5.2. Copulation

After a chalimus female has been located, the male assumes a pre-copula position with its antennae gripping the female in the vicinity of its frontal attachment apparatus (Boxshall, 1990a), often penetrating the gill filament to grip the frontal filament of the female (Kabata, 1958). The male remains in this position until the female moults into the adult. Immediately after the female's ecdysis, the male crawls backwards on the dorsal side of the female using its antennae while the maxillipeds are moved back and forth over the ventral surface of the female (Anstensrud, 1989). Once the male reaches the copula position, it grasps the female's trunk just behind the last leg-bearing thoracic segment with its antennae. Transfer of spermatophores takes place within 1 h of the establishment of the copula position. The spermatophores are expelled after a series of contractions of the male's genital complex and are pushed anteriorly towards the ventral surface of the male's genital complex, being retained by several rows of hook-like processes just anterior to the genital orifices. The male then bends its posterior segments under the female's genital complex and positions the spermatophores on the female's genital opening. The spermatophores are held in position by the long setae of the male's first swimming legs until the male's trunk slowly straightens out. Copulation is complete within 2–3 min. Males take 3 days to produce the first pair of spermatophores and 17 h to produce each of the following ones (Anstensrud, 1990b).

Anstensrud (1989) found that males without maxillipeds are not usually able to fertilise females successfully. In the Pennellidae, only males have maxillipeds (Kabata, 1979), and this suggests that they may be connected with reproduction, possibly being used for mechano-reception of the female's genital opening, orientation on the female, determination of the maturity of the female (Anstensrud, 1989) and relaxation of the female's genital opening.

5.3. Male competition

The females can be inseminated more than once and often the first male to assume pre-copula with the female will inseminate her repeatedly (Anstensrud, 1990c). This may be to reduce sperm competition from subsequent males by filling up the *receptaculum seminis* of the female, as sperm masses from different males become mixed within the receptaculum seminis (Anstensrud, 1990b,c). Egg production starts 7–9 months after copulation (Anstensrud, 1990c; Khan, 1988), allowing plenty of time for sperm masses to become mixed. In experiments, 97% of females were inseminated at least twice and the number of males copulating with the first female more than three times increased significantly in the presence of a male competitor (Anstensrud, 1990c). Given other options, mature males avoid inseminated females, preferring virgin females or immature stages, again indicating the likelihood of sperm competition within the receptaculum seminis (Heuch and Schram, 1996). Based on spermatophore volumes and quantities of sperm in the receptaculum seminis of females, the maximum number of inseminations per female was five (Anstensrud, 1990c).

The amount of time spent on the intermediate host is reduced as a female is inseminated more. Virgin females tend to spend around 8 days on the host, whereas females that have been copulated once or twice spend around 4 and 3 days, respectively (Anstensrud, 1990a). This suggests that one copulation is not sufficient and that lingering on the intermediate host is an adaptation to secure sufficient insemination. Due to the short period of time that females spend on the intermediate host, the males remain and tend to accumulate as they do not leave the intermediate host, leading to high male:female ratios. The 50% survival time of adult males is ~5 weeks (Anstensrud, 1990c).

5.4. Fecundity

Once a pair of egg strings has been extruded, embryonic development of *L. branchialis* takes around 13 days at 10 °C (Whitfield *et al.*, 1988). The number of eggs in a single pair of egg strings is high for parasitic copepods. Whitfield *et al.* (1988) found a mean of 1445 and a maximum in excess of 3,000 eggs. Under laboratory conditions, hatching continued for a maximum of 12 days and followed an exponential decline. Less than half (44.2%) reached the infective copepodid stage, the rest failing to hatch or develop to the nauplius II or copepodid stages. However, if the egg strings are removed, the female extrudes another set within 48 h, demonstrating the iteroparous reproductive capacity of *L. branchialis*. It is likely that the female can rapidly produce several egg strings, maybe one pair every 2 weeks, depending on water temperature (Anstensrud, 1990c). Given that oviposition may continue for a year (Khan, 1988), the maximum fecundity of a single *L. branchialis* female is estimated to be around 36,400 eggs (Heuch and Schram, 1996).

5.5. Female fitness

The high fecundity of *L. branchialis* females may be a result of the two-host life cycle (Anstensrud, 1990c). Females probably have relatively high post-copula mortality once they leave the intermediate host and as a consequence their fitness may be reduced. This is compensated for by high fecundity. The need to accommodate this high fecundity of females may be a selective mechanism resulting in a relatively high volume and rapid production of spermatophores by males (Anstensrud, 1990b). This high spermatophore volume may also serve to benefit the female nutritionally during the pelagic phase, as the seminal constituents may be absorbed from the receptaculum seminis after insemination (Anstensrud, 1990b). The presence of seminal fluid in the *receptaculum seminis* is also essential for the metamorphosis of the female as Anstensrud (1990a) found that significantly fewer virgin females reached the penella stage and none metamorphosed past the U-stage on the final host.

5.6. Egg strings and egg-string attachment

The egg strings of *L. branchialis* are long, narrow and possess individually stacked eggs, that is, uniseriate. Each irregularly coiled egg string is extruded from a gonopore situated just above the constriction which divides the abdomen from the genital complex. Branching tree-like structures, developed from the vitellarium, form the central axis of each egg string and are held to them by thin membranes, giving them an interwoven appearance (Schram and Heuch, 2001). This structure is unique to the genus *Lernaeocera* (Kabata, 1979).

Like many other parasitic copepods, the egg strings are mechanically secured by a hook inside the genital complex, which is found just outside the atrium on each side (Schram and Heuch, 2001). The cupulate hook bases are attached to the body wall with connective tissue and muscle, with five bands of muscle on each hook for suspension and movement, between 1 and 1.8 mm long. A suture in the atrium allows the hook tip to enter and protractor and retractor muscles allow the hook to act as a lever and swing across the atrium. Where the tip of the hook makes contact with the proximal end of the egg string, a concave depression is formed in the egg string as the hook enters a notch in the opposite side of the atrium, locking the egg string in place.

5.7. The male reproductive system and spermatozoon ultrastructure

In the mature male, the reproductive system consists of two linearly arranged sets of reproductive organs, each comprising a testis, a *vas deferens*, a seminal vesicle and a spermatophore sac (Capart, 1948). The testes are located in the cephalothorax and are bilobed, each consisting of a main lobe attached to the vas deferens and a smaller central lobe (Grant and Whitfield, 1988). Spermiogenesis begins with the spermatocytes which are located more dorsally and proceeds ventrally in the testis and along the *vas deferens* where the spherical spermatids, each about 4 μm diameter, are transformed into mature spermatozoa. The mature spermatozoa are packed in ovoid spermatophores in the spermatophore sacs.

The mature spermatozoa are thread-like cells about 30 μm in length and 1 μm in diameter (Grant and Whitfield, 1988). The cells are immotile in seawater. SEM analysis reveals that the cell is javelin-shaped, tapering at both ends and displaying a four-lobed helical twist along its surface. The spermatozoon does not have a defined nucleus enclosed by a nuclear envelope, but the nuclear chromatin is a finely granular/filamentous material associated with a pseudomembranous structure that extends along the central core of the cell. The cell membrane consists of longitudinally orientated microfilaments and some microtubules, although there are no mitochondria or acrosome present in the cell. The characteristics possessed by the spermatozoa of *L. branchialis* are not shared by any other copepod species, although each individual characteristic is possessed by at least one other species (Grant and Whitfield, 1988).

6. PHYSIOLOGY

6.1. Feeding and maintenance of the internal environment

The cuticle of *L. branchialis* is highly chitinised (Sproston and Hartley, 1941b), meaning that the parasite is relatively impermeable. Sproston and Hartley (1941b) found that unlike most marine invertebrates, *L. branchialis* is able to keep its body hypotonic in its marine environment as long as it remains attached to the host. Feeding on the hypotonic blood of the host compensates for slow osmosis of water through the body wall. However, once removed from the host, isotonicity is established with the external medium due to water being taken in through the mouth in place of the blood of the host (Sproston and Hartley, 1941b). Since the parasite is hypotonic to its external environment, it is likely that there is no intake of water by the anus. Sproston and Hartley (1941b) observed no movement of fine particles in suspension around the anus and microscopical examination shows a funnel-like depression at the posterior extremity of the parasite but it does not perforate the cuticle. This is not surprising, as in many blood feeding parasites the digestion is slow but complete. However, free iron liberated from haemoglobin catabolism is toxic and blood feeding parasites must develop mechanisms to prevent iron reaching toxic levels. In the pennellid, *Cardiodectes medusaeus* (Wilson), free iron is converted to non-toxic ferritin and stored in the frontal attachment organ as ferritin crystals (Perkins, 1985). As *L. branchialis* is unable to excrete dietary waste products, it is likely that a similar mechanism is present and a non-toxic form of iron is stored in the parasite.

Although *L. branchialis* can maintain its haemolymph hypotonic to sea water, it is dependent on the osmolarity of the surrounding medium. In experiments, the osmolarity of the parasite haemolymph decreased asymptotically in 50% seawater such that *L. branchialis* cannot survive below a salinity of about 16‰ (Knaus Knudsen and Sundnes, 1998). This is in line with a survey in Baltic waters that reported that *L. branchialis* was not found in waters of salinities below 18‰ (Sundnes *et al.*, 1997).

Feeding is discontinuous and the mouth is opened only at well-spaced intervals to admit a meal of blood which is retained in the gut for a long period (Sproston and Hartley, 1941b). The blood pressure of the host is increased around the site of injury by the parasite due to the fibrotic thickening of the ventral aorta and bulbus arteriosus (Smith *et al.*, 2007) so the parasite controls the intake of blood by keeping the mouth closed and opening it only briefly to feed at intervals. These feeding periods are suggested to be triggered by a drop in hydrostatic pressure due to the slow osmosis of water through the cuticle in response to the hypotonicity of the parasite to sea water (Sproston and Hartley, 1941b). Once the hydrostatic pressure drops below a limiting value, the mouth is opened to admit host blood and the hypotonicity is restored. This discontinuous feeding is supported by measurements of osmotic pressure of the body fluid of *L. branchialis* attached to the host, where a considerable variation in values was observed (Panikkar and Sproston, 1941).

Although *L. branchialis* generally has a red colouration, there is no open connection between the vascular system of the host and the digestive system of the parasite (Sundnes, 1970). The haemocoel has a red colouration, but this is distinct from host haemoglobin and the digestive system of the parasite has no red colouration with no haemoglobin present, which indicates that haemoglobin is separated from the blood serum before it enters the parasite (Sundnes, 1970).

Although it has not been demonstrated in *L. branchialis*, immunosupression of the host is common in ectoparasites and it is possible that *L. branchialis* uses some form of immunosupression to counteract the host's immune response. Fast *et al.* (2002, 2004) found evidence that the caligid *L. salmonis* immunoregulates its host at the sites of attachment and feeding. Two substances amongst other undescribed substances were found, prostaglandin E_2 and trypsin, which were shown to immunomodulate the host. In other arthropod parasites, prostaglandin E_2 is known to play a variety of roles, including vasodilation, which would be useful in maintaining blood flow to the site of feeding since blood constitutes a component of the diet of *L. salmonis* (Brandal *et al.*, 1976; Bricknell *et al.*, 2003). As *L. branchialis* is entirely a blood feeder, prostaglandin E_2 could have a role in maintaining a blood flow. Prostaglandin E_2 could also adversely affect site-specific leucocyte recruitment and activity (Papadogiannakis and Johnsen, 1987; Papadogiannakis *et al.*, 1984; To and Schrieber, 1990), which could help prevent rejection of *L. branchialis* by its host. As trypsin is found in the guts and saliva of some arthropod parasites (Kerlin and Hughes, 1992) and also inhibits phagocytosis in monocytes (Huber *et al.*, 1968), Fast *et al.* (2003) suggested that trypsin derived from *L. salmonis* may decrease host phagocytic activity and immune responses after infection and it is possible that *L. branchialis* employs a similar mechanism to locally immunomodulate the host. Although immunosuppression of the host is most likely to occur at the site of attachment, an accumulation of several parasites may eventually cause immunosuppression of the entire host (Fast *et al.*, 2003) and may be the reason why fish harbouring secondary infections concurrent with *L. branchialis* are more likely to succumb to the infection (Khan and Lacey, 1986).

6.2. Respiration

In many parasitic copepods, respiration occurs principally through the cuticle and/or the anus but Sproston and Hartley (1941b) suggest that in *L. branchialis*, cuticular respiration is minimal due to heavy chitinisation and that oxygen enters the parasite via the mouth in the form of the blood of the host. However, Sundnes (1970) propose that this is not the case as the usual location of *L. branchialis* is on the venous side of the gills where oxygen content is low. Diffusion of oxygen through the cuticle may be sufficient for a sluggish animal with the limited size of *L. branchialis* and osmotic experiments indicated that gas diffusion is evident. A two-way peristalsis causes turbulence within the haemocoel and may be the mechanism of internal oxygen distribution (Sproston and Hartley, 1941b).

7. DISTRIBUTION

7.1. Geographic distribution

The geographic distribution of *L. branchialis* is circumscribed by the distribution of its intermediate and final hosts and restricted by salinity and temperature. *L. branchialis* is therefore restricted to the North Atlantic and adjacent seas (Kabata, 1979) (Fig. 5.5). As the swimming abilities of the pre-metamorphosed adult are suggested to be limited (Sproston, 1942) and the timing of this stage is short (Schuurmans Stekhoven and Punt, 1937), infection of the definitive host is suggested to be possible only where its distribution directly overlaps that of the intermediate host.

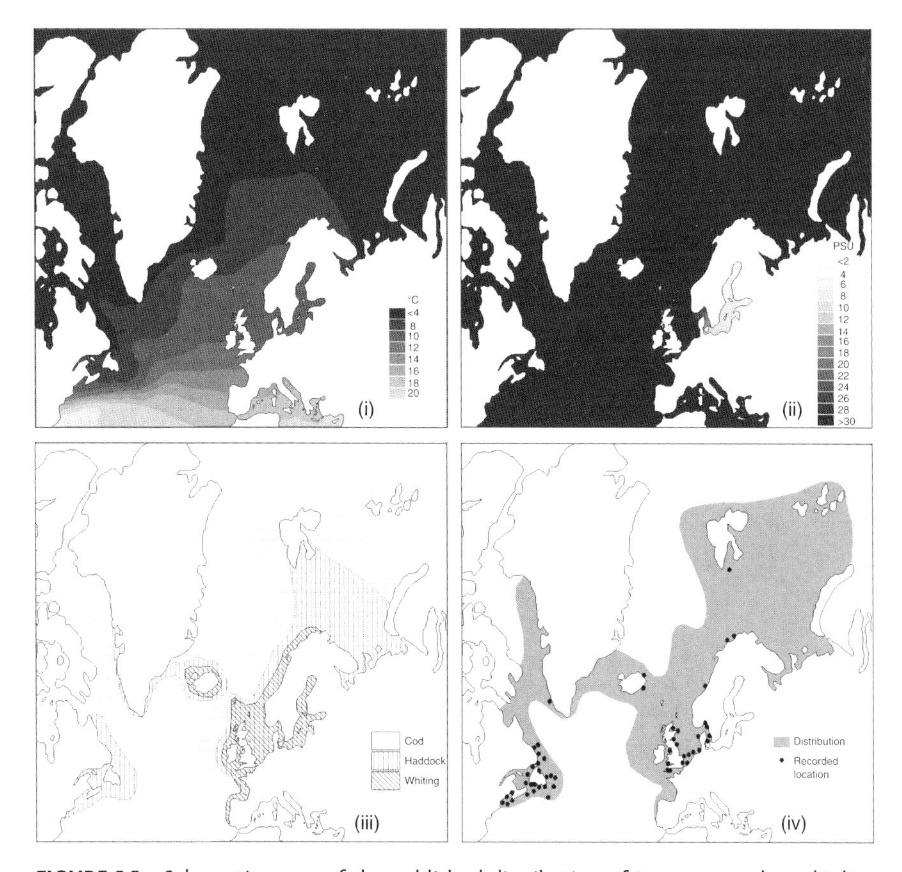

FIGURE 5.5 Schematic maps of the published distribution of *Lernaeocera branchialis* and the environmental boundaries that may determine the distribution. (i) Mean sea surface temperature, (ii) salinity, (iii) definitive host distribution, (iv) theoretical and observed distribution of *L. branchialis*.

Around the majority of the United Kingdom, the life cycle of *L. branchialis* alternates between *P. flesus* (flounder) as the intermediate host and *M. merlangus* (whiting) or *G. morhua* (cod) as the definitive host. As the flounder is predominantly an inshore and estuarine species, the habitats of these species only overlap when juvenile whiting and cod migrate inshore to nursery grounds and this is where the greatest infection takes place (Pilcher *et al.*, 1989; Potter *et al.*, 1988; Shotter, 1973a,b; Sproston and Hartley, 1941a; Van den Broek, 1979a). Whiting caught in deeper, offshore waters have very little infection as they have moved out of the range of the intermediate host (Pilcher *et al.*, 1989; Potter *et al.*, 1988; Shotter, 1973a,b; Sproston and Hartley, 1941a), although fish that have previously been infected and outlived the parasite can be identified by anchor wounds in the gill arch (Pilcher *et al.*, 1989). Around the Isle of Man, *L. branchialis* was found to be more prevalent in inshore locations than in offshore deeper water, although the parasite was more prevalent on the east coast than on the west coast (Shotter, 1973a,b). This is thought to be due to flounder being more common on the east coast than the west coast. In the northern North Sea, lemon sole (*M. kitt*) are the predominant intermediate host for *L. branchialis* (Kabata, 1957, 1958). As lemon sole is not restricted to shallow inshore waters, maximum prevalence of *L. branchialis* is usually found where there is a high abundance of lemon sole and haddock or cod, the most common definitive hosts in this area.

On the Northwest Atlantic coast, *Cyclopterus lumpus* L. (lumpfish) have been found to be the most common intermediate host of *L. branchialis* (Fleming and Templeman, 1951; Templeman *et al.*, 1976). In the waters around Newfoundland, the prevalence of *L. branchialis* often reaches 100% on lumpfish (Templeman *et al.*, 1976), although the prevalence of the parasite decreases in southern Newfoundland waters. The parasite is most abundant on cod from inshore waters, where they inhabit similar areas to the lumpfish, than in deeper offshore waters. Templeman and Fleming (1963) noted that the highest prevalence on cod occurs in the warmer, southern waters of Newfoundland, rather than cooler, northern waters. They suggest that higher growth rates in cod in warmer waters may reduce the pathogenic effects of *L. branchialis*, allowing southern Newfoundland cod populations to support a higher level of infection. Further north around Labrador, prevalence on cod decreases with increasing latitude (Jones and Taggart, 1998). This pattern correlates with the distribution of the intermediate host, lumpfish, which is found less frequently in more northern waters (Stevenson and Baird, 1988). Conversely, on the Norwegian coast, the prevalence on cod was found to decrease from north to south (Sundnes, 1970). The prevalence in cod was very low (<1%) in catches from the Spitzbergen Bank, which may be due to the distance from the coast (and the habitat of the intermediate host)

(Sanchez-Lizaso and Vasquez, 1987; Vazquez *et al.*, 1988). However, there may also be other factors affecting infected fish, which could explain the low prevalence of infection in offshore waters, such as mortality caused by the parasite, an inability to swim into deeper waters due to retention of excessive air in the swim bladder and predation of weakened infected fish during migration (Khan, 1988). Compared with studies in European waters, the prevalence on the definitive host is relatively low, around 10–20%, which may reflect the fact that the habitats of the intermediate host and definitive host often do not overlap directly.

L. branchialis cannot survive in salinities below 16‰ (Knaus Knudsen and Sundnes, 1998) (see Section 6.1). In a study of *L. branchialis* on cod in Baltic waters, no parasites were found in salinities below 20‰ (Sundnes *et al.*, 1997). There is only one record of *L. branchialis* being found in the Mediterranean (Brian, 1906), although this may have been a misidentification. Sherman and Wise (1961) found that the prevalence of *L. branchialis* decreases with decreasing latitude on the West Atlantic coast. At 44°N, where ocean surface temperatures can range from 2–5°C in the winter to 16–20°C in the summer, the prevalence reached 18% but decreased to only 1% at 41°N, where ocean surface temperatures can range from 8–10°C in the winter to 24–26°C in the summer (Fig. 5.5i). As the cod extends its range south of 41°N in sufficient numbers to support a fishery, it appears that *L. branchialis* may be limited by warmer waters.

7.2. Prevalence and intensity of host infection

In species such as cod, whiting and haddock, juvenile fish migrate inshore to nursery grounds until they mature and migrate to deeper offshore waters. As the highest prevalence of *L. branchialis* is usually found in inshore waters, where the habitats of the intermediate and definitive hosts overlap, juvenile fish are the most common definitive host (Table 5.4). Sproston and Hartley (1941a) found that juvenile whiting less than 10 cm in length entering the Tamar Estuary, United Kingdom, were free from infection. Prevalence increased the longer the fish remained in the estuary. Prevalence was 7.5% for fish 10–15 cm in length and 22.3% for fish 15–25 cm in length. Similarly, juvenile whiting less than 7.5 cm in length entering the Medway Estuary, United Kingdom, were almost free from infection with a prevalence of 1.8%, but infection was rapid and peaked soon after their inshore migration (Van den Broek, 1977). In fish 12.5–14.9 cm in length, prevalence was 56.1%. By the time the whiting mature and migrate offshore out of the range of the intermediate host, most fish have outlived the parasite and the prevalence of *L. branchialis* in these older fish is low (Pilcher *et al.*, 1989). Prevalence on flounder was very high (up to 100% in some cases) in all age groups,

TABLE 5.4 Studies regarding the prevalence and intensity of *Lernaeocera branchialis* infections in varying sizes of different fish hosts

Location	Species	Size range	Year class	Prevalence (%)	Intensity	References
Tamar Estuary, United Kingdom	*Merlangius merlangus*	10–15 cm	–	7.5	–	Sproston and Hartley, 1941a
		15–25 cm	–	22.3	–	
North Norway	*Gadus morhua*	–	>5+	0	–	Sundnes, 1970
		–	6+	5	–	
		–	>7+	10	–	
Borgenfjord, Norway	*Gadus morhua*	–	2+	24	–	
		–	3+	28	–	
		–	4+	32	–	
		–	5+	26	–	
Isle of Man, United Kingdom	*Merlangius merlangus*	13–28 cm	–	19.6	–	Shotter, 1973a
		28–38 cm	–	36.8	–	
Isle of Man, United Kingdom (inshore)	*Merlangius merlangus*	–	0+	2.5	1.0	Shotter, 1973b
		–	2+	6.7	1.0	
Isle of Man, United Kingdom (offshore)	*Merlangius merlangus*	–	0+	0	0	
		–	2+	0.8	1.0	

Newfoundland	*Cyclopterus lumpus*	30–34 cm	–	–	231.25	Templeman *et al.*, 1976
		35–39 cm	–	–	122.7	
		40–44 cm	–	–	93.38	
		45–49 cm	–	–	168.40	
		50–54 cm	–	–	191.8	
	Gadus morhua	11–30 cm	–	2.6	–	
		31–50 cm	–	4.9	–	
		51–90 cm	–	2.2	–	
Medway Estuary, United Kingdom	*Platichthys flesus*	–	0+	95.7	20.7	Van den Broek, 1979a
		–	1+	95.7	46.3	
		–	2+	97.6	92.1	
		–	3+	100	149.8	
		–	4/5+	100	111.6	
	Merlangius merlangus	<7.5 cm	–	1.8	–	
		12.5–14.9 cm	–	56.1	–	
Spitzbergen Bank, North Atlantic	*Gadus morhua*	<50 cm	–	0.175	–	Sanchez-Lizaso and Vasquez, 1987
		51–59 cm	–	0.610	–	
		60–68 cm	–	1.533	–	
		>69 cm	–	0.633	–	

(continued)

TABLE 5.4 *(continued)*

Location	Species	Size range	Year class	Prevalence (%)	Intensity	References
Severn Estuary, United Kingdom	*Merlangius merlangus*	–	0+	2	–	Potter *et al.*, 1988
		–	1+	10.5	–	
Spitzbergen Bank, North Atlantic	*Gadus morhua*	–	3+	0.85	–	Vazquez *et al.*, 1988
		–	4+	0.75	–	
		–	5+	1.01	–	
		–	6+	1.08	–	
Newfoundland/ Labrador	*Gadus morhua*	<36 cm	–	0	–	Jones and Taggart, 1998
		36–49 cm	–	9.48	–	
		50–63 cm	–	6.46	–	
		64–77 cm	–	3.3	–	
		78–91 cm	–	1.61	–	
		>91 cm	–	0	–	
Balsfjord, Norway	*Gadus morhua*		2 + 3	31	–	Hemmingsen *et al.*, 2000
			4	20.5	–	
			5	19	–	
			6	35	–	
			7 + 8	35.5	–	

although the intensity of infection increased from 20.7 in 0+ fish to 149.8 in 3+ fish (Van den Broek, 1979a). As flounder remain inshore throughout the year, where infection is more intense, this result is not unexpected. Around the Isle of Man, the prevalence of *L. branchialis* in different age groups of whiting was similar to other areas (Table 5.4), with younger 0+ fish having a lower prevalence than older 2+ fish, although the prevalence was generally lower in offshore populations (Shotter, 1973b). The pattern of prevalence is different in the northern North Sea where older haddock are the most common definitive host as they are found in the same areas as the intermediate host, lemon sole (Kabata, 1958).

In Norwegian waters, two different populations of cod showed different infection dynamics (Sundnes, 1970). 2+ fish from Bjorgenfjord showed considerable infection prevalence of 27.3%, whereas fish from Arcto-Norwegian populations were not infected until they were at least 5 years old. 0+ Arcto-Norwegian cod leave coastal waters to feed in the Barents Sea where they are out of the range of infected intermediate hosts and are not exposed to *L. branchialis* until they begin their spawning migration. Prevalence decreased in older Bjorgenfjord fish, suggesting that infection often caused mortality, whereas a slight increase in prevalence in older cod was seen in Arcto-Norwegian populations, indicating that infection did not cause mortality in these older fish and also that there was continuous exposure to infection.

A similar pattern exists on the Northwest Atlantic coast where juvenile cod from inshore waters are more likely to be infected than older cod, due to their proximity to the major intermediate host, the lumpfish. In the waters around Newfoundland, Templeman *et al.* (1976) found that the maximum prevalence of 4.9% was found in intermediate length cod of 31–50 cm in length, but declined to 2.2% at fish lengths of 51–90 cm, as these older cod are usually found offshore. There was no distinct pattern in the intensity of infection of lumpfish of different ages, as they tend to inhabit coastal habitats where transmission of infection occurs readily, so infection is high on all ages of fish. This is comparable to the pattern of infection found in flounder in United Kingdom waters. Jones and Taggart (1998) found a similar pattern of prevalence in cod around the Newfoundland and Labrador coasts, with the greatest prevalence of 9.48% occurring in fish 36–49 cm in length and prevalence decreasing at lower and higher fish lengths. However, these authors attribute the highest prevalence at lower fish lengths to a reduction in growth rates of infected cod and not to the migratory behaviour of the fish.

7.3. Seasonality

The seasonal distribution of *L. branchialis* is influenced by the migrations of the host populations. In general, prevalence and intensity of *L. branchialis* on the definitive host is higher in winter than in summer as this is when the definitive hosts migrate inshore and occupy the same habitat as the intermediate hosts (Tables 5.5 and 5.6).

In the southern North Sea, the highest prevalence in whiting occurs in winter due to the increased overlap between the whiting and flounder populations in shallow coastal areas (Pilcher *et al.*, 1989). However, this pattern of infection of flounder and whiting is not apparent in the northern North Sea at latitudes higher than 57°N where the cycle alternates between lemon sole and haddock (Kabata, 1958; Van den Broek, 1979a). Recruitment on the definitive host increases between May and June, with maximum prevalence occurring around September.

Other European studies have been more localised, examining fish mainly from inshore waters or estuaries. In the Medway Estuary, United Kingdom, the prevalence on flounder was high throughout the year with a mean of 96.7% (Van den Broek, 1979a). Infection intensity was also high with a mean intensity of 64.9 parasites per infected fish although this did vary throughout the year with peaks in June and December. Whiting entering the estuary in the autumn from offshore waters were generally uninfected, but the infection reached a peak soon after their migration inshore, with the maximum intensity of infection in flounder being recorded 1–3 months after the highest prevalence of mature adult parasites was recorded on whiting (Van den Broek, 1979b). A similar pattern was seen in the Bristol Channel and Severn Estuary, United Kingdom, where the greatest prevalence on whiting was found in January and February, several months after their migration from offshore (Potter *et al.*, 1988). The greatest prevalence on whiting in the Tamar Estuary was seen in July and August, several months after the year class 0 fish moved into the estuary (Sproston and Hartley, 1941a). Once these fish migrated offshore in the spring, the prevalence of *L. branchialis* on whiting in the estuary was 100% although fish that migrated were virtually free from infection. Sproston and Hartley (1941a) concluded that the infected fish tend to linger in the estuary allowing *L. branchialis* to complete its life cycle and shed its eggs in the inshore waters, increasing the probability of encounter of copepodids with the intermediate host following hatching and development. No significant difference was found in the prevalence on flounder throughout the year. This could be attributed to the ability of the adult parasites to produce more than one pair of egg strings and the extended incubation of around 2 weeks (Whitfield *et al.*, 1988).

In the Oosterschelde (SW Netherlands), transmission to whiting can only occur in the spring and autumn when these fish are present in

TABLE 5.5 Studies regarding seasonality in the prevalence of *Lernaeocera branchialis* on different hosts

Location	Species	Habitat	Prevalence	Seasonality	References
Tamar Estuary, United Kingdom	*Platichthys flesus*	Inshore	Low	No	Sproston and Hartley, 1941a
Tamar Estuary, United Kingdom	*Merlangius merlangus*	Inshore	Low–high	Yes	Sproston and Hartley, 1941a
Tamar Estuary, United Kingdom	*Pollachius pollachius*	Inshore	Low	Yes	Sproston and Hartley, 1941a
Plymouth, United Kingdom	*Merlangius merlangus*	Offshore	Low	No	Sproston and Hartley, 1941a
Borgenfjord, Norway	*Gadus morhua*	Inshore	Moderate	No	Sundnes, 1970
North Norway	*Gadus morhua*	Inshore	Low	No	Sundnes, 1970
Isle of Man, United Kingdom	*Merlangius merlangus*	Inshore	Low–moderate	Yes	Shotter, 1973a,b
Isle of Man, United Kingdom	*Merlangius merlangus*	Offshore	None	No	Shotter, 1973a,b
Newfoundland	*Cyclopterus lumpus*	Inshore	High	No	Templeman *et al.*, 1976
Newfoundland	*Cyclopterus lumpus*	Offshore	Low–moderate	Yes	Templeman *et al.*, 1976

(*continued*)

TABLE 5.5 *(continued)*

Location	Species	Habitat	Prevalence	Seasonality	References
Newfoundland	*Gadus morhua*	Inshore	Low	Yes	Templeman *et al.*, 1976
Medway Estuary, United Kingdom	*Platichthys flesus*	Inshore	High	No	Van den Broek, 1979a
Medway Estuary, United Kingdom	*Merlangius merlangus*	Inshore	Moderate–high	Yes	Van den Broek, 1979a
Sweden, south and west coasts	*Gadus morhua*	Inshore	Low–moderate	Yes	Linderby and Thulin, 1983
Severn Estuary, United Kingdom	*Merlangius merlangus*	Inshore	Low–moderate	Yes	Potter *et al.*, 1988
Balsfjord, Norway	*Gadus morhua*	Deep, inshore	Moderate	No	Hemmingsen *et al.*, 1995
Oosterschelde, The Netherlands	*Merlangius merlangus*	Inshore	Moderate–high	Yes	Van Damme and Hamerlynck, 1992; Van Damme *et al.*, 1997

TABLE 5.6 Studies regarding seasonality in the intensity of *Lernaeocera branchialis* on different hosts

Location	Species	Habitat	Intensity	Seasonality	References
Newfoundland	*Cyclopterus lumpus*	Inshore	Low–high	Yes	Templeman *et al.*, 1976
Newfoundland	*Cyclopterus lumpus*	Offshore	Low	No	Templeman *et al.*, 1976
Medway Estuary, United Kingdom	*Platichthys flesus*	Inshore	Moderate–high	Yes	Van den Broek, 1979a
Medway Estuary, United Kingdom	*Merlangius merlangus*	Inshore	Moderate	Yes	Van den Broek, 1979a
North Sea	*Merlangius merlangus*	Inshore and offshore	Low	Yes	Pilcher *et al.*, 1989
Balsfjord, Norway	*Gadus morhua*	Deep, inshore	Moderate	No	Hemmingsen *et al.*, 1995
Oosterschelde, The Netherlands	*Merlangius merlangus*	Inshore	Low–high	Yes	Van Damme and Hamerlynck, 1992; Van Damme *et al.*, 1997

shallow gullies with the intermediate host, flounder (Van Damme and Hamerlynck, 1992; Van Damme *et al.*, 1997). During the hot summer months, whiting occupy the deeper gullies and in the winter they migrate offshore. The prevalence of infection in whiting peaked at 30–50% in May and 70–90% in December, with a mean infection intensity of 1.6 in December. This high prevalence in whiting is indicative of an area where parasite transmission is highly successful due to the high abundance of flounder as is the case for the Tamar and the Medway Estuaries, United Kingdom (Sproston and Hartley, 1941a; Van den Broek, 1979a).

In Balsfjord, Norway, no seasonality was found in *L. branchialis* on cod in samples taken from deep water (ranging from 110 to 180 m) (Hemmingsen *et al.*, 1995). The fjord is relatively isolated with a sill at the entrance and it is likely that the cod have formed a distinct population and that transmission of parasites occurs throughout the year. The lack of seasonality might also be attributed to the lack of major change in temperature throughout the year at these depths.

On the Northwest Atlantic coast in the Newfoundland area, lumpfish are common in coastal waters and during the spring and summer a prevalence of 100% was generally found, although in southern waters the prevalence was much lower, around 0–20% (Templeman *et al.*, 1976). The intensity of infection in inshore waters was often very high, sometimes as many as 5,000–6,000 parasites per fish and on one individual 26,700 parasites were estimated (Templeman *et al.*, 1976). The intensity peaked in the summer and tailed off to minimal numbers during winter. Transmission to cod occurs in the summer and autumn when the smaller cod are found close inshore and the prevalence is highest on cod during these months. In offshore waters, the prevalence on lumpfish was much lower, usually around 20–50% and intensity was around 0.1–0.5 parasites per fish, but with no apparent seasonality.

8. PATHOGENICITY

Infection of the definitive host by adult female *L. branchialis* has been observed to induce anaemia and loss of weight in its host; reduces fat content, liver somatic index, haematocrit levels and reproductive capacity; and causes localised sores, scarring and mortality (Hislop and Shanks, 1981; Kabata, 1958; Khan, 1988; Khan and Lacey, 1986; Khan *et al.*, 1990; Mann, 1970; Moller, 1983; Smith *et al.*, 2007; Van den Broek, 1978; Van Damme *et al.*, 1994). In comparison, *L. lusci* is suggested to be less pathogenic towards its definitive host (Van Damme *et al.*, 1994), whilst Eiras (1986) and Brinkworth *et al.* (2000) failed to find evidence for the pathogenicity of *L. lusci*. Mortality of infected definitive hosts is greatest amongst the smallest size classes and Khan (1988) reported a mortality of 43% in infected cod 36–41 cm, compared with 31% in fish >54 cm. Most deaths occurred in the first 4 months after infection and were largely associated with multiple infections. Death of young fish was associated with emaciation, haemorrhage of the tissue adjacent to the holdfast, open lesions, severe necrosis and low haemoglobin levels. However, in older fish, death probably resulted from occlusion of the ventral aorta and/or branchial arteries. Sundnes (1970) found that fish <40 cm in length showed a significant decrease in condition factor and haematocrit values, whereas fish >40 cm showed no significant difference in condition between infected and uninfected fish, indicating that older fish are more able to cope with the infection.

Mature parasites often protrude from the gills, preventing opercular closure and putting pressure on the surrounding area, leading to atrophy of sections of gill (Van den Broek, 1978). Adverse effects in whiting have been shown to include small localised sores, decreased liver lipid content, complete emaciation and death of the host (Van den Broek, 1978, 1979a). In cod, the early stages of *L. branchialis* infection induce local gill hyperplasia, large intravascular thrombus formation and a cellular immune response in cardiac and branchial tissues (Smith *et al.*, 2007). It has been suggested by some authors that a decrease in haematocrit value and liver lipid content is an indication of a lower metabolic rate, which in turn may result in reduced swimming speed, increased susceptibility to predation and lower prey uptake (Van den Broek, 1978; Van Damme *et al.*, 1994).

Generally, cod parasitised with *L. branchialis* have a lower condition factor and growth rate than uninfected fish, especially in younger fish and those with multiple infections. Khan (1988) found that young cod around 33 cm in length and harbouring three or more parasites had a significantly lower condition factor than control fish after 4 weeks. In a different study (Khan and Lee, 1989), cod showed lower increments in body length and weight over a period of 10 months, with infected fish weighing up to 28% less than controls. However, after the first 10 months, infected fish gained weight at a comparable rate to controls and at 16 months, after the parasites had reproduced, no significant difference in weight was apparent. This indicates that although *L. branchialis* can significantly reduce condition and growth in cod, those fish that outlive the parasite may eventually catch up with uninfected fish. Adult cod infected with one or two parasites initially showed an increase in food consumption and weight gain compared with controls, although condition factor and food conversion efficiency were lower throughout the period of study (Khan, 1988; Khan and Lee, 1989). Presumably this compensates for the effect of the parasites, resulting in a transitory weight gain. Another study showed that this compensatory growth occurs through the consumption of more food in the autumn months rather than during the winter (Khan *et al.*, 1993). A further explanation may be that resistance to *L. branchialis* infection is costly, in terms of growth and maturation, as Lysne *et al.* (2006) found that caged cod infected with one *L. branchialis* had a higher growth rate than fish that remained uninfected throughout the study period.

There are several reports of abnormal behaviour of fish infected with *L. branchialis*. Khan (1988) noted that infected fish were hyperactive and swam in an erratic manner, remaining at the surface because of the excess air in the swim bladders. Sproston and Hartley (1941a) suggested that the parasites cause metabolic alterations resulting in abnormal uptake of fluid and that these fish may remain in low salinity estuarine environments. These behavioural abnormalities may increase the likelihood of predation or capture by fishing and can therefore have an impact on fish populations.

Infection with *L. branchialis* has also been shown to affect gonadal maturation. According to Hislop and Shanks (1981), fecundity was 21% lower in infected haddock than in uninfected fish and Kabata (1958) found that infected female haddock had lower gonadal weights. Templeman *et al.* (1976) and Khan (1988) showed evidence that infection with *L. branchialis* delayed gonadal maturation in cod. In a study by Khan (1988), only 19% of fish infected with *L. branchialis* were sexually mature and only those fish with single infections were mature. In comparison, 64% of control fish were mature. In the wild populations, this delayed maturation and reduced fecundity could lead to changes in the population dynamics of the infected fish populations over time. This may be particularly pertinent for species such as cod which are under considerable current pressure as a result of overfishing.

Although most studies have shown considerable pathogenicity of *L. branchialis*, some authors have reported only minimal changes (Khan and Lacey, 1986; Sherman and Wise, 1961; Sproston and Hartley, 1941a). It is likely that these differences can be attributed to the time of year that sampling took place. Older infected fish examined in the summer, around 10 months after infection, will have had time to adjust to the infection and any effects may not be readily apparent.

It appears that infected fish show indications of chronic stress. Khan (1988) found that most infected fish 30–60 cm in length, transported in tank trucks during the summer, died within 24 h. Mann (1952) reported that infected fish are less able to utilise oxygen and it could be assumed that these fish would be more likely to die than uninfected fish if oxygen was limited as their oxygen requirements would be higher. Sub-adult cod harbouring infections and exposed to low hydrocarbon concentrations were also more likely to die than uninfected fish (Khan, 1988). Additionally, it has been shown that fish harbouring secondary infections concurrent with *L. branchialis* are more likely to succumb (Khan and Lacey, 1986). Khan (1988) and Khan and Lacey (1986) infected sub-adult cod with dual infections of *L. branchialis* and a blood protozoan, *Trypanosoma murmanensis* Nikitin. About 60% of the sub-adult fish died and the surviving fish were emaciated to such an extent that their survival in a natural environment would be unlikely. Surviving fish had pale gills, significantly lower haemoglobin concentrations, condition factors and liver somatic indices than uninfected fish or those infected only with *L. branchialis*. With the chronic stress associated with *L. branchialis* infection, fish subjected to any form of secondary stress, such as disease infection or poor water quality, are unlikely to survive.

Although adult *L. branchialis* have been shown to have a major effect on their definitive hosts, little or no adverse effects have been seen in hosts infected with the juvenile and pre-metamorphic stages of the parasite. Juvenile *L. branchialis* feed on the gill tips of the host, but do not penetrate the heart or blood vessels. Up to 700 larvae on flounder and over 26,000 larvae on lumpfish have been reported, but with no apparent ill-effects (Kabata, 1979; Templeman *et al.*, 1976).

9. CONCLUDING REMARKS

It is clear from the literature that *L. branchialis* is capable of severely affecting the health of gadoid fish, particularly juveniles, and is therefore likely to pose a considerable risk to the health of pressured and recovering wild gadoid populations and cultured gadoids. Factors that increase these risks include its high fecundity, wide distribution in northern coastal waters, and the considerable host impact of individual adult parasites. Despite the potential impact of this pathogen, there are numerous research areas that require more detailed examination. These include (1) host–parasite interactions, both in terms of host immune responses and the immunomodulation of such responses by the parasite; (2) pathology and pathogenesis in intermediate and final hosts; (3) epidemiological studies, which are key to assessing the risks to fish populations associated with this pathogen and developing improved tools for management and control; (4) parasite behaviour and in particular host location and settlement behaviour and the cues that drive them; (5) population genetics studies, which may allow for the recognition of different populations and thereby reveal any existing differences in host preference, environmental optima and pathogenicity between genotypes; (6) physiological parameters, for example, feeding, digestion, respiration, etc.; (7) the development of management and control strategies including assessment of the efficacy of existing chemotherapeutants, for example, emamectin benzoate. In a situation where many wild gadoid fisheries are under increasing pressure, demand for these fish products remains high. As a result, the emerging gadoid culture industry is expanding rapidly to meet demand. Current technologies, however, largely favour the use of coastal sites where the likelihood of infection by this parasite is highest. There is therefore an urgent need to anticipate the health problems that infection by this parasite may engender and to develop contingency plans for the management and control of this wide-spread and potentially devastating pathogen.

REFERENCES

Anon (2005). Final report of the Aquaculture Health Joint Working Group sub-group on disease risks and interactions between farmed salmonids and emerging marine aquaculture species. 54 p.

Anstensrud, M. (1989). Experimental studies of the reproductive behaviour of the parasitic copepod *Lernaeocera branchialis* (Pennellidae). *J. Mar. Biol. Assoc. UK* **69,** 465–476.

Anstensrud, M. (1990a). Effects of mating on female behaviour and allometric growth in the two parasitic copepods *Lernaeocera branchialis* (L., 1767) (Pennellidae) and *Lepeophtheirus pectoralis* (Muller, 1776) (Caligidae). *Crustaceana* **59,** 245–258.

Anstensrud, M. (1990b). Male reproductive characteristics of two parasitic copepods, *Lernaeocera branchialis* (L.) (Pennellidae) and *Lepeophtheirus pectoralis* (Muller) (Caligidae). *J. Crust. Biol.* **10,** 627–638.

Anstensrud, M. (1990c). Mating strategies of two parasitic copepods [*Lernaeocera branchialis* (L.) (Pennellidae) and *Lepeophtheirus pectoralis* (Muller) (Caligidae)] on flounder: Polygamy, sex-specific age at maturity and sex ratio. *J. Exp. Mar. Biol. Ecol.* **136,** 141–158.

Anstensrud, M. (1992). Mate guarding and mate choice in two copepods, *Lernaeocera branchialis* (L.) (Pennellidae) and *Lepeophtheirus pectoralis* (Müller) (Caligidae), parasites on flounder. *J. Crust. Biol.* **12,** 31–40.

Begg, G. S., and Bruno, D. W. (1999). The common dab as definitive host for the pennellid copepods *Lernaeocera branchialis* and *Haemobaphes cyclopterina*. *J. Fish Biol.* **55,** 655–657.

Boxshall, G. A. (1974). Infections with parasitic copepods in North Sea marine fishes. *J. Mar. Biol. Assoc. UK* **54,** 355–372.

Boxshall, G. A. (1990a). Precopulatory mate guarding in copepods. *Bijdragen tot de Dierkunde* **60,** 209–213.

Boxshall, G. A. (1990b). The skeletomusculature of siphonostomatoid copepods, with an analysis of adaptive radiation in structure of the oral cone. *Philos. Trans. R. Soc. Lond. (B)* **328**(1246), 167–212.

Brandal, P. O., Egidius, E., and Romslo, I. (1976). Host blood: A major food component for the parasitic copepod *Lepeophtheirus salmonis* Kröyer 1838 (Crustacea: Caligidae). *Norwegian J. Zool.* **24,** 431–443.

Brian, A. (1906). "Copepodi parassiti dei pesci d'Italia," 190 p. Stabilimento Tipo-Litografico Reale Instituto Sordomuti Genova.

Bricknell, I. R., Bron, J. E., Cook, P., and Adamson, K. (2003). Is blood an important component of the diet of sea lice? *In* "Proceedings of the 6th International Conference on Biology and Control of Sea Lice, St. Andrews, NB, July 1–4."

Brinkworth, R. I., Harrop, S. A., Prociv, P., and Brindley, P. J. (2000). Host specificity in blood feeding parasites: A defining contribution by haemoglobin-degrading enzymes? *Int. J. Parasitol.* **30,** 785–790.

Bron, J. E., Sommerville, C., Jones, M., and Rae, G. H. (1991). The settlement and attachment of early stages of the salmon louse, *Lepeoptheirus salmonis* (Copepoda: Caligidae) on the salmon host, *Salmo salar*. *J. Zool.* **224,** 201–212.

Bron, J. E., Sommerville, C., and Rae, G. H. (1993). Aspects of the behaviour of copepodid larvae of the salmon louse *Lepeophtheirus salmonis* (Krøyer, 1837). *In* "Pathogens of Wild and Farmed Fish: Sea Lice'' (G. A. Boxshall, and D. Defaye, eds.), pp. 125–142. Ellis Horwood, Chichester, UK.

Capart, A. (1948). Lernaeocera branchialis. *Cellulae* **52,** 159–212.

Claus, C. (1868a). Beobachtungen uber *Lernaeocera, Peniculus* und *Lernaea.* Ein Beitrag zur Naturgeschichte der Lernaeen. *Shriften der Gesellschaft Naturforschenden zu Marburg* **9,** 1–32.

Claus, C. (1868b). Über die Metamorphose und systematische Stellung der Lernaeen. *Sitzungsberichte der Gesellschaft zur Beforderung der Gesammten Naturwissenschaften zu Marburg* **2,** 5–13.

Desbrosses, P. (1945). Le merlan (*Gadus merlangus* L.) de la côte française de l'Atlantique. *Revue des Travaux de l'Office de Peches Maritimes* **13,** 177–195.

Devine, G., Ingvarsdottir, A., Mordue, W., Pike, A., Pickett, J., Duce, I., and Mordue, A. (2000). Salmon lice, *Lepeophtheirus salmonis,* exhibit specific chemotactic responses to semiochemicals originating from the salmonid, *Salmo salar. J. Chem. Ecol.* **26,** 1833–1848.

Edwards, J. A. (1984). The utilisation of the gills of flounder (*Platichthys flesus*) by the parasitic copepod *Lernaeocera branchialis. Parasitology* **89**,p v.

Eiras, J. C. (1986). Some aspects of the infection of bib, *Trisopterus luscus* (L.), by the parasitic copepod *Lernaeocera lusci* (Bassett-Smith, 1896) in Portuguese waters. *J. Fish Biol.* **28,** 141–145.

Fast, M. D., Ross, N. W., Mustafa, A., Sims, D. E., Johnson, S. C., Conboy, G. A., Speare, D. J., Johnson, G., and Burka, J. F. (2002). Susceptibility of rainbow trout *Oncorhynchus mykiss,* Atlantic salmon *Salmo salar* and coho salmon *Oncorhynchus kisutch* to experimental infection with sea lice *Lepeophtheirus salmonis. Dis. Aquat. Org.* **52,** 57–68.

Fast, M. D., Burka, J. F., Johnson, S. C., and Ross, N. W. (2003). Enzymes released from *Lepeophtheirus salmonis* in response to mucus from different salmonids. *J. Parasitol.* **89**(1), 7–13.

Fast, M. D., Ross, N. W., Craft, C. A., Locke, S. J., MacKinnon, S. L., and Johnson, S. C. (2004). *Lepeophtheirus salmonis*: A characterization of prostaglandin E_2 in secretory products of the salmon louse by RP-HPLC and mass spectrometry. *Exp. Parasitol.* **107,** 5–13.

Fleming, A. M., and Templeman, W. (1951). Discovery of larval forms of *Lernaeocera branchialis* on lumpfish. *Fisheries Research Board Canadian Biological Station St. John's, Newfoundland Annual Report* 47.

Fox, M. H. (1945). Haemoglobin in blood-sucking parasites. *Nature* **156,** 475–476.

Fraile, L., Escoufier, Y., and Raibaut, A. (1993). Analyse des Correspondances de Données Planifiées: Etude de la Chémotaxie de la Larvae Infestante d'un Parasite. *Biometrics* **49**(4), 1142–1153.

Gouillart, M. (1937). Recherches sur les copepodes parasites (biologie, spermatogenese, ovogenese). *Travaux de la Station Zoologique de Wimereux* **12,** 308–457.

Grant, H. J., and Whitfield, P. J. (1988). The ultrastructure of the spermatozoon of *Lernaeocera branchialis* (Copepoda, Pennellidae). *Hydrobiologia* **167,** 607–616.

Hansen, H. J. (1923). Crustacea Copepoda II. Copepoda parasita and hemiparasita. *Danish Ingolf Expedition* **7,** 1–92.

Hemmingsen, W., Lile, N., and Halvorsen, O. (1995). Search for seasonality in occurrence of parasites of cod, *Gadus morhua* L. in a fjord at 70°N. *Polar Biol.* **15,** 517–522.

Hemmingsen, W., Halvorsen, O., and MacKenzie, K. (2000). The occurrence of some metazoan parasites of Atlantic cod, *Gadus morhua* L., in relation to age and sex of the host in Balsfjord (70°N), North Norway. *Polar Biol.* **23,** 368–372.

Heuch, P. A., and Karlsen, E. (1997). Detection of infrasonic water oscillations by copepodids of *Lepeophtheirus salmonis* (Copepoda Caligida). *J. Plankton Res.* **19,** 735–747.

Heuch, P. A., and Schram, T. A. (1996). Male mate choice in a natural population of the parasitic copepod *Lernaeocera branchialis* (Copepoda: Pennellidae). *Behaviour* **133,** 221–239.

Hislop, J. R. G., and Shanks, A. M. (1981). Recent investigations on the reproductive biology of the haddock, *Melanogrammus aeglefinus,* of the northern North Sea and the effects on

fecundity of infection with the copepod parasite *Lernaeocera branchialis*. *Journal du Conseil International par l'Exploration de la Mer* **39,** 244–251.
Huber, H., Pokey, M. J., Linscott, W. D., Fudenburg, H. H., and Muller-Eberhard, H. J. (1968). Human monocytes: Distinct receptor sites for the 3rd component of complement and for IgG. *Science* **162,** 1281–1283.
Hull, M. Q., Pike, A. J., Mordue (Luntz), A. J., and Rae, G. H. (1998). Patterns of pair formation and mating in an ectoparasitic caligid copepod *Lepeophtheirus salmonis* (Kroyer 1837): Implications for its sensory and mating biology. *Philos. Trans. R. Soc. London (B)* **353,** 753–764.
Huys, R., and Boxshall, G. A. (1991). ''Copepod Evolution,'' 468 p. The Ray Society, London.
Ingvarsdottir, A., Birkett, M., Duce, I., Genna, R., Mordue, W., Pickett, J., Wadhams, L., and Mordue, A. (2002). Semiochemical strategies for sea louse control: Host location cues. *Pest Manag. Sci.* **58,** 537–545.
Jones, M. E. B., and Taggart, C. T. (1998). Distribution of gill parasites (*Lernaeocera branchialis*) infection in Northwest Atlantic cod (*Gadus morhua*) and parasite-induced host mortality: Inferences from tagging data. *Can. J. Fish. Aquat. Sci.* **55,** 364–375.
Kabata, Z. (1957). *Lernaeocera obtusa* n. sp. a hitherto undescribed parasite of the haddock (*Gadua aeglefinus*). *J. Mar. Biol. Assoc. UK* **36,** 569–592.
Kabata, Z. (1958). *Lernaeocera obtusa*. n. sp. its biology and its effects on the haddock. *Mar. Res.* **3,** 26.
Kabata, Z. (1961). *Lernaeocera branchialis* (L.), a parasitic copepod from the European and American shores of the Atlantic. *Crustaceana* **2,** 243–249.
Kabata, Z. (1962). The mouth and mouthparts of *Lernaeocera branchialis* L. *Crustaceana* **3,** 311–317.
Kabata, Z. (1965). *Lernaeocera* (Copepoda) parasitic on ling (*Molva elongata* Otto.). *Crustaceana* **9,** 104–105.
Kabata, Z. (1970). Crustacea as enemies of fishes. *In* ''Diseases of Fishes'' (S. F. Snieszko and H. R. Axelrod, eds.), 171 p. Book 1. T.F.H Publications, Jersey City, NJ.
Kabata, Z. (1979). ''Parasitic Copepoda of British Fishes,'' 468 p. Ray Society, London.
Kerlin, R. L., and Hughes, S. (1992). Enzymes in saliva from four parasitic arthropods. *Med. Vet. Entomol.* **6,** 121–126.
Khan, R. A. (1988). Experimental transmission, development, and effects of a parasitic copepod, *Lernaeocera branchialis*, on Atlantic cod, *Gadus morhua*. *J. Parasitol.* **74,** 586–599.
Khan, R. A., and Lacey, D. (1986). Effect of concurrent infections of *Lernaeocera branchialis* (Copepoda) and *Trypanosoma murmanensis* (Protozoa) on Atlantic cod, *Gadus morhua*. *J. Wildl. Dis.* **22,** 201–208.
Khan, R. A., and Lee, E. M. (1989). Influence of *Lernaeocera branchialis* (Crustacea: Copepoda) on growth rate of Atlantic cod, *Gadus morhua*. *J. Parasitol.* **75,** 449–454.
Khan, R. A., Lee, E. M., and Barker, D. (1990). *Lernaeocera branchialis*: A potential pathogen to cod ranching. *J. Parasitol.* **76,** 913–917.
Khan, R. A., Barker, D. E., and Lee, E. M. (1993). Effect of a single *Lernaeocera branchialis* (copepoda) on growth of Atlantic cod. *J. Parasitol.* **79,** 954–958.
Knaus Knudsen, K., and Sundnes, G. (1998). Effects of salinity on infection with *Lernaeocera branchialis* (L.) (Copepoda: Pennellidae). *J. Parasitol.* **84,** 700–704.
Køie, M. (1999). Metazoan parasites of flounder *Platichthys flesus* (L.) along a transect from the southwestern to the northwestern Baltic Sea. *ICES J. Mar. Sci.* **56,** 157–163.
Kurlansky, M. (1998). Cod. A biography of the fish that changed the world. 294 p.
Lee, E. M., and Khan, R. A. (2000). Length-weight-age relationships, food and parasites of Atlantic cod (*Gadus morhua*) off coastal Labrador within NAFO Divisions 2H and 2J-3K. *Fish. Res.* **45,** 65–72.
Linderby, E., and Thulin, J. (1983). Occurrence and abundance of parasitic copepods on cod along the west and south coast of Sweden. *Proc. 11th Scand. Symp. Parasitol.* **17,** 79.

Lysne, D. A., Hemmingsen, W., and Skorping, A. (2006). Is reduced body growth of cod exposed to the gill parasite *Lernaeocera branchialis* a cost of resistance? *J. Fish Biol.* **69,** 1281–1287.

Machado Cruz, J. A. (1959). *Lernaeocera caparti* sp. nov., copépode parasite de *Merluccius merluccius* (Linné). *Publicacoes do Instituto de zoologia "Dr. Augusto Nobre" Faculdade de Ciencias, Universidade do Porto* **64,** 7–12.

Mann, H. (1952). *Lernaeocera branchialis* (Copepoda parasitica) und seine Schadwirkung bei einigen Gadiden. *Archiv fur Fischereiwissenschaft* **4,** 133–144.

Mann, H. (1970). Copepods and isopods as parasites of marine fishes. *In* "A Symposium on Diseases of Fishes and Shell Fishes" (S. F. Snieszko, ed.), pp. 177–188. American Fisheries Society, Washington, DC, Special Publication 5.

Markevich, A. P. (1956). Parraziticheskie veslonogie ryb SSSR [Parasitic copepods of fishes of the USSR]. *Izvestiya Akademii Nauk Ukraine SSR* **1,** 259.

Metzger, A. (1868). Ueber das Männchen und Weibchen der Gattung *Lernaea* vor den Eintritt der sogenannten rüchschreitenden Metamorphose. *Transl. Ann. Mag. Nat. Hist.* **4,** 154–157.

Moller, H. (1983). The effects of *Lernaeocera* infestation on cod. *Bull. Eur. Assoc. Fish Pathol.* **3,** 21–22.

Olsson, P. (1869). Prodromus faunae copepodorum parasitantium Scandinaviae. *Acta Univ. Lund* **5,** 1–49.

Osorio, B. (1892). Appendice ao catalogo dos crustaceos de Portugal existentes no Museu Nacional de Lisboa. *J. Sci. Math. Phys. Nat. Hist. Lisbon* **2,** 233–241.

Panikkar, N. K., and Sproston, N. G. (1941). Osmotic relations of some metazoan parasites. *Parasitology* **33,** 214–223.

Papadogiannakis, N., and Johnsen, S. A. (1987). Motogenic action of phorbol ester TPA and calcium ionophore A23187 on human cord and maternal/adult peripheral lymphocytes: Regulation by PGE_2. *Clin. Exp. Immunol.* **70,** 173–181.

Papadogiannakis, N., Johnsen, S. A., and Olding, L. B. (1984). Strong prostaglandin associated suppression of the proliferation of human maternal lymphocytes by neonatal lymphocytes linked to T versus T cell interactions and differential PGE_2 sensitivity. *Clin. Exp. Immunol.* **61,** 125–134.

Pedashenko, D. D. (1898). Die Embryonalentwickelung und Metamorphose von *Lernaea branchialis. Travaux de la Societe Imperiale de Naturalistes de St. Petersbourg* **26,** 247–307.

Perkins, P. S. (1985). Iron crystals in the attachment organ of the erythrophagous copepod *Cadiodectes medusaeus* (Pennellidae). *J. Crust. Biol.* **5**(4), 591–605.

Piasecki, W. (1996). The developmental stages of *Caligus elongatus* von Nordmann, 1832 (Copepoda: Caligidae). *Can. J. Zool.* **74,** 1459–1478.

Piasecki, W., and MacKinnon, B. M. (1995). Life cycle of a sea louse, *Caligus elongatus* von Nordmann, 1832 (Copepoda: Siphonostomatoida, Caligidae). *Can. J. Zool.* **73,** 74–82.

Pike, A. W., MacKenzie, K., and Rowand, A. (1993). Ultrastructure of the frontal filament in chalimus larvae of *Caligus elongatus* and *Lepeoptheirus salmonis* from Atlantic salmon, *Salmo salar*. *In* "Pathogens of Wild and Farmed Fish: Sea Lice" (G. A. Boxshall, and D. Defaye, eds.), pp. 99–113. Ellis Horwood, Chichester, UK.

Pilcher, M. W., Whitfield, P. J., and Riley, J. D. (1989). Seasonal and regional infestation characteristics of three ectoparasites of whiting, *Merlangius merlangus* L. in the North Sea. *J. Fish Biol.* **35,** 97–110.

Polyanski, Yu. I. (1955). Materialy po parazitologii ryb severnykh morey SSSR. Parazity ryb Barentsova Morya [Data on parasitology of fishes of the northern sea of the USSR. Parasites of fishes of the Barents Sea]. *Trudy Zoologicheskogo Instituta. Akademiya Nauk SSSR* **19,** 5–170.

Potter, I. C., Gardner, D. C., and Claridge, P. N. (1988). Age composition, growth, movements, meristics and parasties of the whiting, *Merlangius merlangus*, in the Severn estuary and Bristol channel. *J. Mar. Biol. Assoc. UK* **68,** 295–313.

Radulescu, I. I., Nalbant, T. T., and Angelescu, N. (1972). Nei contributii la cunoasterea parasitofaunei pestilar din Oceane Atlantic. *Buletinul de Cercetari Piscicole* **31,** 71–78.

Reed, G. B., and Dymond, J. R. (1951). Report Newfoundland Fisheries Station. *Report of the Fisheries Research Board of Canada* **1951,** 43–65.

Sanchez-Lizaso, J. L., and Vasquez, J. (1987). Infestation of cod by *Lernaeocera branchialis* in ICES Division IIb. *ICES C.M. 1987/ G:6* 6 p.

Schram, T. A., and Heuch, P. A. (2001). The egg string attachment mechanism of selected pennellid copepods. *J. Mar. Biol. Assoc. UK* **81,** 23–32.

Schuurmans Stekhoven, J. H. (1936a). Beobachtungen zur Morphologie und Physiologie der *Lernaeocera branchialis* L. und der *Lernaeocera lusci* Bassett-Smith (Crustacea parasitica). *Zeitschrift fur Parasitenkunde* **8,** 659–698.

Schuurmans Stekhoven, J. H., Jr. (1936b). Copepoda parasitica from the Belgian coast, II. (Included some habitats in the North Sea). *Memoires du Musêe Royal D'Histoire Naturelle de Belgique* **74,** 1–20.

Schuurmans Stekhoven, J. H. Jr., and Punt, A. (1937). Weitere Beitrage zur Morphologie und Physiologie der *Lernaeocera branchialis* L. *Zeitung fur Parasitenkunde* **9,** 648–668.

Scott, A. (1901). *Lepeophtheirus* and *Lernaea*. *Liverpool Marine Biology Committee Memoirs* **6,** 54.

Sherman, K., and Wise, J. P. (1961). Incidence of the cod parasite *Lernaeocera branchialis* L. in the New England area, and its possible use as an indicator of cod populations. *Limnol. Oceanogr.* **6,** 61–67.

Shotter, R. A. (1973a). Changes in the parasite fauna of whiting, *Odontogadus merlangus* (L.) with age and sex of host, season, and from different areas in the vicinity of the Isle of Man. *J. Fish Biol.* **5,** 559–573.

Shotter, R. A. (1973b). A comparison of the parasite fauna of the young whiting, *Odontogadus merlangus* (L.) (Gadidae) from an inshore and an offshore location off the Isle of Man. *J. Fish Biol.* **5,** 185–195.

Shotter, R. A. (1976). The distribution of some helminth and copepod parasites in tissues of whiting, *Merlangius merlangus* L., from Manx waters. *J. Fish Biol.* **8,** 101–117.

Shulman, S. S., and Shulman-Albova, R. E. (1953). Parazity ryb Belogo Morya [Parasites of fishes of the White Sea]. *Izvestiya Akademii Nauk SSSR: Moscow and Leningrad* 199 P.

Smith, J. W. (1969). The distribution of one monogenean and two copepod parasites of whiting, *Merlangius merlangus* (L.), caught in British waters. *Nytt Magasin for Zoologi* **17,** 57–63.

Smith, J. A., and Whitfield, P. J. (1988). Ultrastructural studies on the early cuticular metamorphosis of adult female *Lernaeocera branchialis* (L.) (Copepoda, Pennellidae). *Hydrobiologia* **167,** 607–616.

Smith, J. L., Wotten, R., and Sommerville, C. (2007). The pathology of the early stages of the crustacean parasite, *Lernaeocera branchialis* (L.), on Atlantic cod, *Gadus morhua* L. *J. Fish Dis.* **30**(1), 1–11.

Sproston, N. G. (1942). The developmental stages of *Lernaeocera branchialis*. *J. Mar. Biol. Assoc. UK* **25,** 441–446.

Sproston, N. G., and Hartley, P. H. T. (1941a). The ecology of some parasitic copepods of gadoids and other fishes. *J. Mar. Biol. Assoc. UK* **25,** 361–392.

Sproston, N. G., and Hartley, P. H. T. (1941b). Observations on the bionomics and physiology of *Trebius caudatus* and *Lernaeocera branchialis* (Copepoda). *J. Mar. Biol. Assoc. UK* **25,** 393–417.

Stevenson, S. C., and Baird, J. W. (1988). The fishery for lumpfish (*Cyclopterus lumpus*) in Newfoundland waters. *Canadian Technical Report of Fisheries and Aquatic Science No. 1595.*

Sundnes, G. (1970). "*Lernaeocera branchialis* (L.) on Cod (*Gadus morhua* L.) in Norwegian Waters," 48 p. Institute of Marine Research Bergen.

Sundnes, G., Mork, J., Solemdal, P., and Solemdal, K. (1997). *Lernaeocera branchialis* (L., 1767) on cod in Baltic waters. *Helgolander Meeresuntersuchungen* **51,** 191–196.
Templeman, W., and Fleming, A. M. (1963). Distribution of *Lernaeocera branchialis* (L.) on cod as an indicator of cod movements in the Newfoundland area. *ICNFA Spec. Publ.* **4,** 318–322.
Templeman, W., Hodder, V. M., and Fleming, A. M. (1976). Infection of lumpfish (*Cyclopterus lumpus*) with larvae and of Atlantic cod (*Gadus morhua*) with adults of the copepod, *Lernaeocera branchialis,* in and adjacent to the Newfoundland area, and inferences thereof on inshore-offshore migrations of cod. *J. Fish. Res. Board Can.* **33,** 711–731.
Tirard, C., Berrebi, P., Raibaut, A., and Renaud, F. (1993). Biodiversity and biogeography in heterospecific teleastean (Gadidae)–copepod (*Lernaeocera*) associations. *Can. J. Zool.* **71,** 1639–1645.
To, S. S. T., and Schrieber, L. (1990). Effect of leukotriene B_4 and prostaglandin E_2 on the adhesion of lymphocytes to endothelial cells. *Clin. Exp. Immunol.* **81,** 160–165.
Van Damme, P. A., and Hamerlynck, O. (1992). The infection dynamics and dispersion pattern of *Lernaeocera branchialis* L. on 0+ whiting (*Merlangius merlangus* L.) in the Oosterschelde (SW Netherlands). *J. Fish Biol.* **41,** 265–275.
Van Damme, P. A., and Ollevier, F. (1995). Morphological and morphometric study of crustacean parasites within the genus *Lernaeocera. Int. J. Parasitol.* **25,** 1401–1411.
Van Damme, P. A., Ollevier, F., and Hamerlynck, O. (1994). Pathogenicity of *Lernaeocera lusci* and *L. branchialis* in bib and whiting in the North Sea. *Dis. Aquat. Org.* **19,** 61–65.
Van Damme, P. A., Geets, A., and Hamerlynck, O. (1997). The suprapopulation dynamics of *Lernaeocera branchialis* and *L. lusci* in the Oosterschelde: Seasonal abundance on three definitive host species. *ICES J. Mar. Sci.* **54,** 24–31.
Van den Broek, W. L. F. (1977). Aspects of the biology of fish populations from the Medway Estuary, based on power station intake sampling, with special reference to parasitism and pollution. Ph.D. thesis, University of London.
Van den Broek, W. L. F. (1978). The effects of *Lernaeocera branchialis* on the *Merlangius merlangus* population in the Medway estuary. *J. Fish Biol.* **13,** 709–715.
Van den Broek, W. L. F. (1979a). Copepod ectoparasites of *Merlangius merlangus* and *Platichthys flesus. J. Fish Biol.* **14,** 371–380.
Van den Broek, W. L. F. (1979b). Infection of estuarine populations by *Lernaeocera branchialis* and its effect on whiting *Merlangius merlangus. N.Z. J. Zool.* **6,** 646–647.
Van Oared-de Lint, G. M., and Schuurmans Stekhoven, J. H. (1936). Copepoda parasitica. *Tier welt der Nerd- und Osteen* **31,** 73–197.
Vazquez, J., Paz Caballero, X., and Escalante J. L. G. (1988). Rates of infestation of cod by *Lernaeocera branchialis* in ICES Division IIb. *ICES C. M. 1988/G:2? Session O,* 9 p.
Whitfield, P. J., Pilcher, M. W., Grant, H. J., and Riley, J. (1988). Experimental studies on the development of *Lernaeocera branchialis* (Copepoda, Pennellidae)—Population processes from egg production to maturation on the flatfish host. *Hydrobiologia* **167,** 579–586.
Wilson, C. B. (1917). North American parasitic copepods belonging to Lernaeidae, with a revision of the entire family. *Proc. US Natl. Museum* **53,** 1–150.

INDEX

A

R

S

T

CONTENTS OF VOLUMES IN THIS SERIES

Volume 45

Volume 46

Volume 47

Volume 48

Volume 55

Volume 56

Volume 57

Volume 58

Volume 59

Volume 60

Volume 61

Volume 62

Volume 63

Volume 64

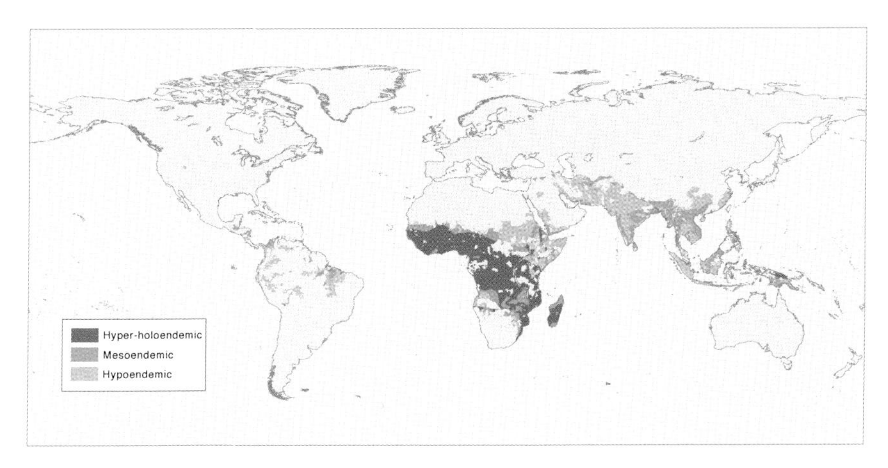

Plate 1.2 Spatial limits and potential endemic levels of *P. falciparum* in 2005. *Source:* The Malaria Atlas Project (http://www.map.ox.ac.uk). Several sources of information on malaria risk (notably international travel health guidelines on malaria chemoprophylaxis, altitude limits for dominant vectors, climate limits for malaria transmission and human population density thresholds) have been combined in a geographic information system to generate this map. See Guerra *et al.* (2006a,b) for details of the methodology. The method for defining the endemic levels within these limits can be found in Snow *et al.* (2005). This map is provided only as a guide and the Malaria Atlas Project is committed to assembling the medical intelligence required to improve future iterations.

Plate 3.6 Prediction map describing the geographic distribution of *Bancroftian filariasis* microfilaria (mf) prevalence in Tanzania obtained by ordinary krigging (OK) of mf point prevalence data (symbols). The krigged surface (filled contours) was produced using parameter values from the best-fitting theoretical model for the spatial correlation among these data, which was given by a spherical semivariogram model (Cressie, 1993) with values of 66.106 for the nugget, 129.216 for the sill and 0.807 degrees for the range. The contours depicted on the krigged surface show areas of similar autocorrelated prevalence of infection with darker fills indicating higher prevalence. Methods for fitting semivariogram models, including those used for detrending data, are as given in Srividya *et al.* (2002), while the data used in the analysis were collected under the auspices of the Tanzania National Lymphatic Filariasis programme (M.N.M.-L., personal communication). The blowout shows an overlay of the krigged mf probability surface, categorized into three zones of low, medium and high filariasis endemicity, on a village location map for Lindi district (circles denote individual human habitations). Random sampling of habitations within each zone will give rise to the stratified random samples of sentinel sites for monitoring filariasis control in the district.

(*See Plate on next page*)

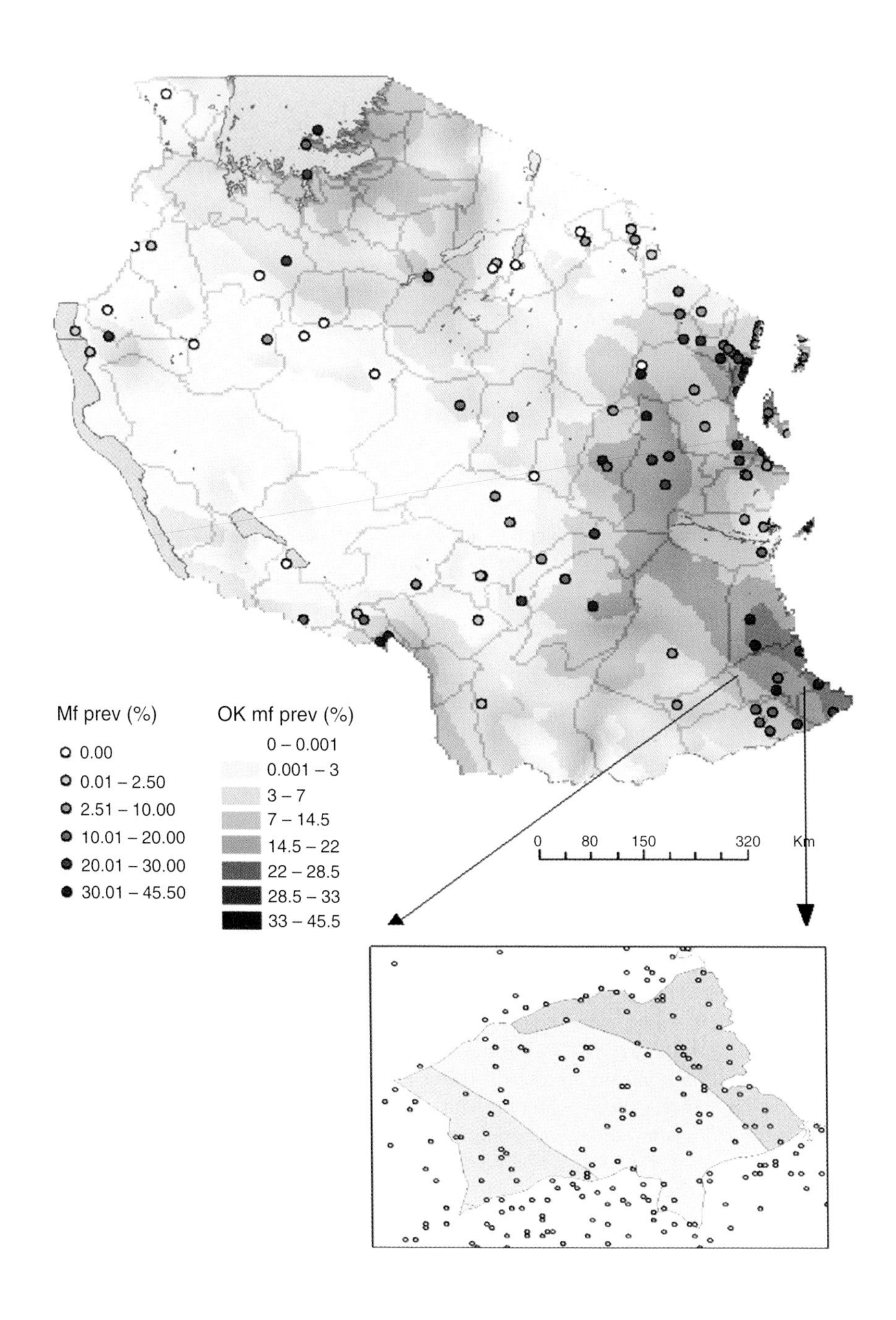

Mf prev (%)
0.00
0.01 – 2.50
2.51 – 10.00
10.01 – 20.00
20.01 – 30.00
30.01 – 45.50
OK mf prev (%)
0 – 0.001
0.001 – 3
3 – 7
7 – 14.5
14.5 – 22
22 – 28.5
28.5 – 33
33 – 45.5
0
80
150
320
Km